4th International Conference on Precision Machinery and Manufacturing Technology (ICPMMT 2019)

IOP Conference Series: Materials Science and Engineering
Volume 644

Kenting, Taiwan
24 – 26 May 2019

ISBN: 978-1-5108-9843-1
ISSN: 1757-8981

Printed from e-media with permission by:

Curran Associates, Inc.
57 Morehouse Lane
Red Hook, NY 12571

Some format issues inherent in the e-media version may also appear in this print version.

Institute of Physics
Dirac House, Temple Back
Bristol BS1 6BE UK

Phone: 44 1 17 929 7481
Fax: 44 1 17 920 0979

techtracking@iop.org

Additional copies of this publication are available from:

Curran Associates, Inc.
57 Morehouse Lane
Red Hook, NY 12571 USA
Phone: 845-758-0400
Fax: 845-758-2633
Email: curran@proceedings.com
Web: www.proceedings.com

Table of contents

Volume 644

4th International Conference on Precision Machinery and Manufacturing Technology

24–26 May 2019, Kenting, Taiwan

Accepted papers received: 13 September 2019
Published online: 24 October 2019

Preface

Papers

IOP Conf. Series: Materials Science and Engineering **644** (2019) 011001 doi:10.1088/1757-899X/644/1/011001

Foreword

The 1st International Conference on Precision Machinery and Manufacturing Technology 2015 (ICPMMT) was held at Howard Beach Resort, Kenting, Taiwan on May 22-24, 2015. The papers submitted to ICPMMT 2015 were recommended to KEM (EI Compendex) through the peer reviewing process.

The 2nd ICPMMT 2017 was held at Howard Plaza Resort, Kenting, Taiwan on May 19-21, 2017. **And the 3rd ICPMMT 2018 has been taken place in Auckland, New Zealand on 4-8 February, 2018**. After conference, all papers submitted to ICPMMT 2017 and 2018 were recommended to MATEC Web of Conferences, Open-access Journal in Materials science, Engineering and Chemistry (Google Scholar, CAS, EI Compendex, Inspec, DOAJ, Web of Science, Scopus). All papers through the peer reviewing process can be published in MATEC Web of Conferences. This year the 4th ICPMMT 2019 is back to Howard Plaza Resort, Kenting, Taiwan on May 24-26, 2019. All papers through the peer reviewing process can be published Key Engineering Materials with EI Compendex.

The topics are as follows:

1. Design & Manufacturing
   Machinery, Aerospace, Fasteners, Electronics, Plant, Robot, EV, Hybrid Car, CAD, CAE, CAM, PLM, SCM, Optimization, Reverse Engineering
2. Processing Technology
   Machining, Forming, Grinding, Casting, Forging, Molding, Welding, Joining, Micro· Nano, Lase, Energy-Beam, Printed Electronics, 3D Printing, RP&M, R2R, Electro-Chemical, Non-traditional, Re-manufacturing, Recycle, MQL, Smart Manufacturing
3. Green Manufacturing and Materials Application
   Light·Eco-friendly Material & Structure, Natural, Bio-inspired, Light-weight, Flexible, Powder, Multifunctional, Metals, Ceramics, Polymers, Composites, FGM, Graphene, Smart Material & Structure, Photonic, Display
4. Automation and Intelligent Robot
5. Information Integration and Communication Technology
6. Industry 4.0 Technology
7. Energy and Nano, Micro Technology

IOP Conf. Series: Materials Science and Engineering **644** (2019) 011001 doi:10.1088/1757-899X/644/1/011001

8. Educations and Training
9. Related Topics to Precision Machinery and Manufacturing Technology
10. Other Topics

In ICPMMT 2019, Prof. Cheng Lu from Wollongong University, Australia (The topic: Numerical Simulations of Crystallographic Orientation Evolution during Accumulative Roll Bonding), Prof. Man-Soo Joun from Gyeongsang National University, Korea (The topic: Accuracy of Metal Forming Simulation), and Prof. Jong-Ning Aoh from National Chung Cheng University (The topic: Friction Stir Additive Manufacturing (FSAM)-- Review from Shape Welding and Laser Additive Manufacturing) are invited as keynote speakers**.** They are very famous in innovation materials, manufactures and metal forming research field, and once again to thank them for strong supports to ICPMMT 2019. I hope all of you can enjoy this lovely place, Kenting, Pingtung, Taiwan, and to absorb excellent knowledge and technologies from these wonderful speeches and oral presentations in ICPMMT 2019.

On behalf of organization committee, I would like to thank Cheng Shiu University, Chung Cheng University, and Changhua University to assist ICPMMT 2019 being successful by way of financial supports.

Many thanks to lots of authors provide your excellent contributions to ICPMMT 2019. Next conference (ICPMMT 2020) will be held on 3~7 Feb 2019, in Auckland, New Zealand.

Gow-Yi Tzou

Guest Editor of IOP:MSE
Conference Chair of ICPMMT
Chair Prof. Gow-Yi Tzou
Dean of Academic Affairs Office and College of Engineering,
Chung-Chou University of Science and Technology

Committee

Conference Chair:

Chair Prof. Gow-Yi Tzou (Chung-Chou University of Science and Technology, Taiwan)

Steering Committee:

- Chair Prof. Gow-Yi Tzou (Chung-Chou University of Science and Technology, Taiwan)
- Prof. Thomas Neitzert (Auckland University of Technology, New Zealand)
- Prof. Jeong-Whan Yoon (Deakin University, Australia; KAIST Korea)

Scientific Committee:

Gow-Yi Tzou (Chung-Chou University of Science and Technology, Taiwan)

Ho-Chi Cheng (Chung-Chou University of Science and Technology, Taiwan)

Thomas Neitzert (Auckland University of Technology, New Zealand)

Timotius Pasang (Auckland University of Technology, New Zealand)

Guy Littlefair (Auckland University of Technology, New Zealand)

Maziar Ramezani (Auckland University of Technology, New Zealand)

Richard Lin (The University of Auckland, New Zealand)

Jeong-Whan Yoon (Deakin University, Australia; KAIST, Korea)

Mansoo Joun (Gyeongsang National University, Korea)

Ming Yang (Graduate School of System Design, Tokyo Metropolitan University, Japan)

Cheng Lu (Wollongong University, Australia)

Zengyi Jiang (Wollongong University, Australia)

IOP Conf. Series: Materials Science and Engineering **644** (2019) 011001 doi:10.1088/1757-899X/644/1/011001

Jing-Jing Li (Penn State University, USA)

Scott Miller (Hawaii University, USA)

Peter Groche (TU Darmstadt, Germany)

Ying-Hua Liu (Tsinghua University Beijing, China)

Fuh-Kuo Chen (Taiwan University, Taiwan)

Jen-Yuan Chang (Tsing Hua University, Taiwan)

De-Shin Liu (Chung-Cheng University, Taiwan)

Jong-Ning Aoh (Chung-Cheng University, Taiwan)

Yung-Chou Kao (Chung-Cheng University, Taiwan)

You-Min Huang (Taiwan University of Science & Technology, Taiwan)

Yeong-Maw Hwang (Sun Yat-Sen University, Taiwan)

Cheng-Tang Pan (Sun Yat-sen University, Taiwan)

Jinn-Jong Sheu (Kaohsiung University of Applies Science, Taiwan)

Chao-Cheng Chang (Kaohsiung University of Applies Science, Taiwan)

Cho-Pei Jiang (Formosa University, Taiwan)

Dyi-Cheng Chen (Changhua University of Education, Taiwan)

Yeong-Lin Lai (Changhua University of Education, Taiwan)

Shao-Yi Hsia (Kao-Yuan University of Science and Technology, Taiwan, Taiwan)

Chih-Liang Chu (Southern Taiwan University of Science and Technology, Taiwan)

Cheng-Chien Wang (Southern Taiwan University of Science and Technology, Taiwan)

Yen-Ming Tseng (Fuzhou Polytechnic, China)

IOP Conf. Series: Materials Science and Engineering **644** (2019) 011002 doi:10.1088/1757-899X/644/1/011002

Peer review statement

All papers published in this volume of *IOP Conference Series: Materials Science and Engineering* have been peer reviewed through processes administered by the proceedings Editors. Reviews were conducted by expert referees to the professional and scientific standards expected of a proceedings journal published by IOP Publishing.

IOP Conf. Series: Materials Science and Engineering **644** (2019) 012001 doi:10.1088/1757-899X/644/1/012001

The Carbon Footprints and Equipments Energy Consumption Assessment for Bicycle Rubber-Tire

Tse-Hung Lin[1], Yu-Shu Chien[2], and Wei-Ming Chiu[1]

1 Graduate Institute of Precision Manufacturing,National Chin-Yi University of Technology, Taiwan
2 Department of Chemical and Materials engineering, National Chin-Yi University of Technology, Taiwan
E-mail: iesony88@gmail.com

Abstract. High-priced rubber tires are important key components in the bicycle industry chain. The annual bicycle tire output is about 150,000 product lines in Taiwan. Relevant carbon emission inventory data analysis is the focus of the industry's sustainable development and circular economy. Most researches study paper did not completely disclose supply chain inventory data. Relative comparison this bicycle rubber-tire case industry chain is more complete and the inventory data is more detailed.

This tire product case study is original and compiles exclusive Taiwan local activity data and carbon emission database. The important investigation results are as follows: (1)A single product is a functional unit of 561g bicycle rubber-tire carbon emissions of 1.64kgCO2e. Internationally verified BSI certification.(2)The analysis result meets the Monte Carlo analysis requirements. (3)This important practical case study completed two major central factories and five satellite suppliers to check the local activity data, equipment energy consumption and carbon footprint emissions, confirmed by a third party international verification unit BSI. (4)The application and review of the B to B product carbon-label were completed, it can be used for international marketing. The database and practical results published in this paper can be used as a reference for academics and business circles.

1. Introduction

In order to cope with the green product policy promotion, combined with the phased joint efforts of the indicator bicycle industry chain, the practical focus is to check the carbon footprint of the functional units of the bicycle key components rubber tires. The future includes carbon footprints of aluminum rims, chains, saddles, complete cycles, etc., and can build a complete carbon footprint database of key components of bicycles in Taiwan.

According to Taiwan's Ministry of Economics product category rule (PCR), the classification of tire products into petrochemical rubber products is a high carbon emission industry. Its carbon footprint inventory needs to be put into record for a long time and requires considerable funds to support, and its related research is rarely published. Recently, Lin, et al (2017) released that the carbon footprint of tire products is concentrated in the raw material stage, and graphene materials can replace carbon black to reduce carbon emissions and environmental impact. The above research objects are small center factories, which cover rubber and yarn manufacturing process is outsourced, and the raw material activity data check is provided by the 1st-tier supplier, so the source data quality attribute

IOP Conf. Series: Materials Science and Engineering **644** (2019) 012001 doi:10.1088/1757-899X/644/1/012001

DQI(data quality Indicator) is the estimated value belonging to the third level. This paper hopes to conduct a more complete activity data check on the suppliers of the raw material stage. The data quality attribute DQI is the calculated value (level II) and the measured value (level I). The inventory data is more objective and confirmed by the third party BSI International.
The research object is a large-scale and 100% self-made processing rubber factory in central Taiwan, and it has a large international scale. Its manufacturing process covers 2 major factories. All tire processes are concentrated in the former engineer in Yunlin-factory (F1) and the post-engineer in Yuanlin-factory (F2) and other five 1st-tier raw material plants (S1, S2, S3, S4, S5) practice check. Table 1 shows the general information on participating suppliers including number of employees and main products.

Table 1. Participating companies' size and raw material.

Supplier	No. of Employees	Raw material	Location
S1	2200	Butadiene Rubber (PR040)	Tainan City
S2	620	Synthetic Rubber (BR,SBR)	Kaohsiung City
S3	5000	Tire cord and fabrics	Yunlin County
S4	42	Heavy paraffinic distillate	Hsinchu County
S5	80	Filler (Zinc oxide)	Changhua County

2. Methods

Practical case study a life cycle assessment model for a rubber tire inventory database, The main materials, auxiliary materials, energy consumption and waste air, waste water and wastes discharge of the whole process of the tire process were analyzed. In-depth study of the main activity data, different factories have different system boundaries, complete the seven-factory system boundary inventory, and optimize the DQI, complete the eight-checklist database, and provide more objective applications.

2.1. Functional unit

The functional unit is the production methodology for a single piece bicycle rubber-tire with a stand weight of 561 g. The methodologies under analysis in this provide similar guidance for the establishment of the functional unit, which includes the input/output material consumption and energy resource consumption.
Five 1st-tier raw material plants (S1, S2, S3, S4, S5) functional unit. Table 2 shows the suppliers including raw material weight and LCI no.

Table 2. 1st-tier suppliers functional unit weight and LCI no.

Raw material	Functional unit	LCI data-base **
S1_Butadiene Rubber (PR040)	35 kg	LCI-4
S2_Synthetic Rubber (BR57%,SBR43%)	35 kg	LCI-5
S3_Tire cord and fabrics	1 kg	LCI-6
S4_Heavy paraffinic distillate	1 ton	LCI-7
S5. Filler (Zinc oxide)	1 kg	LCI-8
F1a*_Compound (S1,S2,S4,S5..etc.)	201.66 kg	LCI-1
F1b*_Yarn agglutination(S3..etc.)	261.4 kg	LCI-2
F2_Sulfur molding	565.54 g	LCI-3

* F1 factory=F1a+F1b
**Consider the text space layout is unable to disclosed LCI report

3. System boundary

3.1. Stages included

Follow the PAS 2050 system and propose cradle to gate assessments. The ISO14040/ISO14044 standards refer that the system boundary should be consistent with the goal of this study. The standards allow consideration of only parts of the life cycle.

As shown in Fig.1 the following stages are generically considered.

1. Main materials product: Consists of cradle-to-gate production using the main materials, including the Butadiene rubber, Synthetic Rubber (BR,SBR), carbon black, zinc sulphate, sodium hydroxide, yarn, steel wire, distances and energy consumption for transportation etc.
2. Auxiliary production materials: Auxiliary materials are used during the manufacturing processes, but the solvent, diesel, paraffin, natural gas, sulphuric acid, gasoline, lubricating oil etc are not contained in the products.
3. Energy consumption: Energy includes electricity, oil, water, LNG, fuel oil etc.
4. Emission to air: Waste air includes the sodium oxalate, nitrogen oxides, sulfur oxides, chemical oxygen demand (COD), volatile organic compounds (VOC) etc. Refer to the registry documents (According to local environmental protection laws and regulations).
5. Emission to water: Waste water refers to the registry documents (According to local environmental protection laws and regulations).
6. Wastes: Disposal includes the solvents mixture, textiles, digester sludge, rubber etc. and refers to the registry documents (According to local environmental protection laws and regulations).

Figure 1 shows Stages considered within the Rubber-tire carbon footprint system boundary.

IOP Conf. Series: Materials Science and Engineering **644** (2019) 012001 doi:10.1088/1757-899X/644/1/012001

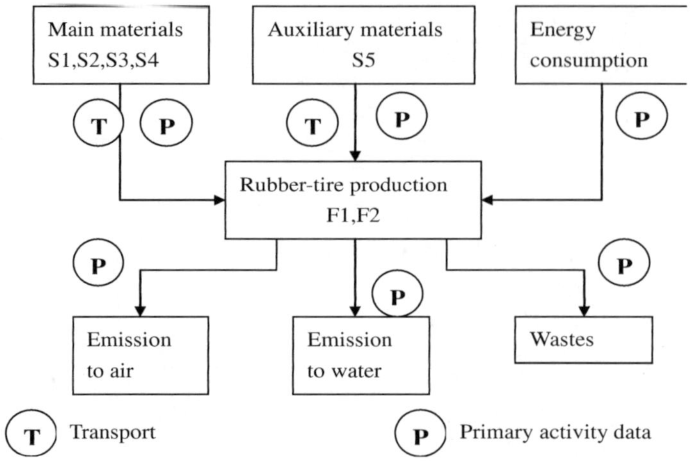

Figure1. Stages considered within the Rubber-tire carbon footprint system boundary.

3.2. Cut-off criteria

The PAS2050 recommends the inclusion of all emission sources that contribute to more than 1% of the anticipated life cycle GHG emissions from the functional unit. Additionally, it includes at least 95% of the anticipated life cycle GHG emissions from the functional unit and shall not exceed 5%. Components and materials omitted from the LCA shall be documented.

Where less than 95% of the anticipated life cycle GHG emissions have been determined, the assessed emissions should be scaled to represent 100% of the GHG emissions associated with the functional unit.

Under this methodology all of the processes generating GHG emissions are included inside the boundary in order to comply with the 95% criterion.

3.2.1. Other exclusions

The production of capital goods (Packaging materials, buildings, machinery and equipment), Auxiliary materials transport and the transport of employees to and from their place of work are excluded from the system boundary. This is in compliance with PAS2050:2011.

4. Data assessment

4.1. Activity data assessment

(1).Inventory the input data: List the main materials that make-up the product. Auxiliary materials are used during the manufacturing processes, but are not contained in the products. Energy and Resources includes electricity, oil, natural gas, water, etc. (2). Inventory the manufacturing process: Energy and resources used by the equipment and machinery. (3). Inventory the output data: Emissions to the air/water/wastes refers to the registry documents, according to local environmental protection laws and regulations.

Table 3 shows Single piece bicycle-tire Input/Output inventory data matrix.

Table 3. Single piece bicycle-tire Input/Output inventory data matrix.

Input			Bicycle-tire	Output		
Main Materials	Auxiliary Material	Energy& Resources	**Process Flow(F1,F2)**	Emission to Air	Emission to Water	Wastes
	LNG	Water, electric	Steam			
V	V	Water, electric	Compound(F1a)	Little	Recycle	
V		Water, electric	Yarn Agglutination (F1b)		Recycle	Recycle
		F1 ➔ Transportation(ton-km) ➔				
V		Electric Power	Blend			V
		Water, electric	pressurize		Recycle	Recycle
V		Electric Power	Wire mold			Recycle
		Electric Power	Mold			Recycle
	V	Water, electric	vulcanization			

4.2. Equipments energy consumption

In the field, the F1 and F2 factories were inspected, and the three major process procedures were inspected. The number of employees in this case was about 900, and the annual bicycle-tire main products output was about 140,000.

F1 factory rubber tire manufacturing process, major equipment energy consumption inventory process with independent electricity meter as the basic requirements for inventory, functional unit inventory to compound process(LCI-1), yarn agglutination process (LCI-2) two independent plants.

Inventory the main process in the latter part of the F2 factory is the project (LCI-3), including the blend project, the pressurize project, the wire mold project, the molding project and the vulcanization project. Table 4 shows eight factories equipment energy consumption data-base.

Table 4. eight factories equipment energy consumption data-base.

Life Cycle Inventory	Factorys	Functional unit	KWH
LCI-4	S1.	1 kg	0.77
LCI-5	S2.	1 kg	0.459
LCI-6	S3.	1 kg	2.2019
LCI-7	S4.	1 ton	0.0029
LCI-8	S5.	1 kg	0.391
LCI-1	F1a.	201.66 kg	9.7946
LCI-2	F1b.	261.4 kg	0.25
LCI-3	F2.	565.54 g	0.2389

4.3. Ghgs and global warming potentials

The GHGs considered in this study, whenever they were available, are those for which the last assessed report of the Inter-governmental Panel on Climate Change (IPCC, 2007) assigns a global warming potential (GWP), as recommend by PAS 2050. The GWPs adopted in this study are also from IPCC (2007) for a time period for assessment of GHG emissions of 100 years, as show Table 5.

IOP Conf. Series: Materials Science and Engineering **644** (2019) 012001 doi:10.1088/1757-899X/644/1/012001

Table 5. IPCC (2007) fourth assessment report: global warming potential.

GWP	Carbon dioxide (CO_2)	Methane (CH_4)	Nitrous oxide (N_2O)
100 years	1	25	298

4.4. Data collection

Completeness, Relevance, Consistency, Accuracy and Transparency are required. Consider the full life cycle assessment including Taiwan power line losses, the energy consumption of GHG Emission is 0.7 KgCO2e/KWH for Consistency. The verification was performed based on PAS 2050:2011(ISO/DIS 14067) and ISO14064-1:2006 Technical Guideline, given to provide for consistent product life cycle GHG emission identification, calculation, monitoring and reporting.
Consider environmentally extended input-output (EEIO) analysis. Age of data and minimum length of time over which data are collected, data that are time-specific to the product being assessed shall be preferred. Follow the third party verification statement results. Analysis validity is certified by an independent third party certification in accordance with PAS 2050:2011(ISO/DIS 14067). Table 6 shows the carbon footprints for the bicycle-tire CO2e emission.

Table 6. Total CO2e emission for single product.

Inventory IN/OUT	Rubber type	single piece Bicycle-tire	
	Factory area	two factory F1(LCI-1,LCI-2) F2(LCI-3)	
	Includable (Primary activity data %)	five engaging suppliers (22%)	
Company Profile	Functional unit / tire	561 g	
	Employee no.	900	
	Data period	12 month	
	Throughput/year	146,868 tires	
IN	main materials (transport)	1.27 kgCO2e (1.61E-04)	77.43% (lower%)
	Auxiliary material	7.41E-03	0.45%
	Energy consumption	0.3617 kgCO2e	22.05%
OUT	Emission to air	Lower	Lower
	Emission to water	3.98E-04	0.02%
	Wastes	Lower	Lower
Emission carbon footprint/Functional unit		1.64 kgCO2e	100.00 wt%
Level of assurance (Third party verification)		*Reasonable level of assurance(BSI)	
Carbon footprint logo *cradle-to-gate		TEEMA Empowered Certification N0. CFP00030	

5. Calculation of inventory analysis

5.1. Emission summary

Total carbon emissions from 561 g bicycle-tire is 1.64 KgCO2e which is composed of the carbon footprint emission 1.27 Kg CO2e from raw materials (ratio 77.43%), the carbon footprint emission 0.3617 Kg CO2e from the energy consumption(ratio 22.05%), and another lower emission.

The unit process emissions comprising the product system are grouped into life-cycle stages, that is, raw material stage, manufacturing stage, transportation stage, the PCF calculation is based on the product life cycle. Table 7 shows the product carbon footprint calculation output.

Table 7. Product carbon footprint calculation.

Category	Raw material	manufacturing	Transportation
Carbon emission per unit (%)	1.347 kgCO2e/unit (82%)	0.295 kgCO2e/unit (18%)	Lower
Total	1.64 kgCO2e/ functional unit (100%)		

5.2. *Inventory analysis*

The carbon footprint has been determined from cradle to gate. The life cycle assessment software Simapro7.3 (as Figure 2 & Figure 3) was used to simulate the GWP factors used in GHG inventories for a 100-year time horizon.

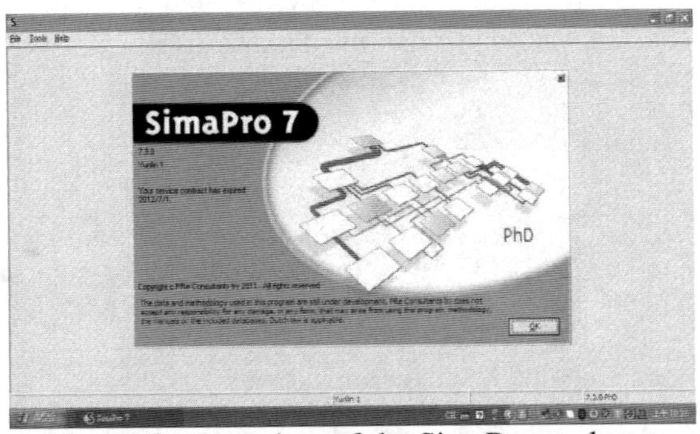

Figure 2. Overview of the SimaPro explorer.

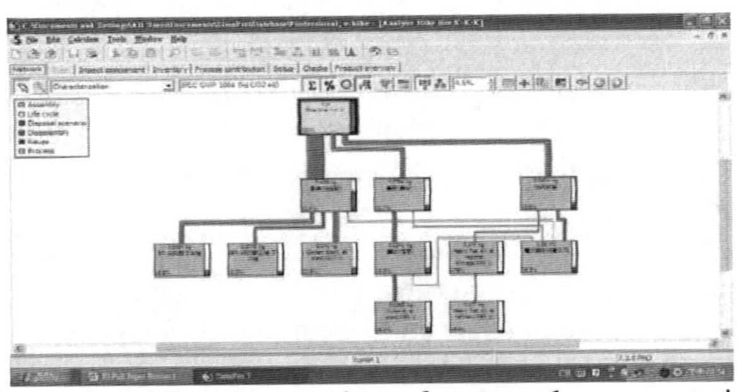

Figure 3. Schematic overview of a network representation

A 561 g bicycle-tire product carbon footprint, total emission is 1.64 KgCO2e, by international certified organization (The British Standard Institute, BSI). The primary activity data includes related facilities from its own processes under the operational control of the organization and its suppliers. The secondary data includes the relative factors from public sources and LCA software Simapro 7.3. TEEMA empowered a carbon footprint logo to certification. This label states that 1.64 KgCO2e is emission per use for the one piece tire. The value of the standard error of mean is especially important to monitor. Method: IPCC2007 GWP 100a (confidence intervals: 95%; unit: kgCO2e). Figure 4 shows bicycle-tire rubber type the Monte Carlo analysis in SimaPro.

IOP Conf. Series: Materials Science and Engineering **644** (2019) 012001 doi:10.1088/1757-899X/644/1/012001

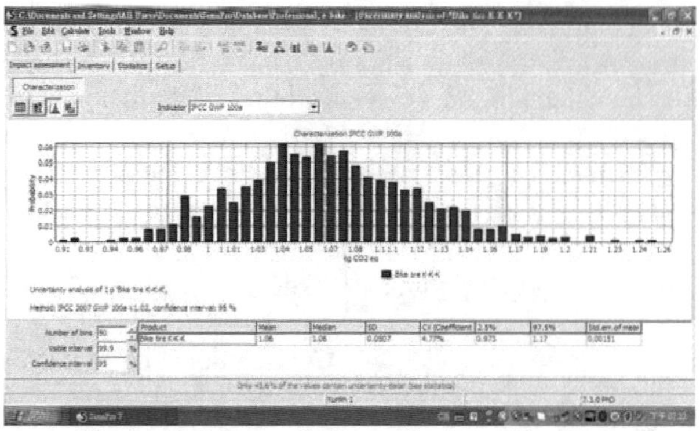

Figure 4. Monte Carlo analysis in SimaPro.

The simulation use over 1000 runs in SimaPro. You are asked to specify the 2XSD value. This is useful as the 95% confidence interval lies between -2xSD and +2xSD. In practice this means that only 2.5% of the data points lie above or below these points, and 95% lie between these points. For obtaining a quick impression, a standard error of mean below 0.01 is quite acceptable. Table 8 shows the simulation standard error data.

Table 8. Standard error data.

95% 1000 runs	Median valuc	Mean	S.D.	Coefficient Variability	2.5%	97.5%	Standard error of mean <0.01
bicycle-tire rubber	1.06	1.06	0.0507	4.77%	0.973	1.17	0.00151

6. Results and discussion

There is very little LCA inventory data on rubber factories in the scientific literature. The results are complete and reveal the single piece rubber-tire of different function units from five factories and 1st-tier suppliers in Taiwan. This paper evaluated the potential impact on the carbon footprint of raw materials, energy consumption and pollutant emissions in the throughput process for 1kg tire at its output life cycle from the "cradle to gate" using the Simapro 7.3 software.

The overall majority of these results carry out product life cycle greenhouse gas emissions verification. These are the opinions of BSI with reasonable assurance.

Certified and Use carbon footprint labels on products are currently being used by a limited number of companies. It is expected that more businesses will seek carbon labeling as Climate Change and Corporate Social Responsibility continues to receive more attention.

Identifying and measuring carbon footprint can be a useful source of carbon competitiveness. This paper is of benefit to academics and managers by providing a detail way to integrate carbon emissions in supply chain management. Managing the carbon footprint of a product across the supply chain is another strategically important next step for businesses to reduce carbon emissions and mitigate climate risks.

More than 1000 SimaPro simulation runs were conducted to obtain a quick impression. A standard error of mean below 0.01 is quite acceptable. The calculation and analysis results show that the material consumption and energy consumption are the main factors that produce the environmental load.

The obtained carbon footprint for the rubber-tire has been 1.64 KgCO2e per bicycle, with the PAS2050:2011(ISO/DIS 14067) standards. That is an international standard for the carbon foot-printing of goods and services across the full product life cycle (BSI,2011). As shows Figure 5.

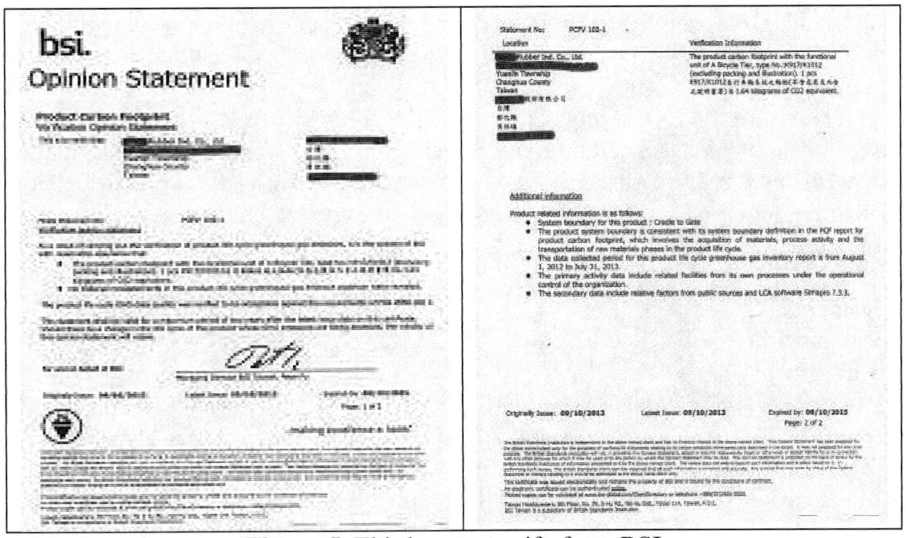

bsi.
Opinion Statement

Statement No: PCFV 102-1

Location	Verification Information
[redacted] Rubber Ind. Co., Ltd. [redacted] Yuanlin Township Changhua County Taiwan [redacted]	The product carbon footprint with the functional unit of A Bicycle Tier, type No. K917/K1012 (excluding packing and illustration). 1 pcs ... is 1.64 kilograms of CO2 equivalent.

Additional information

Product related information is as follows:

- System boundary for this product : Cradle to Gate
- The product system boundary is consistent with its system boundary definition in the PCF report for product carbon footprint, which involves the acquisition of materials, process activity and the transportation of raw materials phases in the product life cycle.
- The data collected period for this product life cycle greenhouse gas inventory report is from August 1, 2012 to July 31, 2013.
- The primary activity data include related facilities from its own processes under the operational control of the organization.
- The secondary data include relative factors from public sources and LCA software Simapro 7.3.3.

Originally Issue: 09/10/2013 Latest Issue: 09/10/2013 Expired by: 08/10/2015

Figure 5. Third-party certify from BSI.

Taiwan Electrical & Electronic Manufacturers' Association (TEEMA) empowered a carbon footprint logo to certify that the Carbon Footprint of these products was independently verified (as Figure 6) to meet all of the necessary requirements under ISO 14025 standard and BSI PAS 2050, with regard to the corresponding Product Category Rules and Life Cycle Analysis results.

Certificate of Achievement

Carbon Footprint Logo

Taiwan Electrical & Electronic Manufacturers' Association (TEEMA)
do hereby certify that an independent verification has been conducted on behalf of :

[redacted] **CO., LTD.**

For the following product(s):

The product carbon footprint with the functional unit of each Bicycle tire, Type No. K917/K1012 is 1.64 kg of CO_2 equivalent

The Carbon Footprint of this product(s) meets all of the necessary requirements under ISO 14025 standard and BSI PAS 2050, with regard to the corresponding Product Category Rules and Life Cycle Analysis results.

Taiwan Electrical & Electronic Manufacturers' Association (TEEMA)

December 10, 2013
Date

Certification No. CFP00030 Valid Until : December 09, 2015

Figure 6. Carbon label from TEEMA certify.

Overall reviews of the Life cycle inventory report and subsequent follow-up interviews have provided BSI with sufficient evidence to determine the stated criteria is fulfilled. The inventory report correctly complies with the PAS2050:2011(ISO/DIS 14067) requirement. The inventory report results in quantification of product life cycle GHG emissions that are real, transparent and measurable.

7. Conclusions

In future, research an energy and climate action plan will be developed with energy-saving opportunities in heat recovery, utilization and environmental improvement. As an example, the thermo-electric heat-pump and heat exchanger techniques will be used to permit industrial heat recovery applications and environmental improvement.

This tire product case study is original and is exclusive to Taiwan local activity data and carbon emission database. The important investigation results are as follows: (1) The carbon footprint of 561g bicycle rubber-tire carbon emissions of 1.64kgCO2e with independent third parties certified our statement results. (2)The error requirement meets the standard error of mean<0.01, and meets the Monte Carlo analysis requirements. (3)This important practical case study completed two major central factories and five satellite suppliers to check the local activity data, equipments energy consumption and carbon footprint emissions, including eight factories inventory checklists. The primary activity data was confirmed by a third party international verification unit BSI. (4)The application and review of the B to B product carbon-label for this bicycle tire product were completed. It can be used for international marketing and brand value added. Track 48 period study found that regression analysis Y=1,592,775+1.483689X, correlation the throughput(X kg) and LNG(Y kcal) can be applied to Boiler-equipment management and energy management analysis. These database and practical results published in this paper can be used as a reference for academics and business circles.

Acknowledgements
This work was supported in part by the NSC under NSC103-313-E-167-001-CC2 (National Science Council of Taiwan)

8. References

[1] A.C. Dias and L. Arroja, Comparison of methodologies for estimating the carbon footprint – case study of office paper, Journal of Cleaner Production 24,30-35 (2011)
[2] BSI, PAS 2050: *Specification for the assessment of the life cycle greenhouse gas emissions of goods and services*, London (2011)
[3] Carbon Trust, *Carbon footprints in the supply chain: the next step for business*, Carbon Trust, London (2006)
[4] IPCC, *IPCC Fourth Assessment Report: Climate Change 2007, the Physical Science Basis, Working Group I report* (2007)
[5] ISO/CD2 14067-1, *Carbon footprint of product – Part 1:Quantification* (2010)
[6] ISO 14064-1,*Greenhouse gases – Part 1: Specification with guidance at the organization level for quantification and reporting of greenhouse gas emissions and removals* (2005)
[7] ISO 14064-3,*Greenhouse gases – Part 3: Specification with guidance for the validation and verification of greenhouse gas assertions* (2005)
[8] J. Morfeldt, W. Nijs , S. Silveira ,The impact of climate targets on future steel production -an analysis based on a global energy system model, Journal of Cleaner Production 103,469-482 (2015)
[9] K.R. Jegannathan, P.H. Nielsen, Environmental assessment of enzyme use in industrial production, Journal of Cleaner Production 42,228-240 (2013)
[10] M.J. Franchetti, D. Apul, *Carbon Footprint analysis: Concepts, Methods, Implementation, and Case Studies* , CRC Press (2012)
[11] M.M. Pang, M.Y. Pun , W.S. Chow , Z.A.M. Ishak , Carbon footprint calculation for thermoformed starch-filled polypropylene biobased materials , Journal of Cleaner Production 64,602-608 (2014)
[12] Yilmaz , A. Anctil , T. Karanfil, LCA as a decision support tool for evaluation of best available techniques(BATs) for cleaner production of iron casting , Journal of Cleaner Production 105,337-347 (2015)
[13] S.P. Dayaratne, K.D. Gunawardana, Carbon footprint reduction: a critical study of rubber production in small and medium scale enterprises in Sri Lanka , Journal of Cleaner Production 103,87-103 (2015)
[14] T.H. Lin, Y.S. Chien, W.M. Chiu, Rubber Tire life cycle assessment and the effect of reducing carbon footprint by replacing carbon black with graphene , Journal of Green Energy 14,97-104 (2016)
[15] V. Siracusa, C. Ingrao, A.L. Giudice, C. Mbohwa, M.D. Rosa, Environmental assessment of a multilayer polymer bag for food packaging and preservation: An LCA approach, Food Research International 62,151-161 (2014)

IOP Conf. Series: Materials Science and Engineering **644** (2019) 012002 doi:10.1088/1757-899X/644/1/012002

Investigation on Optimization of Baking in Sole Bonding Process

Dyi-Cheng Chen, Hua-Wei Jin, and Jhang-Zong Su

Department of Industrial Education and Technology, National Changhua University of Education, No.1, Jinde Rd., Changhua City, Changhua County 50007, Taiwan
E-mail: dcchen@cc.ncue.edu.tw

Abstract. Nowadays, Taiwan's footwear industry has become more computerized and automated, but there are still places in the manufacturing process that can be optimized. In the general shoe bonding process, baking requires very high power and a long time. This Investigation used Solidworks to build a common shoe model, and used Solidworks simulation and other CAE software to discuss the baking soles and achieve the soles. The heat required for t optimize the parameters he temperature of the work is bonded, and through the TRIZ theory to optimize the parameters such as power and time required for the baking process, it can be heated by asymmetric heating to obtain a surface with relatively uniform heating to obtain a solution that effectively reduces the production cost in the process. It is found that when the heating source is separated by 40 mm, a surface with relatively uniform heat can be obtained.

1. Introduction

Although there are many computer-aided equipment and automated machinery, not only can the shoemaking time be greatly shortened, but also the customized shoemaking process can be developed. However, the materials of the shoes are very diverse, and because of the development of customization, each pair of shoes, the size and thickness of the outsole and the midsole are not the same. At this time, it is difficult to control the temperature and time when the midsole and the outsole are bonded. The problem in this aspect is that the engineer or the field master uses the trial and error method. The test is constantly repeating to find the right temperature and time to solve the problem, so it will waste a considerable amount of time in this part of the test, but also many resources are not easy to reduce the cost.

Common materials for the sole are rubber (RUBBER), thermoplastic rubber (TPR), ethylene (EVA), and synthetic resin (PVC). In 2005, Pern and Jorgensen [1] studied adhesion of EVA laminates to glass substrates and enhanced the adhesion strength of EVA. In 2010, Saboo [2] studied bonding strength of EVA modified mortar was tested. In 2011, Ethylene/Vinyl Acetate Copolymers (EVA) for encapsulation of photovoltaic modules were investigated by the thermal analysis methods of Differential Scanning Calorimetry (DSC) and Dynamic Mechanical Analysis (DMA) in the temperature range from −150 °C to 200 °C [3].

In 2012, Alsaed and Jalham [4] investigated polyvinyl butyral (PVB) and vinyl acetate (EVA) as bonding materials on laminated glass on mechanical behaviour. In 2013, Polansky et al. [5] studied the mechanical behaviour and the thermal stability of an encapsulant based on ethylene-vinyl acetate (EVA). The EVA properties were verified at temperatures ranging from −70 °C to 500 °C. In 2016, Saboo and Kumar [6] studied obtaining the optimum blending temperature, time and shear rate for EVA co-polymer.

2. Research method and procedures

2.1. Research method

In order to achieve the purpose of research, this study uses computer-aided engineering analysis (CAE) technology to simulate the situation when the outsole and midsole of the shoes are bonded, and then use TRIZ theory to find out ways to improve the production capacity of the footwear industry, and to do a series of simulation analysis and optimization of experiments and parameters.

2.2. Research procedures

First set the research process to use the relevant parameters, and then use the heat transfer related formula to calculate the time required for the soles to be baked, so that the bonding surfaces reach the specified temperature, and finally use the Solidworks drawing software to draw the outsole and the midsole model, and open the simulation analysis module of Solidworks to simulate the state of the shoe outsole and midsole heating process, verify that the time required for heat to reach the specified temperature is consistent, and then use TRIZ theory to find a solution to the problem.

2.3. Parameter setting

In order to make the results close to the needs of the site, this study investigated the marketed machines and adhesives on the market, and set parameters that meet the site requirements showing Table 1 for analysis.

Table 1. Various parameter settings.

Item	value
The size of the baking machine	4500*760*730mm
Baking machine thermal power	20kw
Number of soles placed in the baking machine	150 piece
Average maximum thermal power per shoe sole	133w/piece
Hot melt adhesive working temperature	175°C
Sole size	37
Room temperature	27°C
Midsole material	EVA
Outsole material	TPR

2.4. Calculation of heat transfer

Heat transfer is the process of transferring heat from high temperature to low temperature. There are three ways of heat transfer [7]:

2.5. Heat conduction

It is the process of transferring heat from high temperature to low temperature. It is the result of the vibration energy transmitted by one molecule to another. Formula:

$$T = \mu[\Delta v + (\nabla v)^T] - \rho I \tag{1}$$

$$q_h = -k\nabla T \tag{2}$$

2.6. Thermal convection

It refers to the relative displacement between various parts of the fluid caused by the movement of the fluid, and the heat transfer process caused by the mixing of the hot and cold fluids. Average convection formula:

$$q = \bar{h}A_s(T_s - T_\infty) \quad (3)$$

2.7. Thermal radiation

It radiates heat directly through electromagnetic radiation. The conduction speed depends on the absolute temperature of the heat source, and the higher the temperature, the stronger the radiation. Formula:

$$P_{rad} = \in \sigma T^4 \quad (4)$$

The unit of Heat Transfer Coefficient (k) is W/m.K, and the thermal conductivity formula is k = (Q/t) *L/(A*T), which is used to measure the heat energy that can be transmitted per unit time [8]. Using the formula to calculate the temperature of the sole at room temperature after baking in a 20kw oven, how long does it take to make the bonding surface reach 175 °C.

2.8. Computer aided design (CAD) software

CAD (Computer Aided Design) refers to the use of computers for design and drawing, that is, the use of computer power to design products. Therefore, CAD is an important technology in automation, affecting industrial productivity and quality. The introduction of this technology enables people in related industries to predict and verify the design's performance on the computer in advance, reduce the cost and time-consuming waste of actual trial and error on the spot, and point out the potential of design and production in a scientific way. problem.

SolidWorks software is often used in small and medium-sized product development, with strong features, good compatibility, and easy access to the interface. It is used for exterior design, mechanism design, drawing engineering drawings, and structural analysis. It is a software that can combine design end and production end. Software usage is the most powerful indicator of software functionality, user satisfaction, and future viability. SolidWorks has a high market share in the design and manufacturing of mechanical equipment.

2.9. Process of bonding the outsole and midsole of the shoe

The final procedure for making shoes is to glue the outsole and the midsole in a glued manner. Glue the already produced outsole and midsole by spraying or rolling hot melt adhesive. After baking, the outsole and the midsole can be aligned and pressed together to complete the outsole. Bonding action is with the midsole. Common bonding methods are cast rubber (C-PUR) and thermoplastic (T-PUR). Casting gluethe-linear liquid polymer is synthesized by a two-step or one-step method using a compounding agent such as a polyhydric alcohol, an isocyanate or a chain extender, and is poured into a mold during processing, and heated and matured to be converted into a rubbery shape having a certain network structure solid. This method has many advantages and is developing at a faster rate.

Thermoplastic type-the reaction of a polyol and an isocyanate to form a linear polymer, processed into a granular solid, and thermoplastic. Direct production using thermoplastic processing equipment and work procedures finished product. This method is the fastest growing in polyurethane rubber. The disadvantages of this type of rubber are poor oil resistance and heat resistance, and mechanical properties are inferior to C-PUR. Characteristics of urethane rubber included strong anti-wear performance, excellent mechanical and mechanical properties, large hardness range, wide modulus range, good oil and low temperature resistance, excellent radiation and ozone resistance.

2.10. Material of the sole

The sole is the core of the shoe, the sole will directly affect the ability of the shoe, the sole of different materials will have different friction coefficient, hardness, weight, durability and so on. The sole is the only part of the whole shoe that will directly touch the ground. In different occasions and different grounds, it is necessary to match the sole of the material to fully realize the ability of the scorpion. Common materials are:

2.10.1. Rubber

This type of outsole material has excellent anti-slip properties and used to be one of the most commonly used materials for special shoes such as rock climbing shoes. The rubber is divided into natural and synthetic, and the supply of natural rubber is not enough for the shoe factory, so the synthetic rubber is the main. The nature of rubber is wear resistance, good elasticity, heavy weight and high cost. Most of today's shoes use such materials for outsole production. In terms of environmental protection, rubber finished materials have various recycling technologies, such as grinding powder into a part of the rubber sole, which is a bit like the concept of recycled paper pulp. Many brand manufacturers specify certain types of shoes to be filled with a certain proportion. Recycling rubber, in addition to reducing the cost of shoe factories, also contributes to environmental protection.

2.10.2 Thermoplastic rubber (TPR)

The full name of TPR is Thermo, Plastic and Rubber to be a thermoplastic composite material with soft touch and adjustable physical properties and hardness. It also reduces production costs. It has both the easy processing properties of thermoplastics and the characteristics of thermosetting rubber and is non toxic. It is one of the trend materials that save energy and respond to green environmental protection. The resilience, wear resistance, slip resistance and shock absorption of thermoplastic rubber are better than those of ordinary rubber. Therefore, if it is superior to general rubber in terms of comfort, the disadvantage is that the material is inherently stretched and resistant to fatigue. It is not as good as ordinary rubber.

2.10.3 Ethylene (EVA)

It has good flexibility, rubber-like elasticity, and good flexibility at low temperatures. The material itself has high transparency, good surface gloss, good chemical stability, high antioxidant strength, and is as non-toxic as TPR. Because of its light weight, it is the largest amount of material used in sports shoes and casual shoes. Its slip resistance, abrasion resistance and weather resistance can be improved by adding modifiers or other polymer materials.

2.10.4. Synthetic resin (PVC)

It has the advantages of other plastic materials, and it is quite excellent in processing and plasticity. In addition, it is very popular because of its low price, and the production of soles is ecologically friendly. The problem is now rarely used. PVC is generally directly injected into the factory. The texture can be adjusted by pre-injection molding. It can be hard or soft, but the sole of PVC usually has a bit odor and is not popular with consumers.

2.11. Theory of inventive problem solving, TRIZ

TRIZ is derived from the systematic innovation and practice of the Soviet inventor Genrich Altshuller's research on 200,000 patents to solve problems. Emphasis on invention or innovation can be carried out according to certain procedures and steps, rather than just random or imaginative brain stimulation, that is, extracting the principles invented by the predecessors and summarizing them into general rules, thinking direction for solving problems later.

3. Results and discussions

Take the simple size of the store as the object, measure the size of each size of the foot, and complete the model of the outsole and midsole showing Fig. 1, and use Solidworks 2016, the simulation

IOP Conf. Series: Materials Science and Engineering **644** (2019) 012002 doi:10.1088/1757-899X/644/1/012002

analysis module simulates the situation when the outsole and the midsole of the shoe are bonded showing Fig. 2, and the entire bonding surface must reach the working temperature (175 °C) or more. The simulation results show that the center temperature of the outsole and the midsole is 271°C, and the boundary temperature is 176°C showing Fig. 3. The temperature difference between the center and the boundary is 95° C, and the temperature difference is quite large. Heating the boundary temperature above the operating temperature (175°C) must be baked at a heat power of 57 W for 20 seconds.

Figure 1. Model of shoe midsole.

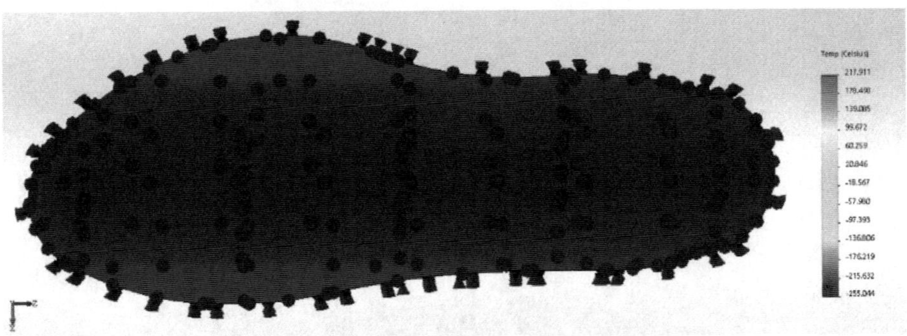

Figure 2. Heat transfer simulation of shoe midsole.

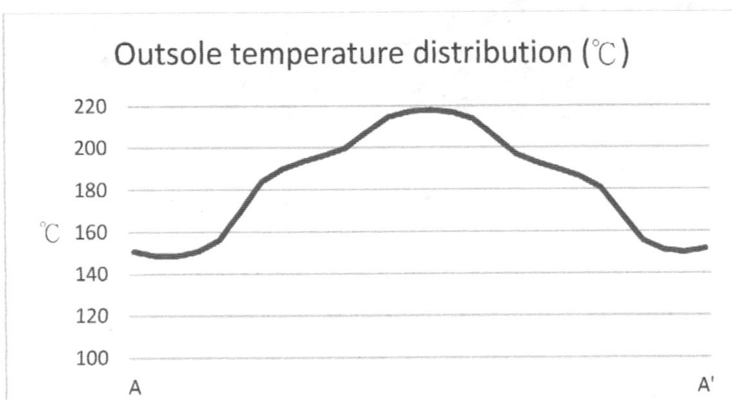

Figure 3. Outsole temperature distribution.

Modify the heating method of the outsole and the midsole, Fig. 4 shows the original heating method, Fig. 5 shows the improved heating method to modify the problem of uneven heating to shorten the baking time and increased bond strength. Fig. 6 shows the results of the modified simulated baking. After modifying the heating mode, the center temperature was 210°C and the boundary temperature was 177°C showing Fig. 7. The temperature difference between the center and the boundary was 33°C.

IOP Conf. Series: Materials Science and Engineering **644** (2019) 012002 doi:10.1088/1757-899X/644/1/012002

And use the optimized design function of Solidworks soft body to find the most suitable parameters such as baking temperature and time in Table 2.

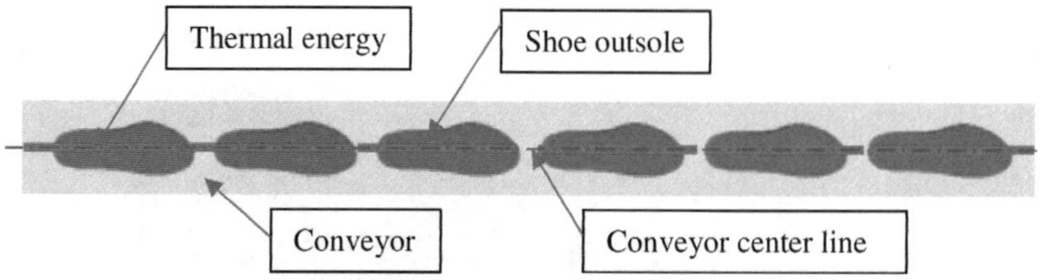

Figure 4. Original heating method.

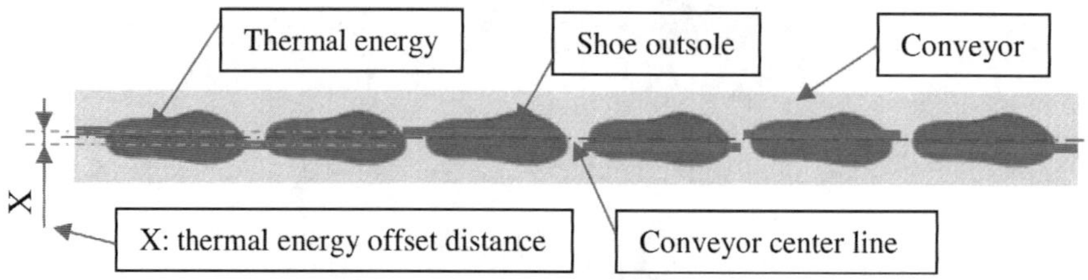

Figure 4. Improved heating method.

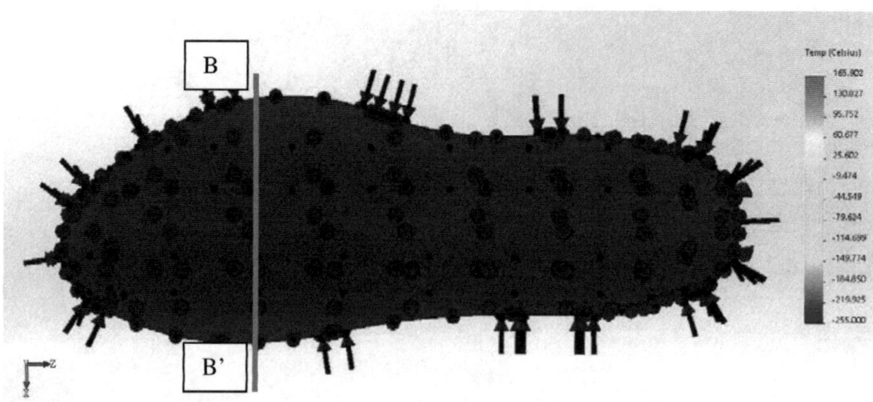

Figure 5. Mid-floor optimized heat transfer analysis.

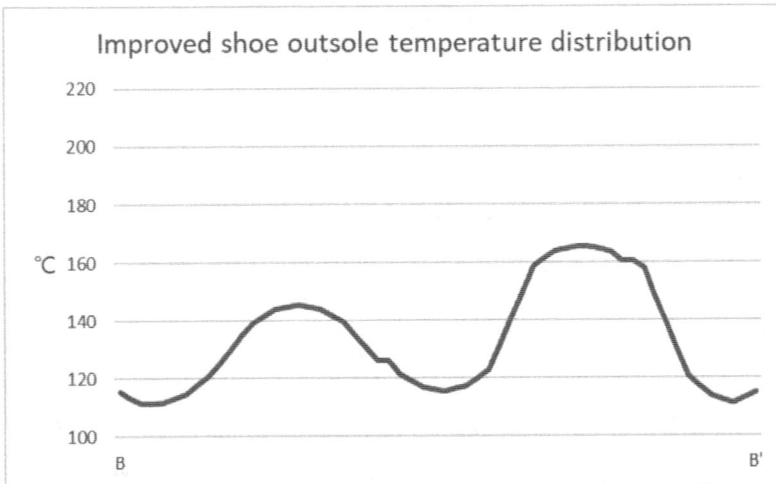

Figure 6. Improved shoe outsole temperature distribution.

Table 2. Optimized design experiment.

Case	Material	Thermal power (w)	Heating method	X value	Maximum temperature (°C)	Minimum temperature (°C)	Temperature difference (°C)
Case 1	EVA	500	original	0	381	108	273
Case 2	EVA	500	Improvement	10	337	108	229
Case 3	EVA	500	Improvement	20	333	143	190
Case 4	EVA	500	Improvement	30	219	147	72
Case 5	EVA	500	Improvement	40	163	133	30
Case 6	EVA	500	Improvement	50	198	144	54
Case 7	EVA	500	Improvement	60	183	137	46

Fig. 8 shows the line graph drawn by the temperature difference of each item in Table 2, when the X value is equal to 40, the obtained temperature difference is the smallest, and a surface having a relatively uniform heat can be obtained.

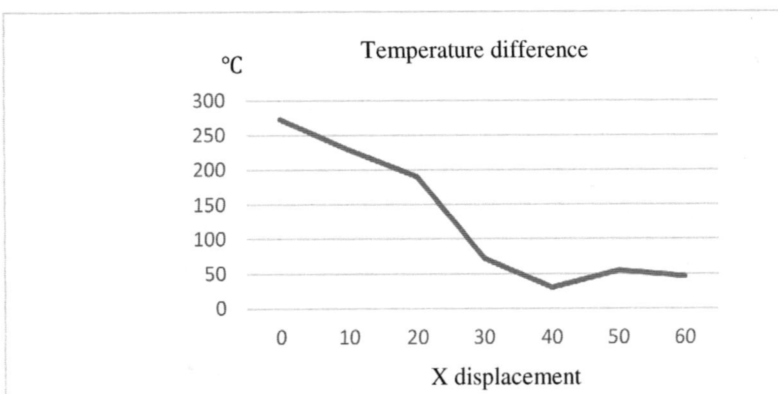

Figure 7. Temperature difference line chart.

4. Conclusions

This Investigation used Solidworks to build a common shoe model, and used Solidworks simulation and other CAE software to discuss the baking soles and achieve the soles. From this simulation

IOP Conf. Series: Materials Science and Engineering **644** (2019) 012002 doi:10.1088/1757-899X/644/1/012002

analysis are found: In order to solve the problem of uneven heating of the sole, it is possible to perform heating by means of asymmetrical heating, and it is found that when the heating source is separated by 40 mm, a surface with relatively uniform heat and increased bond strength can be obtained.

Acknowledgement
The authors gratefully acknowledge the financial support of the Ministry of Science and Technology, Taiwan under Grant No. MOST 107-2813-C-018-019-E.

5. References

[1] F.J. Pern, G.J. Jorgensen, Enhanced adhesion of EVA laminates to primed glass substrates subjected to damp heat exposure, Conference Paper in Conference Record of the IEEE Photovoltaic Specialists Conference, 1, 1-7 (2005)
[2] N. Saboo, P. Kumar, Effect of curing regime on bonding strength of EVA modified mortar to tile, Journal of Wuhan University of Technology-Mater. Sci. Ed. 25, Issue 2, 346–348 (2010)
[3] W. Stark, M. Jaunich, Investigation of ethylene/vinyl acetate copolymer (EVA) by thermal analysis DSC and DMA, Polymer Testing, 30, 236-242 (2011)
[4] O. Alsaed, Issam S. Jalham, Polyvinyl Butyral (PVB) and Ethyl Vinyl Acetate (EVA) as a binding material for laminated glass, JJME, **6**, 127-133 (2012)
[5] R. Polansky, M. Pinkerova, M. Bartůňkova, P. Prosr, The Journal of Slovak University of Technology, 64, 361-365 (2013)
[6] N. Saboo, P. Kumar, Optimum blending requirements for EVA modified binder, Transportation Research Procedia, 17, 98-106 (2016)
[7] D.K. Nguyen and J.Y. San, Heat transfer and energy analysis of a spiral heat exchanger, Heat Transfer Engineering, 37, 1013-1026 (2016)
[8] B.W. Li and Z.R. Chang, Comparisons of solving procedures and computation time for fast cosine transformation and matrix multiplication transformation in spectral methods application, Numerical Heat Transfer, Part A: Applications, 70, 384-398, (2016)

Numerical Optimization of Rectangular Mufflers with Multi-channel Splitters Using the Simulated Annealing Method

Ho-Chih Cheng[1] and Min-Chie Chiu[2]

1 Department of Intelligent Automation Engineering, Chung Chou University of Science and Technology, No. 6, Lane 2, Sec. 3, Shanjiao Rd., Yuanlin City, Changhua County 510, Taiwan
2 Department of Electrical Engineering and Energy Technology, Chung Chou University of Science and Technology, No. 6, Lane 2, Sec. 3, Shanjiao Rd., Yuanlin City, Changhua County 510, Taiwan
E-mail: minchie.chiu@msa.hinet.net

Abstract. Rectangular mufflers internally hybridized with splitters have been extensively applied in industrial noise abatement. However, there has been a palpable lack of academic work directed toward space-constrained mufflers conjugated with multi-channel splitters that disperse venting fluid and reduce secondary noise. That being so, an analysis of the Sound Transmission Loss (STL) of rectangular mufflers internally hybridized with multiple parallel splitters that are optimally designed to perform within a limited space will be considered, here. By using an acoustical lumped method, a four-pole system matrix for evaluating the acoustic performance (STL) emerges. During the optimization process, the simulated annealing (SA) method, which is a robust scheme utilized to search for the global optimum by imitating a physical annealing process, is used. Before dealing with a broadband noise, the STL's maximization relative to a one-tone noise (250Hz) is offered to confirm the SA method's reliability. Moreover, the mathematical model is also checked for accuracy. To appreciate the influence of acoustical efficiency with respect to the design parameters, the sensitivity analysis of six design parameters (dh: the diameter of a perforated hole; W: the width of air channel; R: acoustic flow resistivity of the acoustic fiber; σ: the porosity of the perforated plate; Df: the thickness of the acoustic fiber; L1: the horizontal length of the splitter) is performed. Subsequently, to bring into focus the acoustical interaction with respect to the number of air-channels (between the splitters), three types of mufflers (mufflers A~C) hybridized with one, two, and four air-channels (with parallel splitters) have been surveyed. Results divulge that for a rectangular muffler internally conjugated with a one-channel splitter, the maximal STL is located at the desired tone. The acoustical performance of a rectangular muffler will increase and the induced back pressure will decrease simultaneously if the number of the splitters internally conjugated within the rectangular muffler increases. Consequently, optimally designed rectangular mufflers with multiple parallel splitters that avoid secondary noise induced by high speed flow while simultaneously maximizing acoustical performance within a constrained space are preferable.

1. Nomenclature

This paper is constructed on the basis of the following notations:
c_o: sound speed (m s^{-1})
dh : the diameter of a perforated hole on the front perforated plate (m)

IOP Conf. Series: Materials Science and Engineering **644** (2019) 012003 doi:10.1088/1757-899X/644/1/012003

D_f : the thickness of the acoustic fiber
L_i : diameter of the i-th horizontal segment of the muffler (m)
$d_{1,2}$: diameter of the inlet/outer tube (m)
f : cyclic frequency (Hz)
H : the thickness of the splitter (m)
iter : maximum iteration
j : imaginary unit
k : wave number ($=\omega/c_o$)
k_1 : real part of complex $k_{fiber-i}$
k_2 : image part of complex $k_{fiber-i}$
k_{fiber} : complex propagation constant of the acoustic fiber
kk : cooling rate in *SA*
L_o : total length of the muffler (m)
L_1 : the horizontal length of the splitter (m)
L_2 : the height of the splitter (m)
M : mean flow Mach number
OBJ : objective function (dB)
p_i : acoustic pressure at the i-th node (Pa)
pb(T) : transition probability
Q : volume flow rate of venting gas ($m^3 s^{-1}$)
R : acoustic flow resistivity of the acoustic fiber (MKS rayls m^{-1})
R_{fiber} : real part of the complex Z_{fiber}
S_i : section area at the i-th node(m^2)
STL: sound transmission loss (dB)
SWLO : unsilenced sound power level inside the muffler's inlet (dB)
SWL_T : overall sound power level inside the muffler's output (dB)
q : the thickness of the front perforated plate in the splitter (m)
TC_{ij} : components of four-pole transfer matrices for an acoustical mechanism with sudden-contracted ducts
TE_{ij} : components of four-pole transfer matrices for an acoustical mechanism with sudden-expanded ducts
TS_{ij} : components of four-pole transfer matrices for an acoustical mechanism with straight ducts
TSP_{ij} : components of a four-pole transfer matrix for an acoustical mechanism with a one-channel splitter
T : current temperature (^oC)
T_{ij}^* : components of a four-pole transfer system matrix
T_o : initial temperature (^oC)
u_i : acoustic particle velocity at the i-th node ($m\ s^{-1}$)
W : the width of air channel (m)
Z_i : specific normal impedance at i.
Z_{fiber} : characteristic impedance of the acoustic fiber
Z_p : characteristic impedance of the perforated front plate
X_{fiber} : image part of the complex Z_{fiber}
ρ_o : air density (kg m^{-3})
ρ_i : acoustical density at the i-th node (kg m^{-3})
ν : kinematic viscosity of air ($=15*10^{-6}\ m^2/s$)
σ : the porosity of the perforated plate (m)
δ : viscous boundary layer thickness of the i-th layer of the perforated plate (m)
ω : angular velocity ($=2\pi f$)

2. Introduction

Morse, in 1939 [1], began research on mufflers that reduced high frequency noise using a dissipative duct (a duct lined with sound absorbing material). Later, Scott [2] analyzed the sound transmission of circular and rectangular mufflers internally lined with porous material using bulk reactive model. Ko [3], in 1975, investigated the sound transmission loss in acoustically lined flow ducts separated by porous splitters. The research mentioned above [1-3] focused on the sound attenuation of an infinite duct. Cummings and Chang [4] then, in 1988, analyzed a finite length dissipative flow duct silencer with an internal mean flow in the absorbent using a modal method. Peat [5], addressing the volume modulus, adopted a transfer matrix for evaluating the acoustical performance of an absorption silencer element in 1991. Subsequently, using a one-dimensional analytical method and a three-dimensional boundary element method (BEM), Selamet *et al.* [6, 7] assessed the acoustical attenuation for perforated concentric absorbing silencers and hybrid silencers. In 2003, Munjal [8] analyzed pod silencers using a four-pole transfer matrix. Then, in 2004, Xu *et al.* [9] proposed a characteristic equation to calculate the sound attenuation in dissipative expansion chambers. However, an assessment of a muffler's optimal design within a limited space was seldom addressed. Therefore, Chiu [10] has examined shaped optimization of a circular muffler internally lined with sound absorbing material within a constrained space. Later, Chiu [11] assessed circular mufflers internally equipped with multi-channel splitters using neural networks in conjunction the boundary element method and a genetic algorithm. However, the study is limited in the optimization assessment for pure tone noise. In order to develop a space-constrained rectangular muffler that is internally equipped with multi-channel splitters used in reducing broadband noise, a mathematical mode of a rectangular muffler internally equipped with multiple splitters is proposed using an acoustical lump analysis. Three types of rectangular mufflers linked with multiple splitters (muffler A: a muffler equipped with a one-channel splitter; muffler B: a muffler equipped with two-channel splitters; muffler C: a muffler equipped with four-channel splitters) are presented. It should also be noted that the acoustical lump method used to form a four-pole system matrix is compatible with the simulated annealing method.

3. Theoretical background

Three types of rectangular mufflers internally connected with multiple parallel splitters have been adopted for noise elimination in the steam turbine system shown in Fig. 1. Before the acoustical fields of the mufflers were analyzed, the acoustical elements had been identified. As shown in Fig. 2, four types of muffler components that included four straight ducts (points 1~2, points 3~4, points 7~8, and points 9~10), two sudden expanded ducts (points 2~3, points 6~7), two sudden contracted ducts(points 4~5, points 8~9), and one multiple parallel splitter element (points 5~6) are identified. Additionally, the acoustical field within the muffler is represented by ten points. As derived in previous work [12, 13, 14] and shown in appendices A~C, individual transfer matrices with respect to straight ducts, dissipative ducts, and sudden expanded/contracted ducts are described below.

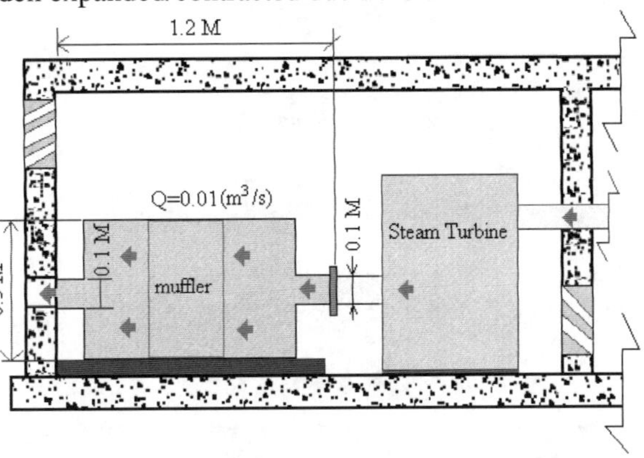

Figure 1. Noise elimination on a steam turbine within a limited space.

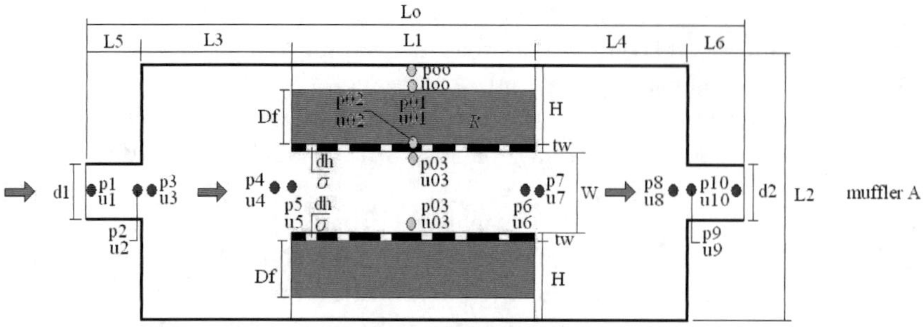

(A) Muffler A

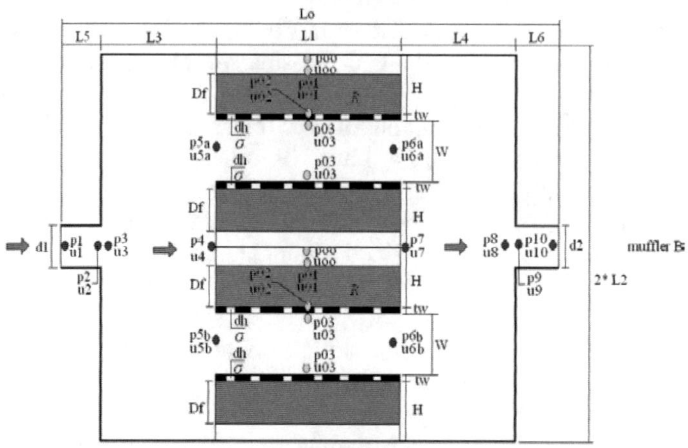

(B) Muffler B

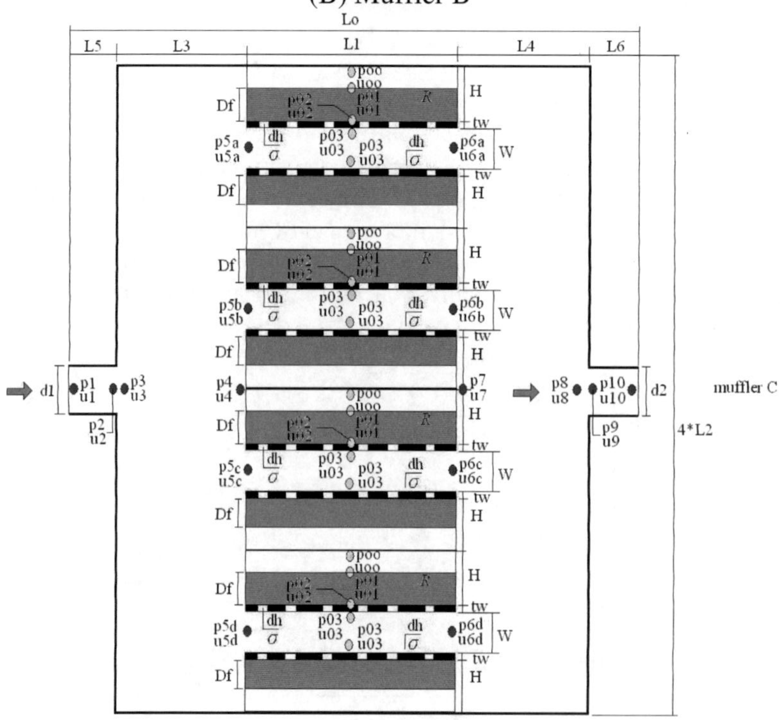

(C) Muffler C

Figure 2. Acoustical elements in three types of rectangular mufflers internally equipped with parallel splitters (muffler A ~ muffler C).

3.1. Muffler A (a rectangular muffler internally connected with a one-channel splitter)

As indicated in Fig. 2, for the acoustical straight element, the four-pole matrix between nodes 1 and 2 is [12, 13, 14]

$$\begin{pmatrix} p_1 \\ \rho_o c_o u_1 \end{pmatrix} = f_1(L_5, d_1, M_1) \begin{bmatrix} TS1_{1,1} & TS1_{1,2} \\ TS1_{2,1} & TS1_{2,2} \end{bmatrix} \begin{pmatrix} p_2 \\ \rho_o c_o u_2 \end{pmatrix} \tag{1}$$

For the acoustical sudden-expanded element, the four-pole matrix between nodes 2 and 3 is [12, 13, 14]

$$\begin{bmatrix} p_2 \\ \rho_o c_o u_2 \end{bmatrix} = \begin{bmatrix} TE1_{1,1} & TE1_{1,2} \\ TE1_{2,1} & TE1_{2,2} \end{bmatrix} \begin{bmatrix} p_3 \\ \rho_o c_o u_3 \end{bmatrix} \tag{2}$$

Similarly, the four-pole matrix between nodes 3 and 4 is

$$\begin{pmatrix} p_3 \\ \rho_o c_o u_3 \end{pmatrix} = f_2(L_3, L_2, M_2) \begin{bmatrix} TS2_{1,1} & TS2_{1,2} \\ TS2_{2,1} & TS2_{2,2} \end{bmatrix} \begin{pmatrix} p_4 \\ \rho_o c_o u_4 \end{pmatrix} \tag{3}$$

For the acoustical sudden-contracted element, the four-pole matrix between nodes 4 and 5 is [12, 13, 14]

$$\begin{bmatrix} p_4 \\ \rho_o c_o u_4 \end{bmatrix} = \begin{bmatrix} TC1_{1,1} & TC1_{1,2} \\ TC1_{2,1} & TC1_{2,2} \end{bmatrix} \begin{bmatrix} p_5 \\ \rho_o c_o u_5 \end{bmatrix} \tag{4}$$

As derived in Appendix A, for a one-channel splitter, the four-pole matrix between nodes 5 and 6 yields

$$\begin{bmatrix} p_5 \\ \rho_o c_o u_5 \end{bmatrix} = \begin{bmatrix} TSP1_{1,1} & TSP1_{1,2} \\ TSP1_{2,1} & TSP1_{2,2} \end{bmatrix} \begin{bmatrix} p_6 \\ \rho_o c_o u_6 \end{bmatrix} \tag{5}$$

Likewise, the four-pole matrix between nodes 6 and 7 for a sudden-expanded duct is

$$\begin{bmatrix} p_6 \\ \rho_o c_o u_6 \end{bmatrix} = \begin{bmatrix} TE2_{1,1} & TE2_{1,2} \\ TE2_{2,1} & TE2_{2,2} \end{bmatrix} \begin{bmatrix} p_7 \\ \rho_o c_o u_7 \end{bmatrix} \tag{6}$$

The four-pole matrix between nodes 7 and 8 in a straight duct is

$$\begin{pmatrix} p_7 \\ \rho_o c_o u_7 \end{pmatrix} = f_3(L_2, L_4, M_3) \begin{bmatrix} TS3_{1,1} & TS3_{1,2} \\ TS3_{2,1} & TS3_{2,2} \end{bmatrix} \begin{pmatrix} p_8 \\ \rho_o c_o u_8 \end{pmatrix} \tag{7}$$

The four-pole matrix between nodes 8 and 9 for a sudden-contracted duct is

$$\begin{bmatrix} p_8 \\ \rho_o c_o u_8 \end{bmatrix} = \begin{bmatrix} TC2_{1,1} & TC2_{1,2} \\ TC2_{2,1} & TC2_{2,2} \end{bmatrix} \begin{bmatrix} p_9 \\ \rho_o c_o u_9 \end{bmatrix} \tag{8}$$

Moreover, the four-pole matrix between nodes 9 and 10 in a straight duct is

$$\begin{pmatrix} p_9 \\ \rho_o c_o u_9 \end{pmatrix} = f_4(L_6, d_2, M_4) \begin{bmatrix} TS4_{1,1} & TS4_{1,2} \\ TS4_{2,1} & TS4_{2,2} \end{bmatrix} \begin{pmatrix} p_{10} \\ \rho_o c_o u_{10} \end{pmatrix} \tag{9}$$

The total transfer matrix assembled by multiplication is

$$\begin{pmatrix} p_1 \\ \rho_o c_o u_1 \end{pmatrix} = f_1(L_5, d_1, M_1) f_2(L_2, L_3, M_2) f_3(L_2, L_4, M_3) f_4(L_6, d_2, M_4)$$

$$\begin{bmatrix} TS1_{1,1} & TS1_{1,2} \\ TS1_{2,1} & TS1_{2,2} \end{bmatrix} \begin{bmatrix} TE1_{1,1} & TE1_{1,2} \\ TE1_{2,1} & TE1_{2,2} \end{bmatrix} \begin{bmatrix} TS2_{1,1} & TS2_{1,2} \\ TS2_{2,1} & TS2_{2,2} \end{bmatrix} \begin{bmatrix} TC1_{1,1} & TC1_{1,2} \\ TC1_{2,1} & TC1_{2,2} \end{bmatrix} \tag{10}$$

$$\begin{bmatrix} TSP1_{1,1} & TSP1_{1,2} \\ TSP1_{2,1} & TSP1_{2,2} \end{bmatrix} \begin{bmatrix} TE2_{1,1} & TE2_{1,2} \\ TE2_{2,1} & TE2_{2,2} \end{bmatrix} \begin{bmatrix} TS3_{1,1} & TS3_{1,2} \\ TS3_{2,1} & TS3_{2,2} \end{bmatrix} \begin{bmatrix} TC2_{1,1} & TC2_{1,2} \\ TC2_{2,1} & TC2_{2,2} \end{bmatrix}$$

$$\begin{bmatrix} TS4_{1,1} & TS4_{1,2} \\ TS4_{2,1} & TS4_{2,2} \end{bmatrix} \begin{pmatrix} p_{10} \\ \rho_o c_o u_{10} \end{pmatrix}$$

A simplified form is expressed in a matrix as

$$\begin{pmatrix} p_1 \\ \rho_o c_o u_1 \end{pmatrix} = \begin{bmatrix} T_{11}^* & T_{12}^* \\ T_{21}^* & T_{22}^* \end{bmatrix} \begin{pmatrix} p_{10} \\ \rho_o c_o u_{10} \end{pmatrix} \tag{11}$$

3.2. Muffler B (a rectangular muffler internally connected with two-channel splitters)

Similarly, the acoustical four-pole matrix between points 1~2, points 2~3, points 3~4, points 4~5, points 6~7, points 7~8, points 8~9, and points 9~10 is the same as Eqs.(1)~(4) and Eqs.(6)~(9).

As derived in Appendix B, for two parallel air-channel splitters, the four-pole matrices between nodes 5a~6a and nodes 5b~6b can be combined into an equivalent matrix

$$\begin{bmatrix} p_5 \\ \rho_o c_o u_5 \end{bmatrix} = \begin{bmatrix} TSP1_{1,1} & \frac{1}{2}\cdot TSP1_{1,2} \\ 2\cdot TSP1_{2,1} & TSP1_{2,2} \end{bmatrix} \begin{bmatrix} p_6 \\ \rho_o c_o u_6 \end{bmatrix} \tag{12}$$

where $[TSP1_{i,j}]$ is the four-pole matrix for a one-channel splitter. Consequently, the total transfer matrix assembled by multiplication is

$$\begin{pmatrix} p_1 \\ \rho_o c_o u_1 \end{pmatrix} = f_1(L_5,d_1,M_1) f_2(L_2,L_3,M_2) f_3(L_2,L_4,M_3) f_4(L_6,d_2,M_4)$$
$$\begin{bmatrix} TS1_{1,1} & TS1_{1,2} \\ TS1_{2,1} & TS1_{2,2} \end{bmatrix} \begin{bmatrix} TE1_{1,1} & TE1_{1,2} \\ TE1_{2,1} & TE1_{2,2} \end{bmatrix} \begin{bmatrix} TS2_{1,1} & TS2_{1,2} \\ TS2_{2,1} & TS2_{2,2} \end{bmatrix} \begin{bmatrix} TC1_{1,1} & TC1_{1,2} \\ TC1_{2,1} & TC1_{2,2} \end{bmatrix} \tag{13}$$
$$\begin{bmatrix} TSP1_{1,1} & \frac{1}{2}\cdot TSP1_{1,2} \\ 2\cdot TSP1_{2,1} & TSP1_{2,2} \end{bmatrix} \begin{bmatrix} TE2_{1,1} & TE2_{1,2} \\ TE2_{2,1} & TE2_{2,2} \end{bmatrix} \begin{bmatrix} TS3_{1,1} & TS3_{1,2} \\ TS3_{2,1} & TS3_{2,2} \end{bmatrix} \begin{bmatrix} TC2_{1,1} & TC2_{1,2} \\ TC2_{2,1} & TC2_{2,2} \end{bmatrix}$$
$$\begin{bmatrix} TS4_{1,1} & TS4_{1,2} \\ TS4_{2,1} & TS4_{2,2} \end{bmatrix} \begin{pmatrix} p_{10} \\ \rho_o c_o u_{10} \end{pmatrix}$$

A simplified form is expressed in a matrix as

$$\begin{pmatrix} p_1 \\ \rho_o c_o u_1 \end{pmatrix} = \begin{bmatrix} T_{11}^{**} & T_{12}^{**} \\ T_{21}^{**} & T_{22}^{**} \end{bmatrix} \begin{pmatrix} p_{10} \\ \rho_o c_o u_{10} \end{pmatrix} \tag{14}$$

3.3. Muffler C (a rectangular muffler internally connected with three-channel splitters)

Likewise, the acoustical four-pole matrix between points 1~2, points 2~3, points 3~4, points 4~5, points 6~7, points 7~8, points 8~9, and points 9~10 is the same as Eqs.(1)~(4) and Eqs.(6)~(9).

As derived in Appendix C, for four parallel air-channel splitters, the four-pole matrices between nodes 5a~6a, 5b~6b, 5c~6c, and 5d~6d nodes can be combined into an equivalent matrix

$$\begin{bmatrix} p_5 \\ \rho_o c_o u_5 \end{bmatrix} = \begin{bmatrix} TSP1_{1,1} & \frac{1}{4}\cdot TSP1_{1,2} \\ 4\cdot TSP1_{2,1} & TSP1_{2,2} \end{bmatrix} \begin{bmatrix} p_6 \\ \rho_o c_o u_6 \end{bmatrix} \tag{15}$$

where $[TSP1_{i,j}]$ is the four-pole matrix for a one-channel splitter. Consequently, the total transfer matrix assembled by multiplication is

$$\begin{pmatrix} p_1 \\ \rho_o c_o u_1 \end{pmatrix} = f_1(L_5,d_1,M_1) f_2(L_2,L_3,M_2) f_3(L_2,L_4,M_3) f_4(L_6,d_2,M_4)$$
$$\begin{bmatrix} TS1_{1,1} & TS1_{1,2} \\ TS1_{2,1} & TS1_{2,2} \end{bmatrix} \begin{bmatrix} TE1_{1,1} & TE1_{1,2} \\ TE1_{2,1} & TE1_{2,2} \end{bmatrix} \begin{bmatrix} TS2_{1,1} & TS2_{1,2} \\ TS2_{2,1} & TS2_{2,2} \end{bmatrix} \begin{bmatrix} TC1_{1,1} & TC1_{1,2} \\ TC1_{2,1} & TC1_{2,2} \end{bmatrix} \tag{16}$$
$$\begin{bmatrix} TSP1_{1,1} & \frac{1}{4}\cdot TSP1_{1,2} \\ 4\cdot TSP1_{2,1} & TSP1_{2,2} \end{bmatrix} \begin{bmatrix} TE2_{1,1} & TE2_{1,2} \\ TE2_{2,1} & TE2_{2,2} \end{bmatrix} \begin{bmatrix} TS3_{1,1} & TS3_{1,2} \\ TS3_{2,1} & TS3_{2,2} \end{bmatrix} \begin{bmatrix} TC2_{1,1} & TC2_{1,2} \\ TC2_{2,1} & TC2_{2,2} \end{bmatrix}$$
$$\begin{bmatrix} TS4_{1,1} & TS4_{1,2} \\ TS4_{2,1} & TS4_{2,2} \end{bmatrix} \begin{pmatrix} p_{10} \\ \rho_o c_o u_{10} \end{pmatrix}$$

A simplified form is expressed in a matrix as

$$\begin{pmatrix} p_1 \\ \rho_o c_o u_1 \end{pmatrix} = \begin{bmatrix} T_{11}^{***} & T_{12}^{***} \\ T_{21}^{***} & T_{22}^{***} \end{bmatrix} \begin{pmatrix} p_{10} \\ \rho_o c_o u_{10} \end{pmatrix} \tag{17}$$

3.4. Overall sound power level

The sound transmission loss (*STL*) of mufflers A is defined as

$$STL_1(Q,f,RT_1,RT_2,RT_3,RT_4,RT_5,RT_6) = 20\log\left(\frac{\left|T_{11}^{*}+T_{12}^{*}+T_{21}^{*}+T_{22}^{*}\right|}{2}\right)+10\log\left(\frac{S_1}{S_{10}}\right) \tag{18a}$$

$$STL_2(Q,f,RT_1,RT_2,RT_3,RT_4,RT_5,RT_6) = 20\log\left(\frac{\left|T_{11}^{**}+T_{12}^{**}+T_{21}^{**}+T_{22}^{**}\right|}{2}\right)+10\log\left(\frac{S_1}{S_{10}}\right) \tag{18b}$$

$$STL_3(Q,f,RT_1,RT_2,RT_3,RT_4,RT_5,RT_6) = 20\log\left(\frac{\left|T_{11}^{***}+T_{12}^{***}+T_{21}^{***}+T_{22}^{***}\right|}{2}\right)+10\log\left(\frac{S_1}{S_{10}}\right) \tag{18c}$$

where

$$RT_1=R;RT_2=\sigma;RT_3=dh;RT_4=D_f/H;RT_5=W;RT_6=L_1; \tag{18d}$$

The silenced octave sound power level emitted from a muffler's outlet is

$$SWL_i = SWLO(f_i) - STL(f_i) \tag{19}$$

where (1) $SWLO(f_i)$ is the original *SWL* at the inlet of a muffler (or pipe outlet), and f_i is the relative octave band frequency.

(2) $STL(f_i)$ is the muffler's *STL* with respect to the relative octave band frequency (f_i).

(3) SWL_i is the silenced *SWL* at the outlet of a muffler with respect to the relative octave band frequency.

Finally, the overall SWL_T silenced by a muffler at the outlet is

$$SWL_{T-K} = 10*\log\{\sum_{i=1}^{n} 10^{SWL_i/10}\} = 10*\log\left\{10^{[SWLO(f_1)-STL_K(f_1)]/10}+10^{[SWLO(f_2)-STL_K(f_2)]/10}+10^{[SWLO(f_3)-STL_K(f_3)]/10}+\ldots+10^{[SWLO(f_n)-STL_K(f_n)]/10}\right\} \tag{20}$$

4. Objective function

4.1 STL maximization for a tone (F) noise

For muffler A (a two-chamber muffler hybridized with one dissipative duct), the objective function in maximizing the *STL* at a pure tone (*f*) is

$$OBJ_{11} = STL_1(Q,f,RT_1,RT_2,RT_3,RT_4,RT_5,RT_6) \tag{21a}$$

The related ranges of the parameters are

Q=0.001 (m^3/s); L_o =1.2(m); L_2=0.3(m); L_5=0.1(m); L_6=0.1(m); d_1=0.1(m); d_3=0.1(m);q=0.008 (m); RT1:[3000, 20000]; RT2:[5, 30]; RT3:[0.003, 0.015]; RT4:[0, 0.95]; RT5:[0.025, 0.15]; RT6:[0.3, 1.0] (21b)

4.2 SWL minimization for a broadband noise

To minimize the overall SWLT, the objective functions for mufflers A~C are

$$OBJ_{21} = SWL_{T-1}(Q,RT_1,RT_2,RT_3,RT_4,RT_5,RT_6) \tag{22a}$$

$$OBJ_{22} = SWL_{T-2}(Q,RT_1,RT_2,RT_3,RT_4,RT_5,RT_6) \tag{22b}$$

$$OBJ_{23} = SWL_{T-3}(Q,RT_1,RT_2,RT_3,RT_4,RT_5,RT_6) \tag{22c}$$

5. Model check

Before performing the SA optimal simulation on the mufflers, an accuracy check of the mathematical model on the acoustical elements of a one-channel splitter is performed using the experimental data. As depicted in Fig. 3, the theoretical and experimental data are in agreement. Therefore, the proposed fundamental mathematical model for the dissipative tube is acceptable. Consequently, the model linked with the numerical method is applied to the shape optimization in the following section.

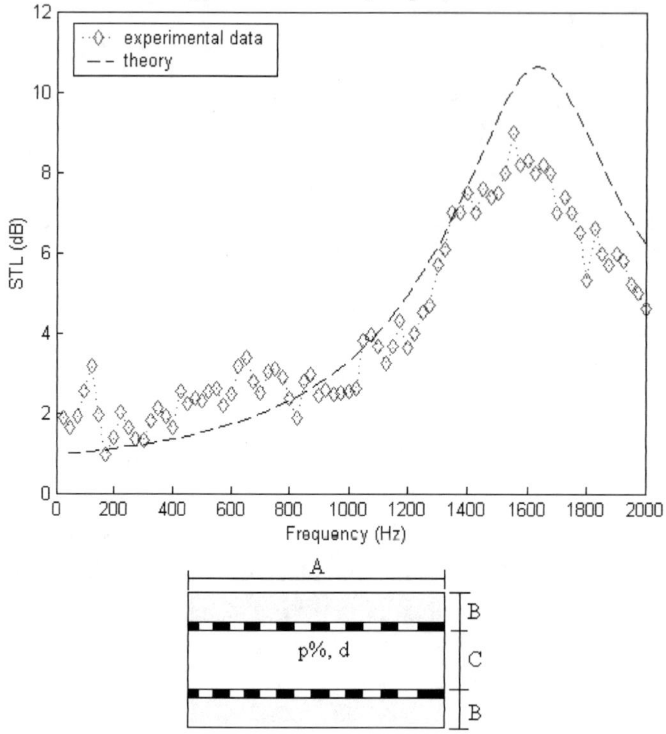

Figure 3. Performance of a rectangular muffler with a one-chamber splitter (A=0.6 M, B=0.05 M, C=0.2 M, p=20%, d=0.005 M, glass fiber (32kg/m^3)).

5.1 Case Study

The noise reduction of a steam turbine system within a space-constrained room is introduced and shown in Fig. 1. The sound power level (*SWL*) inside the venting outlet is shown in Table 1 where the overall *SWL* reaches 127.4 dB. To efficiently reduce the venting noise emitted from the steam turbine system, a rectangular muffler having splitter acoustical elements is necessary. Here, three kinds of rectangular mufflers internally connected with multiple parallel splitters (mufflers A~C) that disperse venting fluid and decrease the secondary flowing noise are considered.

To obtain the best acoustical performance within a fixed space, numerical assessments linked to a *SA* optimizer are applied. Before the minimization of a broadband noise is performed, a reliability check of the *SA* method by maximization of the *STL* at a targeted tone (250 Hz) in muffler A is performed. As shown in Fig. 1, the available space for a muffler is 0.3 m in width (per one channel) and 1.2 m in length. The flow rate (*Q*) is preset at 0.001 (m^3/s). The corresponding *OBJ* functions, space constraints, and ranges of the design parameters are summarized in Eqs.(21)~(22).

Table 1. Unsilenced *SWL* of a fan inside a duct outlet.

f(Hz)		125	250	500	1k	2k	overall
a. -dB(A)	*SWL*	95	110	118	125	122	127.4

5.2. Simulated Annealing Method [13]

Because of the needs of the classical gradient methods *EPFM*, *IPFM* and *FDM* as an appropriate starting point (design data) before optimization is performed, accuracy is limited [15].
Simulated Annealing (*SA*), one of the best stochastic search methods, was first introduced by Metropolis *et al.* [16] and later developed further by Kirkpatrick *et al.* [17]. Because no starting point is required before the optimization is performed, *SA* is then adopted as the optimizer used in the muffler's shape optimization. Because annealing is the process of heating while simultaneously maintaining a metal at a stabilized temperature as it cools, it allows particles to remain close to the minimal energy state. Generating a random initial solution will start the algorithm. The scheme of *SA* is a variation of the hill-climbing algorithm where all downhill movements for improvement are accepted for the decrement of the system's energy. In order to emulate the *SA*'s evolution, a new random solution (X') is selected from the neighborhood of the current solution (X). If it happens that there is a negative change in the objective function, or energy, ($\Delta F \leq 0$), then the resulting solution will be recognized as the new current solution with the transition property (*pb*(X'))of 1. However, if there is not a negative change ($\Delta F > 0$), the probability of transitioning to the new state X' will be the function $pb(\Delta F / CT)$. As shown in Eq. (23), the new transition property (*pb*(X')) varied from 0~1 will be calculated using the Boltzmann factor (pb(X') = $\exp(\Delta F / CT)$) where *C* and *T* are the Boltzmann constant and the current temperature.

$$pb(X') = \begin{cases} 1, \Delta F \leq 0 \\ \exp(\frac{-\Delta F}{CT}), \Delta f > 0 \end{cases} \tag{23a}$$

$$\Delta F = F(X') - F(X) \tag{23b}$$

Each successful substitution of the new current solution will conduct to the decay of the current temperature as

$$T_{new} = kk * T_{old} \tag{24}$$

where kk is the cooling rate.
The process is repeated until the predetermined number (*iter*) of the outer loop is reached.

6. Results and discussion

6.1. Results

The accuracy of the *SA* optimization depends on two types of *SA* parameters that include *kk* (cooling rate) and *iter* (maximum iteration). To achieve good optimization, the following parameters are varied step by step:
kk(0.91, 0.93, 0.95, 0.97, 0.99); *iter* (50, 100, 500, 1000, 2000, 5000).
Two results of optimization (one, pure tone noises used for *SA*'s accuracy check; and the other, a broadband noise occurring in a venting system) are described below.

6.2. Pure tone noise optimization

Before dealing with a broadband noise, the *STL*'s maximization with respect to a one-tone noise (250Hz) of muffler A is introduced for a reliability check on the *SA* method. By using Eq. (21), the maximization of the *STL* with respect to muffler A (a rectangular muffler internally connected with a one-channel splitter) at the specified pure tone (250Hz) was performed first. As indicated in Table 2, ten sets of *SA* parameters are tried in the muffler's optimization. Obviously, the optimal design data can be obtained from the last set of *SA* parameters at (*kk, iter*) = (0.99, 5000). Using the optimal design in a theoretical calculation, the optimal *STL* curves with respect to various *SA* parameters (*kk, iter*) are plotted and depicted in Figs. 4 and 5. As revealed in Figs. 4 and 5, the *STL* is precisely maximized at the desired frequency (250Hz). Consequently, the *SA* optimizer is reliable in the optimization process.

Table 2. Optimal *STL* for splitter muffler A (equipped with one air channel) at various *SA* parameters (targeted tone of 250 Hz).

SA parameter		Design parameters						Result
kk	iter	RT1	RT2	RT3	RT4	RT5	RT6	STL_{250Hz} dB
0.91	50	10060	15.38	0.00798	0.3943	0.07688	0.5905	6.8
0.93	50	5969	9.367	0.005096	0.1659	0.04683	0.4224	8.3
0.95	50	5505	8.685	0.004769	0.14	0.04342	0.4032	9.6
0.97	50	11530	17.54	0.009018	0.4764	0.08769	0.6511	10.7
0.99	50	4739	7.558	0.004228	0.0972	0.03779	0.3716	10.9
0.99	100	12530	19.01	0.009727	0.5326	0.09507	0.6924	12.4
0.99	500	19340	29.03	0.01454	0.9133	0.1452	0.973	13.4
0.99	1000	19780	29.67	0.01484	0.9374	0.1483	0.9908	13.5
0.99	2000	13950	21.11	0.01073	0.6120	0.1055	0.7509	13.8
0.99	5000	15950	24.04	0.01214	0.7235	0.1202	0.8331	14.5

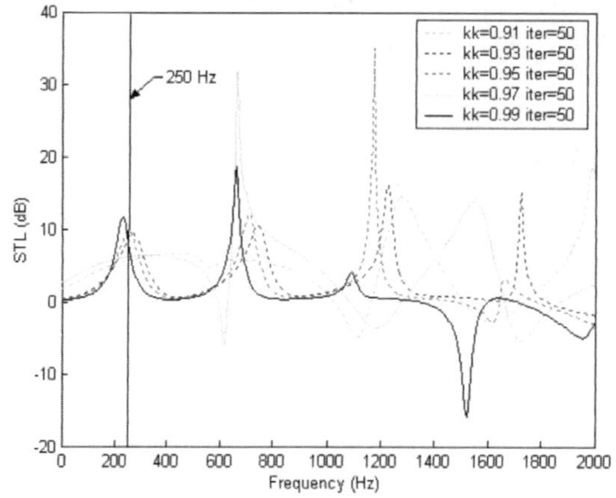

Figure 4. *STL* with respect to various *kk* [muffler A: target tone=250 Hz].

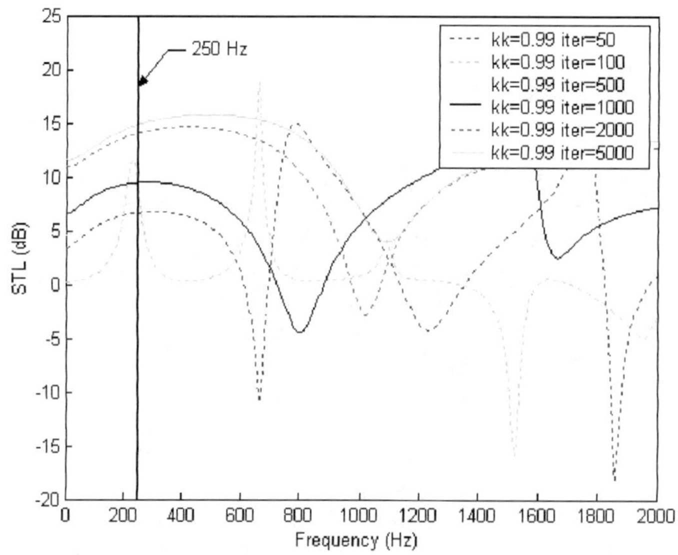

Figure 5. *STL* with respect to various *iter* [muffler A: target tone=250 Hz].

6.3. Broadband noise optimization

Similarly, considering Eq. (22a-c) and using the same *SA* parameters in the broadband optimization process, the minimization of the $SWL_{T\text{-}1}$, $SWL_{T\text{-}2}$, and $SWL_{T\text{-}3}$ with respect to mufflers A~C was performed and shown in Table 3. As illustrated in Table 3, the resultant sound power levels with respect to three types of mufflers have been reduced from 127.4 dB(A) to 113.1 dB(A), 106.5 dB(A), and 99.5 dB dB(A). Using this optimal design in a theoretical calculation, the optimal *STL* curves with respect to various mufflers are plotted and compared with the original *SWL* depicted in Fig. 6.

Table 3. Comparison of the minimized SWL_T of three kinds of mufflers (mufflers A~C) [broadband noise].

Muffler Type	Design parameters						Result
	RT1	RT2	RT3	RT4	RT5	RT6	SWL_T – dB(A)
Muffler A (one channel)	13820	20.91	0.01064	0.6045	0.1045	0.7454	113.1
Muffler B (two channel)	15100	22.8	0.01154	0.6765	0.1140	0.7984	106.5
Muffler C (four channel)	15170	22.9	0.01159	0.6801	0.1145	0.8012	99.5

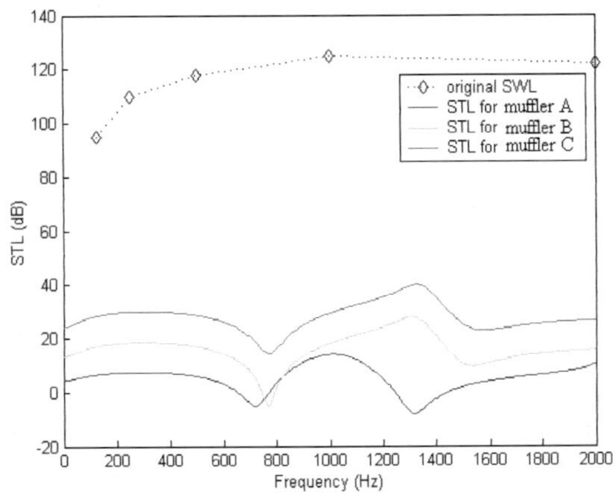

Fig. 6 Comparison of the optimal *STLs* of three kinds of mufflers (mufflers A, B, and C) and the original *SWL*

6.4. Discussion

In order to decrease the secondary flowing noise generated from the higher speed flow, new muffler designs with multiple parallel splitters used to disperse the venting fluid are presented. To achieve a sufficient optimization, the selection of the appropriate *SA* parameter set is essential. As indicated in Table 2, the best *SA* set of muffler A at the targeted pure tone noise of 250 Hz has been shown. The related *STL* curves with respect to various *SA* parameters are plotted in Figs. 4 and 5. Figs. 4 and 5 reveal the predicted maximal value of the *STL* is located at the desired frequency.

In dealing with the broadband noise, the acoustical performance among three types of rectangular mufflers connected with multiple splitters (mufflers A, B, and C) are shown in Table 3 and Fig. 6. As can be observed in Table 3, the overall sound transmission losses with respect to mufflers A~C are 14.3 dB, 20.9 dB, and 27.9 dB. Results shown in Table 3 and Fig. 6 indicate that the rectangular muffler hybridized with more parallel splitters is superior to the other mufflers equipped with fewer splitters. It also can be seen that the passing velocity within the air way will decrease if the number of air-channels increase. With this, the related back pressure will decrease; therefore, the overall back pressure of the mufflers with more parallel splitters will decrease.

To appreciate the influence of the acoustical effect with respect to the design parameters, a sensitivity analysis of six design parameters (*dh*: the diameter of the perforated holes; W: the width of the air channel; *R*: acoustic flow resistivity of the acoustic fiber; σ: the porosity of the perforated plate; D_f: the thickness of the acoustic fiber; L_1: the horizontal length of the splitter) is performed. The influence of acoustical performance with respect to *R* (acoustic flow resistivity of the acoustic fiber) is shown in Fig. 7. As indicated in Fig. 7, the acoustical performance will increase when *R* decreases. In addition, the influence of the acoustical performance with respect to σ (the porosity of the perforated plate) is depicted in Fig. 8. Fig. 8 indicates that the acoustical performance will increase if σ increases. However, as indicated in Fig. 9, there is no influence on the acoustical performance when the *dh* (the diameter of a perforated hole) varies. Also, the influence of acoustical performance with respect to D_f (the thickness of the acoustic fiber) is illustrated in Fig. 10. As indicated in Fig. 10, the acoustical performance at low frequencies will increase when D_f decreases (i.e., the increment of the air layer inside the splitter). This means that the resonating effect at the low frequencies will be remarkable due to the increment of the air layer inside the splitter. Subsequently, the influence of the acoustical performance with respect to W (the width of air channel) is shown in Fig. 11. As indicated in Fig. 11, the acoustical performance at low frequencies (below 600 Hz) will increase when W increases; However, the acoustical performance at higher frequencies (beyond than 1300 Hz) will decrease when W increases. Consequently, the influence of the acoustical performance with respect to L_1 (the horizontal length of the splitter) is depicted in Fig. 12. As can be seen in Fig. 12, the acoustical performance at the frequencies of 0~1000 Hz will increase when L_1 increases.

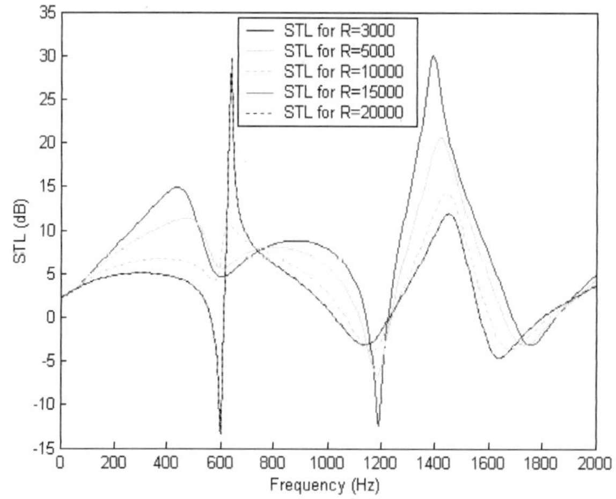

Figure 7. The influence of the acoustical performance with respect to R.

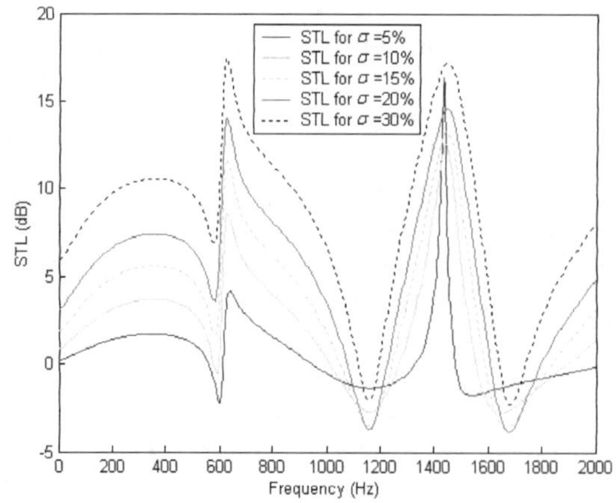

Figure 8. The influence of the acoustical performance with respect to σ .

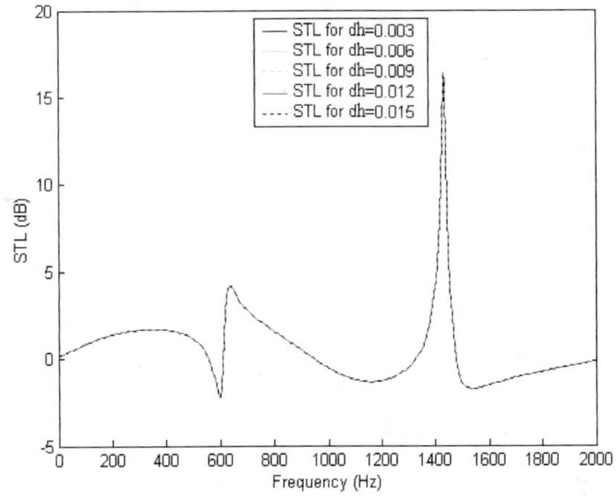

Figure 9. The influence of the acoustical performance with respect to dh.

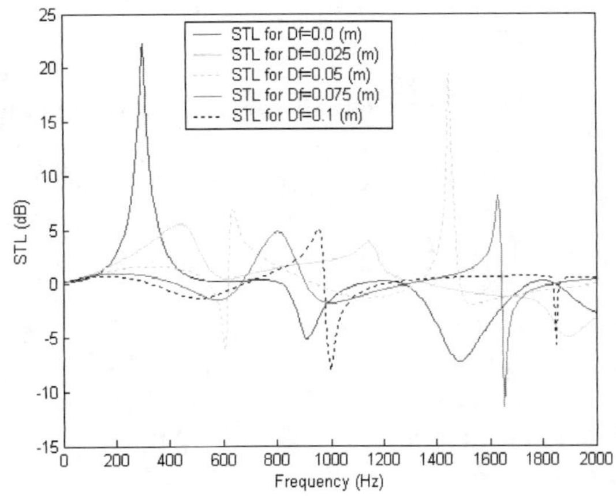

Figure 10. The influence of the acoustical performance with respect to D_f.

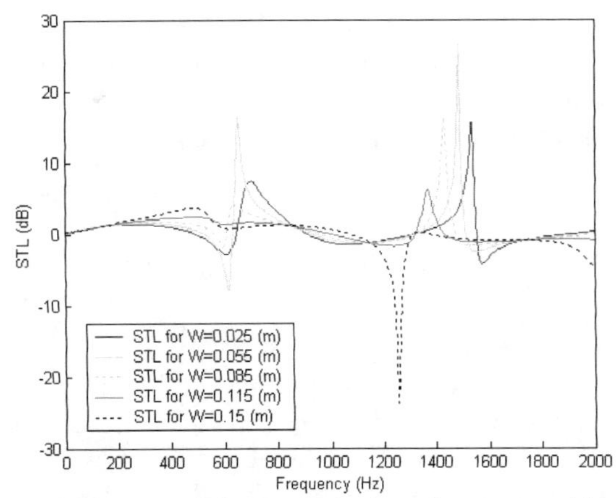

Figure 11. The influence of the acoustical performance with respect to W.

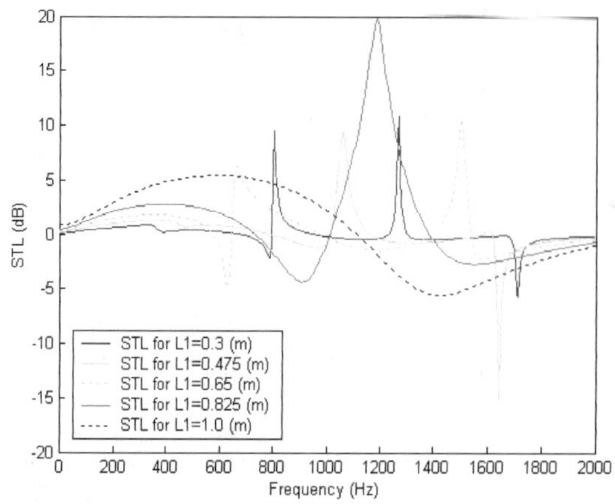

Figure 12. The influence of the acoustical performance with respect to L_1.

7. Conclusion

It has been shown that rectangular mufflers internally connected with multiple parallel splitters can be easily and efficiently optimized within a limited space by using a acoustical lumped technique, a plane wave theory, a four-pole transfer matrix, and a *SA* optimizer. As indicated in Table 2 and Figs. 4~5, two kinds of *SA* parameters (*kk* and *iter*) play essential roles in the solution's accuracy during *SA* optimization. Figs. 4~5 indicate that the tuning ability established by adjusting design parameters of muffler A is reliable.

Additionally, the appropriate acoustical performance curve of three types of rectangular mufflers internally connected with multiple parallel splitters (mufflers A~C) has been assessed. As indicated in Table 3 and Fig. 6, the resultant SWL_T with respect to these mufflers is 113.1 dB(A), 106.5 dB(A), and 99.5 dB(A). Obviously, the muffler hybridized with more parallel splitters is superior to the other mufflers equipped with fewer parallel splitters. It can be seen that more splitters installed inside the rectangular muffler will disperse the venting fluid, avoid the secondary flowing noise, and reduce the back pressure of the muffler. Moreover, as investigated in Section 7.2, the influence of the acoustical performance with respect to five design parameters (W: the width of air channel; *R*: acoustic flow resistivity of the acoustic fiber; σ: the porosity of the perforated plate; D_f: the thickness of the acoustic fiber; L_1: the horizontal length of the splitter) is enormous. Consequently, an appropriate selection of the design parameters will result in a more efficient noise reduction for the mufflers.

8. Acknowledgements

The authors acknowledge the financial support of the Ministry of Science and Technology (MOST 103-2622-E-235 -001 -CC3, ROC)

9. References

[1] P.M. Morse, Transmission of sound inside pipes, J. Acoust. Soc.Am., 11, 205-210 (1939).R.A. Scott, The propagation of sound between walls of porous material, Proceedings of the Physical Society, 58, 358-368 (1946).S.H. Ko, Theoretical analyses of sound attenuation in acoustically lined flow ducts separated by porous splitters (rectangular, annular and circular ducts), propagation of sound between walls of porous material, Journal of Sound and Vibration, 39, 471-487 (1975)

[2] Cummings, I. J. Chang, Sound attenuation of a finite length dissipative flow duct silencer with internal mean flow in the absorbent, Journal of Sound and Vibration, 127, 1-17 (1988)

[3] K.S. Peat, A transfer-matrix for an absorption silencer element, Journal of Sound and Vibration, 146, 353-360 (1991)

[4] Selamet, I.J. Lee, Z.L. Ji, N.T. Huff, Acoustic attenuation performance of perforated concentric absorbing silencers, SAE Noise and Vibration Conference and Exposition, SAE Paper No. 2001-01-1435, Traverse City, MI, April 30-May 3 (2001)
[5] Selamet, I.J. Lee, N.T. Huff, Acoustic attenuation of hybrid silencers, Journal of Sound and Vibration, 262, 509-527 (2003)
[6] M.L. Munjal, Analysis and design of pod silencers, Journal of Sound and Vibration, 262, 497-507 (2003)
[7] M.L. Xu, A. Selamet, I.J. Lee, N.T. Huff, Sound attenuation in dissipative expansion chambers, Journal of Sound and Vibration, 272, 1125-1133 (2004)
[8] M.C. Chiu, Shape optimization of one-chamber perforated mufflers filled with wool using simulated annealing, Journal of Marine Science and Technology, 21(4), 380-390 (2013)
[9] M.C. Chiu, Numerical assessment of multi-Splitter mufflers using neural networks, the boundary element method, and the genetic algorithm, Applied Mechanics and Materials, 58-60, 1049-1055 (2011)
[10] M.C. Chiu, Shape optimisation of multi-chamber mufflers with plug-inlet tube on a venting process by genetic algorithms, Applied Acoustics, 71, 495-505 (2010)
[11] M.C. Chiu, Optimal Design of Multi-chamber Mufflers Hybridized with Perforated Intruding Inlets and Resonated Tube Using Simulated Annealing, ASME J. of Vibration and Acoustics, 132, 1-10 (2010)
[12] M.C. Chiu, Y.C. Chang, Numerical assessment of a venting system with multi-chamber perforated mufflers by GA method, Journal of Marine Science and Technology, 19(5), 483-498 (2011)
[13] Y.C. Chang, L.J. Yeh, M.C. Chiu, G.J. Lai, Shape optimization on constrained single-layer sound absorber by using GA method and mathematical gradient methods, Journal of Sound and Vibration, 286(4-5), 941-961 (2005)
[14] Metropolis, W. Rosenbluth, M.N. Rosenbluth, H. Teller, E. Teller, Equation of static calculations by fast computing machines, The Journal of Chemical Physics, 21, 1087-1092 (1953)
[15] S. Kirkpatrick, C.D. Gelatt, M.P. Vecchi, Optimization by simulated annealing, Science, 220, 671-680 (1983)
[16] M.E. Delany, E. N. Bazley, Acoustical properties of fibrous absorbent materials, Applied Acoustics, 13, 105-116 (1969)
[17] F.C. Lee, W.H. Chen, Acoustic transmission analysis of multi-layer absorbers, Journal of Sound and Vibration, 248, 621-634 (2001)
[18] M. L. Munjal, Acoustics of ducts and mufflers with application to exhaust and ventilation, John Wiley and Sons, New York, (1987)
[19] L.L. Beranek, I.L. Ver, Noise and vibration control engineering, John Wiley and Sons, New York, (1992)

APPENDIX A

Transfer Matrix of a One-Channel Splitter Element

A one-layer sound absorber shown in Fig.13(A) is partitioned with two rigid stiffeners of L_1 and L_2. Based on the plane wave theory, fc < $c_0/2L_{max}$ where L_{max} is the maximum of (L_1, L_2). The acoustic impedance on the perforated front plate is obtained from the bottom wall where the value of the impedance is infinity. As indicated in Fig. 13(B), there exists four points representing the absorbing impedance within the absorbers. The absorber is composed of a "rigid-backing plate +L_i thickness of air + D_{fi} thickness of the acoustic fiber + q_1 thickness of the perforated front plate." As derived in a previous study [15], for a wave propagating normally in a quiescent medium symbolized by "m," the general matrix form between node *i* and node *i+1* is expressed as

IOP Conf. Series: Materials Science and Engineering **644** (2019) 012003 doi:10.1088/1757-899X/644/1/012003

$$\begin{pmatrix} p_{i+1} \\ u_{i+1} \end{pmatrix} = \begin{bmatrix} \cos(k_m L) & jZ_m \sin(k_m L) \\ j\dfrac{1}{Z_m}\sin(k_m L) & \cos(k_m L) \end{bmatrix} \begin{pmatrix} p_i \\ u_i \end{pmatrix} \tag{A1}$$

Therefore, for the upper sound absorber, the relationship of the acoustic pressure p and the acoustic particle velocity u between node 00 and node 01 is expressed as the transfer matrix and shown below.

$$\begin{pmatrix} p_{01} \\ u_{01} \end{pmatrix} = \begin{bmatrix} \cos(\omega(H - D_f)/c_o) & j\rho_o c_o \sin(\omega(H - D_f)/c_o) \\ j\dfrac{\sin(\omega(H - D_f)/c_o)}{\rho_o c_o} & \cos(\omega(H - D_f)/c_o) \end{bmatrix} \begin{pmatrix} p_{oo} \\ u_{oo} \end{pmatrix} \tag{A2}$$

Developing Eq. (A2) yields

$$Z_{01} = -j\rho_o c_o \cot(\frac{\omega(H - D_f)}{c_o}) \tag{A3}$$

The relationship of the acoustic pressure p and the acoustic particle velocity u with respect to node 01 and node 02 is expressed in the transfer matrices below.

$$\begin{pmatrix} p_{02} \\ u_{02} \end{pmatrix} = \begin{bmatrix} \cos(k_{fiber} D_f) & jZ_{fiber} \sin(k_{fiber} D_f) \\ j\dfrac{1}{Z_{fiber}}\sin(k_{fiber} D_f) & \cos(k_{fiber} D_f) \end{bmatrix} \begin{pmatrix} p_{01} \\ u_{01} \end{pmatrix} \tag{A4}$$

By developing Eq. (A4), an alternative form of Eq.(A4) yields

$$Z_{02}\left(= \frac{p_{02}}{u_{02}}\right) = Z_{fiber} \frac{Z_{01}\cosh[k_{fiber} D_f] + Z_{fiber}\sinh[k_{fiber} D_f]}{Z_{01}\sinh[k_{fiber} D_f] + Z_{fiber}\cosh[k_{fiber} D_f]} \tag{A5}$$

By adopting the formula of the specific normal impedance and wave number [18], Eq. (A5) is written as

$$Z_{02} = (R_{fiber} + jX_{fiber})\left[\frac{\sinh(k_2 D_f)\cos(k_1 D_f) - j\sin(k_1 D_f)\cosh(k_2 D_f)}{\cos(k_1 D_f)\cosh(k_2 D_f) - j\sinh(k_2 D_f)\sin(k_1 D_f)}\right] \tag{A6}$$

For sound flowing into the perforated plate, it is assumed that the incident sound passes through the holes of the perforated plate and is immediately transmitted to the porous material behind the perforated plate, the particle velocity is hardly reduced [19, 20]. The continuity of the particle velocity is then applied and expressed as

$$u_{02} = u_2 \tag{A7}$$

Acoustic impedance yields

$$p_2 = Z_p u_{02} + p_{02} \tag{A8}$$

Combining Eqs. (A7)~(A8), the transfer matrix between node 02 and node 2 yields

$$\begin{pmatrix} p_2 \\ u_2 \end{pmatrix} = \begin{bmatrix} 1 & Z_p \\ 0 & 1 \end{bmatrix} \begin{pmatrix} p_{02} \\ u_{02} \end{pmatrix} \tag{A9}$$

Developing Eq. (A9) and substituting Eq. (A7), the specific normal impedance at node 2 is

$$Z_2 = Z_{02} + Z_p \tag{A10}$$

Adopting the formula of the specific normal impedance and the wave number of the perforated plate [21] yields

$$Z_p = \frac{\rho_o}{\sigma}\sqrt{8v\omega}\left(1 + \frac{q}{2dh}\right) + j\frac{\omega\rho_o}{\sigma}\left[\sqrt{\frac{8v}{\omega}}\left(1 + \frac{q}{2dh}\right) + q + \delta\right] \tag{A11a}$$

$$\delta = 0.85(2dh)\left(1 - 1.47\sqrt{\sigma} + 0.47\sqrt{\sigma^3}\right) \tag{A11b}$$

Similarly, for the lower sound absorber, the transfer matrix between node 02 and node 3 yields

IOP Conf. Series: Materials Science and Engineering **644** (2019) 012003 doi:10.1088/1757-899X/644/1/012003

$$\begin{pmatrix} p_3 \\ u_3 \end{pmatrix} = \begin{bmatrix} 1 & Z_p \\ 0 & 1 \end{bmatrix} \begin{pmatrix} p_{02} \\ u_{02} \end{pmatrix} \tag{A12}$$

Developing Eq. (A12), the specific normal impedance at node 3 is

$$Z_3 = Z_{02} + Z_p \tag{A13}$$

A sound wave propagates from the left side to the right side. Considering the acoustical mass conservation and boundary condition yields

$$p_1 = p_4 ; p_2 = p_4 ; p_3 = p_4 ; \tag{A14a}$$

$$v_1 = v_2 + v_3 + v_4 ; \tag{A14b}$$

$$S_1 = S_4 = S_d ; \tag{A14c}$$

$$S_2 = S_3 = L_1 * L_2 * \sigma ; \tag{A14d}$$

Defining the acoustical impedance of node 2 and node 3, the impedance can be expressed as

$$Z_{m=1,n=1,k=1} = Z_2 = \frac{p_2}{u_2} ; Z_{m=1,n=2,k=1} = Z_3 = \frac{p_3}{u_3} \tag{A15}$$

Eq.(A15) can be expressed as

$$u_2 = \frac{p_2}{Z_2} ; u_3 = \frac{p_3}{Z_3} \tag{A16}$$

Rearranging Eqs. (A14)-(A16) yields

$$\begin{cases} p_1 = p_4 \\ \rho_o c_o u_1 = \rho_o c_o u_4 + (\dfrac{1}{Z_{m=1,n=1,k=1}} + \dfrac{1}{Z_{m=1,n=2,k=1}}) p_4 \end{cases} \tag{A17}$$

The matrix form is expressed as

$$\begin{bmatrix} p_1 \\ \rho_o c_o u_1 \end{bmatrix} = \begin{bmatrix} 1 & 0 \\ \dfrac{1}{Z_{m=1,n=1,k=1}} + \dfrac{1}{Z_{m=1,n=2,k=1}} & 1 \end{bmatrix} \begin{bmatrix} p_4 \\ \rho_o c_o u_4 \end{bmatrix} \tag{A18a}$$

An alternative form is

$$\begin{bmatrix} p_1 \\ \rho_o c_o u_1 \end{bmatrix} = \begin{bmatrix} TSP1_{11} & TSP1_{12} \\ TSP1_{21} & TSP1_{22} \end{bmatrix} \begin{bmatrix} p_4 \\ \rho_o c_o u_4 \end{bmatrix} \tag{A18b}$$

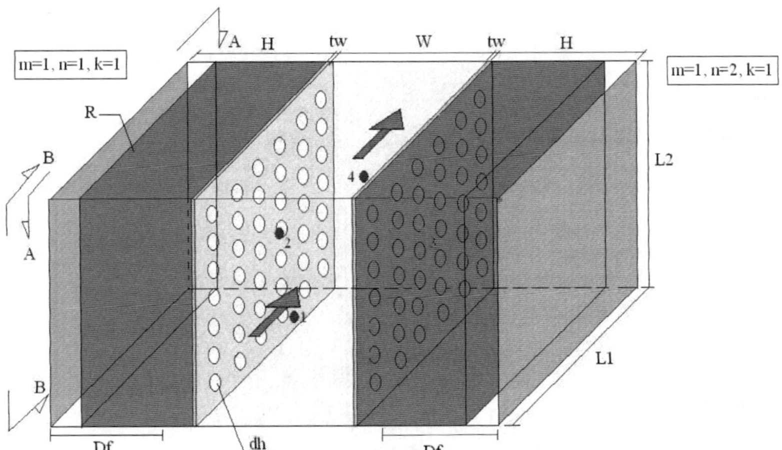

(A) Three-dimensional view of a one-channel splitter muffler composed of two absorbers and an air-channel.

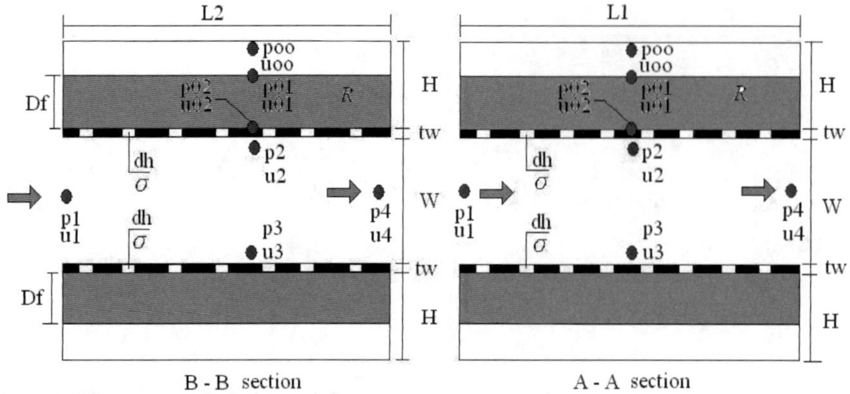

(B) Acoustical field of the three-dimensional view of a one-channel splitter muffler
Fig. 13 A one-channel splitter element.

APPENDIX B

Transfer Matrix of a Two-Channel Splitter Element

As indicated in Fig. 14, two identical one-channel splitters are parallel and combined to form a two-channel splitter element. As derived in Appendix A, the acoustical four-pole matrix between node 5a and 6a for the first one-channel splitter type element is

$$\begin{bmatrix} p_{5a} \\ \rho_o c_o u_{5a} \end{bmatrix} = \begin{bmatrix} TSP1_{1,1} & TSP1_{1,2} \\ TSP1_{2,1} & TSP1_{2,2} \end{bmatrix} \begin{bmatrix} p_{6a} \\ \rho_o c_c u_{6a} \end{bmatrix} \tag{B1}$$

Developing Eq.(B1) yields

$$p_{5a} = TSP1_{1,1} \cdot p_{6a} + TSP1_{1,2} \cdot \rho_o c_c u_{6a} \tag{B2a}$$

$$\rho_o c_o u_{5a} = TSP1_{2,1} \cdot p_{6a} + TSP1_{2,2} \cdot \rho_o c_c u_{6a} \tag{B2b}$$

Similarly, the acoustical four-pole matrix for the second one-channel splitter element between node 5b and 6b is

$$\begin{bmatrix} p_{5b} \\ \rho_o c_o u_{5b} \end{bmatrix} = \begin{bmatrix} TSP1_{1,1} & TSP1_{1,2} \\ TSP1_{2,1} & TSP1_{2,2} \end{bmatrix} \begin{bmatrix} p_{6b} \\ \rho_o c_o u_{6b} \end{bmatrix} \tag{B3}$$

Developing Eq.(B3) yields

$$p_{5b} = TSP1_{1,1} \cdot p_{6b} + TSP1_{1,2} \cdot \rho_o c_c u_{6b} \tag{B4a}$$

$$\rho_o c_o u_{5b} = TSP1_{2,1} \cdot p_{6b} + TSP1_{2,2} \cdot \rho_o c_c u_{6b} \tag{B4b}$$

Combining Eq.(B2a) and Eq.(B4a) yields

$$p_{5a} + p_{5b} = TSP1_{1,1} \cdot [p_{6a} + p_{6b}] + TSP1_{1,2} \cdot \rho_o c_c [u_{6a} + u_{6b}] \tag{B5}$$

Likewise, combining Eq.(B2b) and Eq.(B4b) yields

$$\rho_o c_o [u_{5a} + u_{5b}] = TSP1_{2,1} \cdot [p_{6a} + p_{6b}] + TSP1_{2,2} \cdot \rho_o c_c [u_{6a} + u_{6b}] \tag{B6}$$

where

$$p_5 = p_{5a} = p_{5b};\ p_6 = p_{6a} = p_{6b};\ u_5 = u_{5a} + u_{5b};\ u_6 = u_{6a} + u_{6b} \tag{B7}$$

Plugging Eq.(B7) into Eqs.(B5) and (B6) yields

$$2 \cdot p_5 = 2 \cdot TSP1_{1,1} \cdot p_6 + TSP1_{1,2} \cdot \rho_o c_c u_6 \quad \text{(B8a)}$$

$$\rho_o c_o u_5 = 2 \cdot TSP1_{2,1} \cdot p_6 + TSP1_{2,2} \cdot \rho_o c_c u_6 \tag{B8b}$$

Rearranging Eq.(B8) in a matrix form, the equivalent four-pole matrix between nodes 5 and 6 shown in Fig. 4 is

$$\begin{bmatrix} p_5 \\ \rho_o c_o u_5 \end{bmatrix} = \begin{bmatrix} TSP1_{1,1} & \frac{1}{2} \cdot TSP1_{1,2} \\ 2 \cdot TSP1_{2,1} & TSP1_{2,2} \end{bmatrix} \begin{bmatrix} p_6 \\ \rho_o c_o u_6 \end{bmatrix} \tag{B9}$$

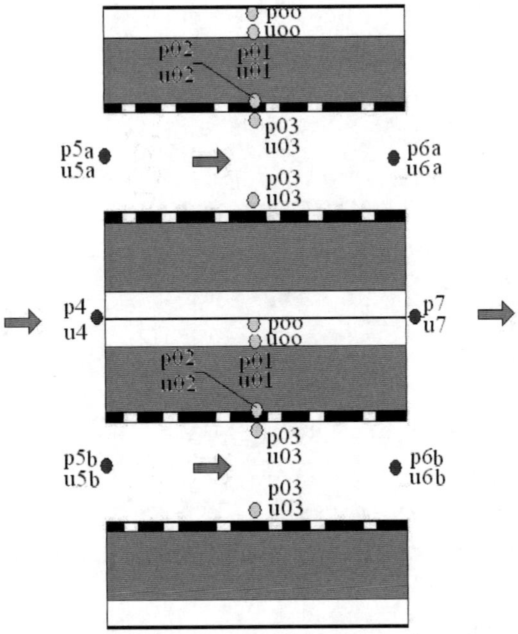

Figure 14. A two-channel splitter element.

APPENDIX C

Transfer Matrix of a Four-Channel Splitter Element

As indicated in Fig. 15, four identical one-channel splitters are parallel and combined to form a four-channel splitter element. As derived in Appendix A, the acoustical four-pole matrix between node 5a and 6a for the first one-channel splitter type element is

$$\begin{bmatrix} p_{5a} \\ \rho_o c_o u_{5a} \end{bmatrix} = \begin{bmatrix} TSP1_{1,1} & TSP1_{1,2} \\ TSP1_{2,1} & TSP1_{2,2} \end{bmatrix} \begin{bmatrix} p_{6a} \\ \rho_o c_c u_{6a} \end{bmatrix} \quad \text{(C1)}$$

Developing Eq.(C1) yields

$$p_{5a} = TSP1_{1,1} \cdot p_{6a} + TSP1_{1,2} \cdot \rho_o c_c u_{6a} \quad \text{(C2a)}$$

$$\rho_o c_o u_{5a} = TSP1_{2,1} \cdot p_{6a} + TSP1_{2,2} \cdot \rho_o c_c u_{6a} \quad \text{(C2b)}$$

Similarly, the acoustical four-pole matrices between node 5b and 6b, node 5c and 6c, and node 5d and 6d are

$$\begin{bmatrix} p_{5b} \\ \rho_o c_o u_{5b} \end{bmatrix} = \begin{bmatrix} TSP1_{1,1} & TSP1_{1,2} \\ TSP1_{2,1} & TSP1_{2,2} \end{bmatrix} \begin{bmatrix} p_{6b} \\ \rho_o c_o u_{6b} \end{bmatrix} \quad \text{(C3a)}$$

$$\begin{bmatrix} p_{5c} \\ \rho_o c_o u_{5c} \end{bmatrix} = \begin{bmatrix} TSP1_{1,1} & TSP1_{1,2} \\ TSP1_{2,1} & TSP1_{2,2} \end{bmatrix} \begin{bmatrix} p_{6c} \\ \rho_o c_o u_{6c} \end{bmatrix} \quad \text{(C3b)}$$

$$\begin{bmatrix} p_{5d} \\ \rho_o c_o u_{5d} \end{bmatrix} = \begin{bmatrix} TSP1_{1,1} & TSP1_{1,2} \\ TSP1_{2,1} & TSP1_{2,2} \end{bmatrix} \begin{bmatrix} p_{6d} \\ \rho_o c_o u_{6d} \end{bmatrix} \quad \text{(C3c)}$$

Developing Eqs.(C3a)~(C3c) yields

$$p_{5b} = TSP1_{1,1} \cdot p_{6b} + TSP1_{1,2} \cdot \rho_o c_c u_{6b} \quad \text{(C4a)}$$

$$\rho_o c_o u_{5b} = TSP1_{2,1} \cdot p_{6b} + TSP1_{2,2} \cdot \rho_o c_c u_{6b} \quad \text{(C4b)}$$

$$p_{5c} = TSP1_{1,1} \cdot p_{6c} + TSP1_{1,2} \cdot \rho_o c_c u_{6c} \quad \text{(C4c)}$$

$$\rho_o c_o u_{5c} = TSP1_{2,1} \cdot p_{6c} + TSP1_{2,2} \cdot \rho_o c_c u_{6c} \quad \text{(C4d)}$$

$$p_{5d} = TSP1_{1,1} \cdot p_{6d} + TSP1_{1,2} \cdot \rho_o c_c u_{6d} \tag{C4e}$$

$$\rho_o c_o u_{5d} = TSP1_{2,1} \cdot p_{6d} + TSP1_{2,2} \cdot \rho_o c_c u_{6d} \tag{C4f}$$

Combining Eq.(C2a) and Eq.(C4a),(C4c)(C4e) yields

$$p_{5a} + p_{5b} + p_{5c} + p_{5d} = TSP1_{1,1} \cdot [p_{6a} + p_{6b} + p_{6c} + p_{6d})] + TSP1_{1,2} \cdot \rho_o c_c [u_{6a} + u_{6b} + u_{6c} + u_{6d}] \tag{C5}$$

Likewise, combining Eq.(C2b) and Eq.(C4b),(C4d), and (C4f) yields

$$\rho_o c_o [u_{5a} + u_{5b} + u_{5c} + u_{5d}] = TSP1_{2,1} \cdot [p_{6a} + p_{6b} + p_{6c} + p_{6d}] + TSP1_{2,2} \cdot \rho_o c_c [u_{6a} + u_{6b} + u_{6c} + u_{6d}] \tag{C6}$$

where

$$p_5 = p_{5a} = p_{5b} = p_{5c} = p_{5d};\ p_6 = p_{6a} = p_{6b} = p_{6c} = p_{6d};$$
$$u_5 = u_{5a} + u_{5b} + u_{5c} + u_{5d};\ u_6 = u_{6a} + u_{6b} + u_{6c} + u_{6d} \tag{C7}$$

Plugging Eq.(C7) into Eqs.(C5) and (C6) yields

$$4 \cdot p_5 = 4 \cdot TSP1_{1,1} \cdot p_6 + TSP1_{1,2} \cdot \rho_o c_c u_6 \tag{C8a}$$

$$\rho_o c_o u_5 = 4 \cdot TSP1_{2,1} \cdot p_6 + TSP1_{2,2} \cdot \rho_o c_c u_6 \tag{C8b}$$

Rearranging Eq.(C8) in a matrix form, the equivalent four-pole matrix between nodes 5 and 6 shown in Fig. 4 is

$$\begin{bmatrix} p_5 \\ \rho_o c_o u_5 \end{bmatrix} = \begin{bmatrix} TSP1_{1,1} & \frac{1}{4} \cdot TSP1_{1,2} \\ 4 \cdot TSP1_{2,1} & TSP1_{2,2} \end{bmatrix} \begin{bmatrix} p_6 \\ \rho_o c_o u_6 \end{bmatrix} \tag{C9}$$

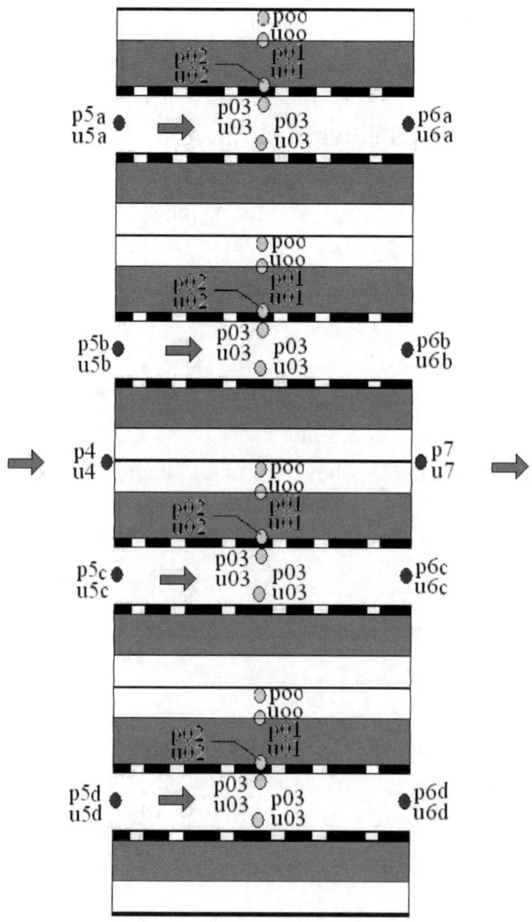

Figure 15. A four-channel splitter element.

IOP Conf. Series: Materials Science and Engineering **644** (2019) 012004 doi:10.1088/1757-899X/644/1/012004

Optimal Design of Mufflers Equipped with Pod and Concentric Acoustical Rings using the SA Method

Min-Chie Chiu[1] and Ho-Chih Cheng[2]

1 Department of Electrical Engineering and Energy Technology, Chung Chou University of Science and Technology, No. 6, Lane 2, Sec. 3, Shanjiao Rd., Yuanlin City, Changhua County 510, Taiwan
2 Department of Intelligent Automation Engineering, Chung Chou University of Science and Technology, No. 6, Lane 2, Sec. 3, Shanjiao Rd., Yuanlin City, Changhua County 510, Taiwan
E-mail: hccheng@gm.ccut.edu.tw

Abstract. There has been, of recently, much research on mufflers lined with sound absorbing material. In dealing with high frequency noise emitted from venting systems, a sound absorbing material inside a tube covered with a perforated plate has been used. However, the acoustical performance of mufflers equipped with one-layer of sound absorbing material has still proved to be insufficient. In order to promote acoustic performance, a design and theoretical assessment on pod mufflers hybridized with multi-layer concentric sound-absorbing rings is proposed. By using the acoustical-lumped-method, a four-pole system matrix for evaluating the acoustic performance (STL) will emerge. During the optimization process, the simulated annealing (SA) method, which is a robust scheme utilized to search for the global optimum by imitating a physical annealing process, is used. Before dealing with a broadband noise, the STL's maximization relative to a one-tone noise (1000Hz) is offered to confirm the SA method's reliability. To appreciate the influence of acoustical efficiency with respect to the design parameters, an analysis of design parameters (dh: the diameter of a perforated hole; W: the width of an air channel; R: acoustic flow resistivity of the acoustic fiber; σ : the porosity of the perforated plate; Df: the thickness of the acoustic fiber; L1: the horizontal length of the sound absorbing ring) is completed. Subsequently, to bring into focus the acoustical interaction with respect to the number of air-channels (between the sound absorbing rings), three types of mufflers (mufflers A~C) hybridized with one, two, and three air-channels have been surveyed. Results reveal that for a rectangular muffler internally conjugated with one air-channel, the maximal STL is located at the desired tone, and the acoustical performance of a cone muffler will increase if the number of air-channels internally conjugated within the cone muffler increases. Consequently, the approach used for the optimal design of the STL on pod mufflers internally equipped with multiple air-channels within a space-constrained situation can be easily assessed.

1. Nomenclature

This paper is constructed on the basis of the following notations:
c_o : sound speed (m s^{-1})
dh : the diameter of a perforated hole on the front perforated plate (m)
D_f : the thickness of the acoustic fiber
L_i : diameter of the i-th horizontal segment of the muffler (m)
D_i : diameter of the concentric sound absorbing ring (m)

IOP Conf. Series: Materials Science and Engineering **644** (2019) 012004 doi:10.1088/1757-899X/644/1/012004

f : cyclic frequency (Hz)
H : the thickness of the sound absorbing ring (m)
iter : maximum iteration
j : imaginary unit
k : wave number ($=\omega/c_o$)
k_1 : real part of complex $k_{fiber-i}$
k_2 : image part of complex $k_{fiber-i}$
k_{fiber} : complex propagation constant of the acoustic fiber
kk : cooling rate in *SA*
L_o : total length of the muffler (m)
L_i : the horizontal length of the i-th segment (m)
M : mean flow Mach number
OBJ : objective function (dB)
p_i : acoustic pressure at the i-th node (Pa)
pb(T) : transition probability
Q : volume flow rate of venting gas (m^3 s^{-1})
R : acoustic flow resistivity of the acoustic fiber (MKS rayls m^{-1})
R_{fiber} : real part of the complex Z_{fiber}
S_i : section area at the i-th node(m^2)
STL: sound transmission loss (dB)
SWLO : unsilenced sound power level inside the muffler's inlet (dB)
SWL_T : overall sound power level inside the muffler's output (dB)
t_w: the thickness of the front perforated plate in the splitter (m)
T: current temperature (^oC)
T^*_{ij} : components of a four-pole transfer system matrix
T_o: initial temperature (^oC)
u_i : acoustic particle velocity at the i-th node (m s^{-1})
W : the width of air channel (m)
Z_i : specific normal impedance at i.
Z_{fiber} : characteristic impedance of the acoustic fiber
Z_p : characteristic impedance of the perforated front plate
X_{fiber} : image part of the complex Z_{fiber}
ρ_o : air density (kg m^{-3})
ρ_i : acoustical density at the i-th node (kg m^{-3})
v : acoustical mass velocity (kg/s)
v v : kinematic viscosity of air ($=15*10^{-6}$ m^2/s)
pp: the porosity of the perforated plate (%)
δ : viscous boundary layer thickness of the i-th layer of the perforated plate (m)
ω : angular velocity ($=2\pi f$)

2. Introduction

Morse [1] started research on mufflers that reduced high frequency noise using a dissipative duct (a duct lined with sound absorbing material) in 1939. Scott [2] predicted the sound transmission of circular and rectangular mufflers internally lined with porous material using a bulk reactive model. Later, in 1975, Ko [3] analyzed the sound transmission loss in acoustically lined flow ducts separated by porous splitters. These researchers [1-3] focused on the sound attenuation of an infinite duct. Cummings and Chang [4] then, in 1988, investigated a finite length dissipative flow duct silencer with an internal mean flow in the absorbent channel using a modal method. Peat [5], in 1991, considering the volume modulus, adopted a transfer matrix for evaluating the acoustical performance of an absorption silencer element. Subsequently, Selamet *et al.* [6, 7], adopting a one-dimensional analytical method and a three-dimensional boundary element method (BEM), explored the acoustical attenuation for perforated concentric absorbing silencers and hybrid silencers. In 2003, Munjal [8] analyzed pod silencers using a four-pole transfer matrix. In 2004, Xu *et al.* [9] proposed a characteristic equation to calculate the sound attenuation in dissipative expansion chambers. However, an exploration of a muffler's optimal design within a limited space was never approached. Therefore, Chiu [10] has presented shaped optimization of a circular muffler internally lined with sound absorbing material within a constrained space. Later, Chiu [11] assessed circular mufflers internally equipped with multi-channel splitters using neural networks in conjunction the boundary element method and a genetic algorithm. However, the study is limited to a one-layer air channel muffler.

Here, then, in order to improve the acoustical performance of space-constrained cylindrical pod mufflers, a mathematical model of a cylindrical pod muffler internally equipped with multi-layer concentric sound- absorbing rings is proposed using an acoustical lump analysis. Three types of cylindrical pod mufflers equipped with multi-layer concentric sound- absorbing rings (muffler A: a muffler equipped with one air channel; muffler B: a muffler equipped with two air channels; muffler C: a muffler equipped with three air channels) are presented. Also note: the acoustical lump method used to form a four-pole matrix is in line with the simulated annealing method.

3. Theoretical background

Three kinds of multi-channel pod mufflers have been adopted for noise elimination in the Nitrogen emergent venting device shown in Fig. 1. Before being analyzed, the acoustical fields and related outline dimensions of the mufflers are shown in Figs. 2(A), (B), and (C). Individual transfer matrices with respect to one-channel, two-channel, and three-channel pod mufflers are described below.

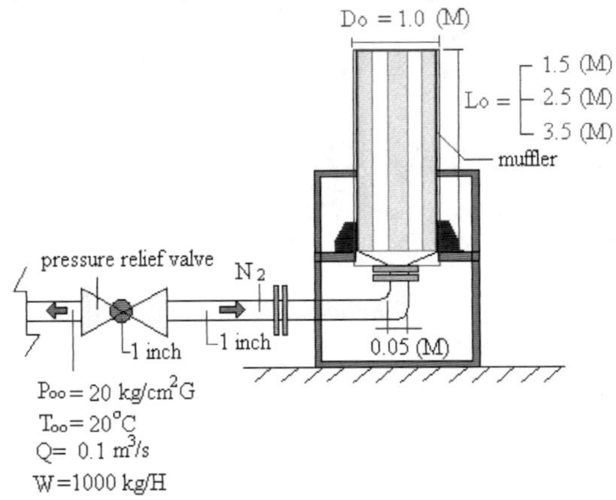

Figure 1. The gas emergent venting device.

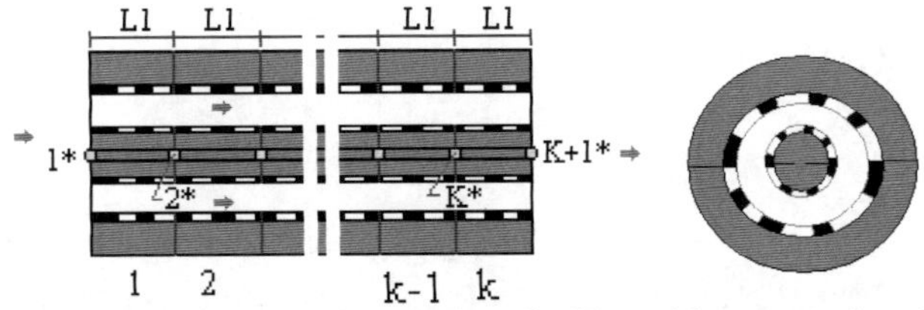

(A) Muffler A with segment number (k=K) and layer number (n = 1)

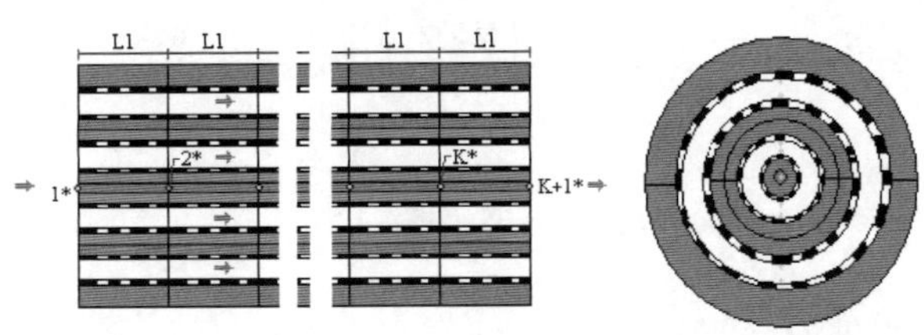

(B) Muffler B with segment number (k=K) and layer number (n = 2)

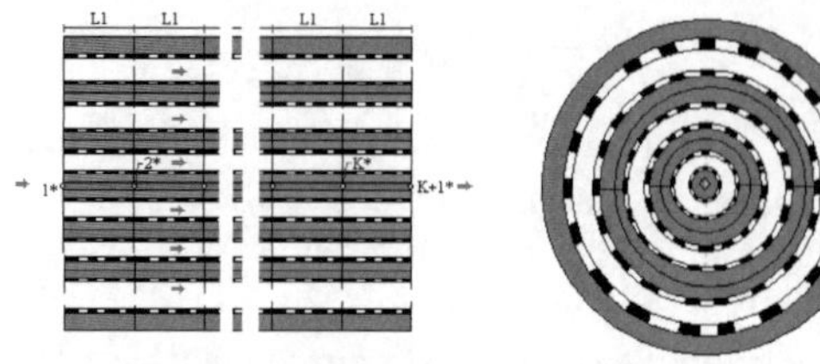

(C) Muffler C with segment number (k=K) and layer number (n = 3)

Figure 2. Three kinds of multi-channel pod mufflers (Muffler A: one-channel; Muffler B: two-channel; Muffler C: three-channel).

3.1. Muffler A (a one-channel pod muffler)

As indicated in Fig.2(A) and Fig. 3(A), the horizontal length of a muffler with 1 layer (n=1) of sound-absorbing rings is divided into k segments, and the cross section of the muffler is divided into 2 sections (m=2). Concerning the acoustical field of the one-segment muffler A shown in Fig. 2(A), the inner perforated plate is partitioned as L_1*L_{11}, and the outer perforated plate is partitioned as L_1*L_{22}. Based on the plane wave theory, the cutoff frequency (fc) is $c_0/2L_{max}$ where L_{max} is the maximum value of (L_1,L_{11}) and (L_1,L_{22}). The related matrix form between node 000 and node 001 is

$$\begin{bmatrix} p_{001} \\ u_{001} \end{bmatrix} = \begin{bmatrix} \cos[K_{fiber}(H_1)] & jZ_{fiber}\sin[K_{fiber}(H_1)] \\ j\dfrac{\sin[K_{fiber}(H_1)]}{Z_{fiber}} & \cos[K_{fiber}(H_1)] \end{bmatrix} \begin{bmatrix} p_{000} \\ u_{000} \end{bmatrix} \quad (1)$$

Developing Eq.(1) yields

$$Z_{001} = \frac{p_{001}}{u_{001}} = \frac{p_{000}\cos[K_{fiber}(H_1)] + ju_{000}Z_{fiber}\sin[K_{fiber}(H_1)]}{jp_{000}\dfrac{\sin[K_{fiber}(H_1)]}{Z_{fiber}} + u_{000}\cos[K_{fiber}(H_1)]} \quad (2)$$

$$= \frac{\dfrac{p_{000}}{u_{000}}\cos[K_{fiber}(H_1)] + jZ_{fiber}\sin[K_{fiber}(H_1)]}{j\dfrac{p_{000}}{u_{000}}\dfrac{\sin[K_{fiber}(H_1)]}{Z_{fiber}} + \cos[K_{fiber}(H_1)]}$$

Because the acoustical particle velocity is zero at the solid boundary, the acoustical impedance at node 000 is

$$Z_{000} = \frac{p_{000}}{u_{000}} = \infty \quad (3)$$

Plugging Eq. (3) into Eq.(2) yields

$$Z_{001} = \frac{\cos[K_{fiber}(H_1)]}{j\sin[K_{fiber}(H_1)]}\cdot Z_{fiber} \quad (4)$$

$$= -j\cot[K_{fiber}(H_1)]\cdot Z_{fiber}$$

where $Z_{fiber} = R_{fiber} + jX_{fiber}$; $K_{fiber} = K_{01A} + jK_{02A}$ (5)

Using the wool's acoustical impedance formula from Delany and Bazley [12] yields

$$K_{01A} = \left\{\left[\frac{\omega\cdot H_1}{c_o}\right]\left[1 + 0.0978\left(\frac{\rho_o f}{R_f}\right)^{-0.700}\right]\right\};\ K_{02A} = \left\{\left[\frac{\omega\cdot H_1}{c_o}\right]\left[-0.189\left(\frac{\rho_o f}{R_f}\right)^{-0.595}\right]\right\} \quad (6a)$$

$$R_{fiber} = \rho_o c_o\left[1 + 0.0571\left(\frac{\rho_o f}{R_f}\right)^{-0.754}\right];\ X_{fiber} = \rho_o c_o\left[-0.087\left(\frac{\rho_o f}{R_f}\right)^{-0.732}\right] \quad (6b)$$

Here, R_f is the flow resistance of the wool.

Assuming that the sound wave quickly passes through node 2 from node 001, the acoustical pressure between node 001 and node 2 can be regarded as the same. Also, adopting the continuity equation yields

$$p_{001} = p_2 ; u_{01} = u_2 \tag{7}$$

According to the definition of acoustical impedance at node 2 and using Eq.(7) yields

$$p_2 = Z_{p1} u_{001} + p_{001} \tag{8}$$

Combining Eq.(7) with Eq.(8), the matrix form between node 001 and node 2a is

$$\begin{pmatrix} p_{2a} \\ u_{2a} \end{pmatrix} = \begin{bmatrix} 1 & Z_{p1} \\ 0 & 1 \end{bmatrix} \begin{pmatrix} p_{001} \\ u_{001} \end{pmatrix} \tag{9}$$

Developing Eq.(9) yields

$$Z_{2a} = Z_{001} + Z_{p1} \tag{10}$$

Using the perforated plate's acoustical impedance formula from Beranek and Ver [13] yields

$$Z_{p1} = \frac{\rho_o}{pp_1}\sqrt{8v v\omega}\left(1+\frac{t_{w1}}{2d_{h1}}\right) + j\frac{\omega\rho_o}{pp_1}\left[\sqrt{\frac{8v}{\omega}}\left(1+\frac{t_{w1}}{2d_{h1}}\right) + t_{w1} + \delta_1\right] \tag{11a}$$

$$\delta_1 = 0.85(2d_{h1})\left(1-1.47\sqrt{pp_1}+0.47\sqrt{pp_1^{\,3}}\right) \tag{11b}$$

Therefore, the acoustical impedance at node 2a is

$$Z_{2a} = -j\cot\left[K_{fiber}(H_1)\right]\cdot Z_{fiber} + \frac{\rho_o}{pp_1}\sqrt{8v\omega}\left(1+\frac{t_{w1}}{2d_{h1}}\right) + j\frac{\omega\rho_o}{pp_1}\left[\sqrt{\frac{8v}{\omega}}\left(1+\frac{t_{w1}}{2d_{h1}}\right) + t_{w1} + \delta_1\right] \tag{12}$$

Similarly, the acoustical impedance at node 2b is

$$Z_{2b} = -j\cot\left[K_{fiber}(H_1)\right]\cdot Z_{fiber} + \frac{\rho_o}{pp_1}\sqrt{8v\omega}\left(1+\frac{t_{w1}}{2d_{h1}}\right) + j\frac{\omega\rho_o}{pp_1}\left[\sqrt{\frac{8v}{\omega}}\left(1+\frac{t_{w1}}{2d_{h1}}\right) + t_{w1} + \delta_1\right] \tag{13}$$

For node 3a, the matrix form between node 010 and node 011 inside the inner perforated plate can be expressed as

$$\begin{bmatrix} p_{011} \\ u_{011} \end{bmatrix} = \begin{bmatrix} \cos\left[K_{fiber}(H_2)\right] & jZ_{fiber}\sin\left[K_{fiber}(H_2)\right] \\ j\dfrac{\sin\left[K_{fiber}(H_2)\right]}{Z_{fiber}} & \cos\left[K_{fiber}(H_2)\right] \end{bmatrix} \begin{bmatrix} p_{010} \\ u_{010} \end{bmatrix} \tag{14}$$

Developing Eq.(14) yields

$$Z_{011} = \frac{p_{011}}{u_{011}} = \frac{p_{010}\cos\left[K_{fiber}(H_2)\right] + ju_{010}Z_{fiber}\sin\left[K_{fiber}(H_2)\right]}{jp_{010}\dfrac{\sin\left[K_{fiber}(H_2)\right]}{Z_{fiber}} + u_{010}\cos\left[K_{fiber}(H_2)\right]} = \frac{\dfrac{p_{010}}{u_{010}}\cos[K_{fiber}(H_2)] + jZ_{fiber}\sin[K_{fiber}(H_2)]}{j\dfrac{p_{010}}{u_{010}}\dfrac{\sin[K_{fiber}(H_2)]}{Z_{fiber}} + \cos[K_{fiber}(H_2)]} \tag{15}$$

Because the acoustical particle velocity is zero at the solid boundary, the acoustical impedance at node 010 is

$$Z_{010} = \frac{p_{010}}{u_{010}} = \infty \tag{16}$$

Plugging Eq. (16) into Eq.(15) yields

$$Z_{011} = \frac{\cos[K_{fiber}(H_2)]}{j\sin[K_{fiber}(H_2)]}\cdot Z_{fiber} = -j\cot\left[K_{fiber}(H_2)\right]\cdot Z_{fiber} \tag{17}$$

where $Z_{fiber} = R_{fiber} + jX_{fiber}$; $K_{fiber} = K_{11A} + jK_{12A}$ (18)

Similarly, using the wool's acoustical impedance formula from Delany and Bazley [12] yields

$$K_{11A} = \left\{\left[\frac{\omega \cdot H_2}{c_o}\right]\left[1+0.0978\left(\frac{\rho_o f}{R_f}\right)^{-0.700}\right]\right\};\ K_{12A} = \left\{\left[\frac{\omega \cdot H_2}{c_o}\right]\left[-0.189\left(\frac{\rho_o f}{R_f}\right)^{-0.595}\right]\right\} \quad (19a)$$

$$R_{fiber} = \rho_o c_o\left[1+0.0571\left(\frac{\rho_o f}{R_f}\right)^{-0.754}\right];\ X_{fiber} = \rho_o c_o\left[-0.087\left(\frac{\rho_o f}{R_f}\right)^{-0.732}\right] \quad (19b)$$

Likewise, assuming that the sound wave quickly passes through node 2 from node 001, the acoustical pressure between node 011 and node 3 can be regarded as the same. Also, adopting the continuity equation yields

$$p_{011} = p_3;\ u_{01} = u_2 \quad (20)$$

According to the definition of the acoustical impedance at node 3 and using Eq.(20) yields

$$p_3 = Z_{p2}u_{011} + p_{011} \quad (21)$$

Combining Eq.(A20) with Eq.(A21), the matrix form between node 011 and node 3a is

$$\begin{pmatrix} p_{3a} \\ u_{3a} \end{pmatrix} = \begin{bmatrix} 1 & Z_{p2} \\ 0 & 1 \end{bmatrix}\begin{pmatrix} p_{011} \\ u_{011} \end{pmatrix} \quad (22)$$

Developing Eq.(22) yields

$$Z_{3a} = Z_{011} + Z_{p2} \quad (23)$$

Similarly, using the perforated plate's acoustical impedance formula from Beranek and Ver [13] yields

$$Z_{p2} = \frac{\rho_o}{pp_2}\sqrt{8v\omega}\left(1+\frac{t_{w2}}{2d_{h2}}\right) + j\frac{\omega\rho_o}{pp_2}\left[\sqrt{\frac{8v}{\omega}}\left(1+\frac{t_{w2}}{2d_{h2}}\right)+t_{w2}+\delta_2\right] \quad (24a)$$

$$\delta_2 = 0.85(2d_{h2})\left(1-1.47\sqrt{pp_2}+0.47\sqrt{pp_2^{\,3}}\right) \quad (24b)$$

Therefore, the acoustical impedance at node 3a is

$$Z_{3a} = -j\cot\left[K_{fiber}(H_2)\right]\cdot Z_{fiber} + \frac{\rho_o}{pp_2}\sqrt{8v\omega}\left(1+\frac{t_{w2}}{2d_{h2}}\right) + j\frac{\omega\rho_o}{pp_2}\left[\sqrt{\frac{8v}{\omega}}\left(1+\frac{t_{w2}}{2d_{h2}}\right)+t_{w2}+\delta_2\right] \quad (25a)$$

Also, the acoustical impedance at node 3b is

$$Z_{3b} = -j\cot\left[K_{fiber}(H_2)\right]\cdot Z_{fiber} + \frac{\rho_o}{pp_2}\sqrt{8v\omega}\left(1+\frac{t_{w2}}{2d_{h2}}\right) + j\frac{\omega\rho_o}{pp_2}\left[\sqrt{\frac{8v}{\omega}}\left(1+\frac{t_{w2}}{2d_{h2}}\right)+t_{w2}+\delta_2\right] \quad (25b)$$

As indicated in Fig. 2, a sound wave propagates from the left side to the right side. According to the continuity equation and geometric data

$$p_1 = p_{2q} = p_{2b} = p_{3a} = p_{3b} = p_4; \quad (26a)$$

$$v_1 = v_{2a} + v_{2b} + v_{3a} + v_{3b} + v_4 \quad (26b)$$

$$S_1 = S_4 = \frac{\pi}{4}(D_3^2 - D_2^2)\cdot\frac{1}{M} = S_M; \quad (26c)$$

$$S_{2a} = S_{2b} = L_1 * L_{11} * pp_1; \quad (26d)$$

$$S_{3a} = S_{3b} = L_1 * L_{22} * pp_2; \quad (26e)$$

For the mass conservation theory $\rho_o S_M u_1 = \rho_o S_{2a}u_{2a} + \rho_o S_{2b}u_{2b} + \rho_o S_{3a}u_{3a} + \rho_o S_{3b}u_{3b} + \rho_o S_M u_4$ (27)

Based on the definition of acoustical impedance

$$Z_{2a} = \frac{p_{2a}}{u_{2a}};\ Z_{2b} = \frac{p_{2b}}{u_{2b}};\ Z_{3a} = \frac{p_{3a}}{u_{3a}};\ Z_{3b} = \frac{p_{3b}}{u_{3b}} \quad (28)$$

Or in the form of

$$u_{2a}=\frac{p_{2a}}{Z_{2a}};u_{2b}=\frac{p_{2b}}{Z_{2b}};u_{3a}=\frac{p_{3a}}{Z_{3a}};u_{3b}=\frac{p_{3b}}{Z_{3b}} \quad (29)$$

Here, $Z_{2a}=Z_{2b}$ and $Z_{3a}=Z_{3b}$

Summing up Eqs. (26)-(29) yields

$$\begin{cases} p_1=p_4 \\ \rho_o c_o u_1=\rho_o c_o\cdot u_4+(\rho_o c_o\cdot 2\cdot\frac{S_{2a}}{S_M}\frac{1}{Z_{2a}}+\rho_o c_o\cdot 2\cdot\frac{S_{3a}}{S_M}\frac{1}{Z_{3a}})p_4 \end{cases} \quad (30)$$

Eq.(30) can be expressed as

$$\begin{bmatrix} p_1 \\ \rho_o c_o u_1 \end{bmatrix}=\begin{bmatrix} 1 & 0 \\ \rho_o c_o\left(2\cdot\frac{S_{2a}}{S_M}\frac{1}{Z_{2a}}+2\frac{S_{3a}}{S_M}\frac{1}{Z_{3a}}\right) & 1 \end{bmatrix}\begin{bmatrix} p_4 \\ \rho_o c_o u_4 \end{bmatrix} \quad (31a)$$

Or

$$=>\begin{bmatrix} p_1 \\ \rho_o c_o u_1 \end{bmatrix}=\begin{bmatrix} T_{11} & T_{12} \\ T_{21} & T_{22} \end{bmatrix}\begin{bmatrix} p_4 \\ \rho_o c_o u_4 \end{bmatrix} \quad (31b)$$

The Sound Transmission Loss (STL) between node 1 and node 4 is

$$STL=10\log\left[\frac{|T_{11}+T_{12}+T_{21}+T_{22}|^2}{4}\right] \quad (32)$$

Similarly, for a cylindrical one-channel muffler with the section number m=M (with the segment number k = 1 and the layer number n = 1), the matrix form between node 1 and node 4 is

$$\begin{bmatrix} p_1 \\ \rho_o c_o u_1 \end{bmatrix}=\begin{bmatrix} 1 & 0 \\ \rho_o c_o\left(M\cdot\frac{S_{2-1}}{S_M}\frac{1}{Z_{2-1}}+M\frac{S_{3-1}}{S_M}\frac{1}{Z_{3-1}}\right) & 1 \end{bmatrix}\begin{bmatrix} p_4 \\ \rho_o c_o u_4 \end{bmatrix} \quad (33a)$$

Or

$$=>\begin{bmatrix} p_1 \\ \rho_o c_o u_1 \end{bmatrix}=\begin{bmatrix} T_{11} & T_{12} \\ T_{21} & T_{22} \end{bmatrix}\begin{bmatrix} p_4 \\ \rho_o c_o u_4 \end{bmatrix} \quad (33b)$$

The STL between node 1 and node 4 is

$$STL=10\log\left[\frac{|T_{11}+T_{12}+T_{21}+T_{22}|^2}{4}\right] \quad (34)$$

Furthermore, as indicated in Fig. 3(A), for a cylindrical one-channel muffler horizontally partitioned into k segments (with the section number m=M and the layer number n = 1), the matrix form from node 1 and node 2 to node k and node K+1 is

$$\begin{bmatrix} p_1^* \\ \rho_o c_o u_1^* \end{bmatrix}=\begin{bmatrix} 1 & 0 \\ \rho_o c_o\left(M\cdot\frac{S_{2-1}}{S_M}\frac{1}{Z_{2-1}}+M\frac{S_{3-1}}{S_M}\frac{1}{Z_{3-1}}\right) & 1 \end{bmatrix}\begin{bmatrix} p_2^* \\ \rho_o c_o u_2^* \end{bmatrix} \quad (35a)$$

$$\begin{bmatrix} p_2^* \\ \rho_o c_o u_2^* \end{bmatrix}=\begin{bmatrix} 1 & 0 \\ \rho_o c_o\left(M\cdot\frac{S_{2-1}}{S_M}\frac{1}{Z_{2-1}}+M\frac{S_{3-1}}{S_M}\frac{1}{Z_{3-1}}\right) & 1 \end{bmatrix}\begin{bmatrix} p_3^* \\ \rho_o c_o u_3^* \end{bmatrix} \quad (35b)$$

...

$$\begin{bmatrix} p_{K-1}^* \\ \rho_o c_o u_{K-1}^* \end{bmatrix}=\begin{bmatrix} 1 & 0 \\ \rho_o c_o\left(M\cdot\frac{S_{2-1}}{S_M}\frac{1}{Z_{2-1}}+M\frac{S_{3-1}}{S_M}\frac{1}{Z_{3-1}}\right) & 1 \end{bmatrix}\begin{bmatrix} p_K^* \\ \rho_o c_o u_K^* \end{bmatrix} \quad (35c)$$

$$\begin{bmatrix} p_K^* \\ \rho_o c_o u_K^* \end{bmatrix}=\begin{bmatrix} 1 & 0 \\ \rho_o c_o\left(M\cdot\frac{S_{2-1}}{S_M}\frac{1}{Z_{2-1}}+M\frac{S_{3-1}}{S_M}\frac{1}{Z_{3-1}}\right) & 1 \end{bmatrix}\begin{bmatrix} p_{K+1}^* \\ \rho_o c_o u_{K+1}^* \end{bmatrix} \quad (35d)$$

The total transfer matrix (between node 1 and node k+1) assembled by multiplication is

$$\begin{bmatrix} p_1^* \\ \rho_o c_o u_1^* \end{bmatrix}=\begin{bmatrix} T_{11} & T_{12} \\ T_{21} & T_{22} \end{bmatrix}\begin{bmatrix} p_{K+1}^* \\ \rho_o c_o u_{K+1}^* \end{bmatrix} \quad (36)$$

The STL between node 1 and node k+1 yields

IOP Conf. Series: Materials Science and Engineering **644** (2019) 012004 doi:10.1088/1757-899X/644/1/012004

$$STL_1 = 10\log\left[\frac{|T_{11} + T_{12} + T_{21} + T_{22}|^2}{4}\right] \tag{37}$$

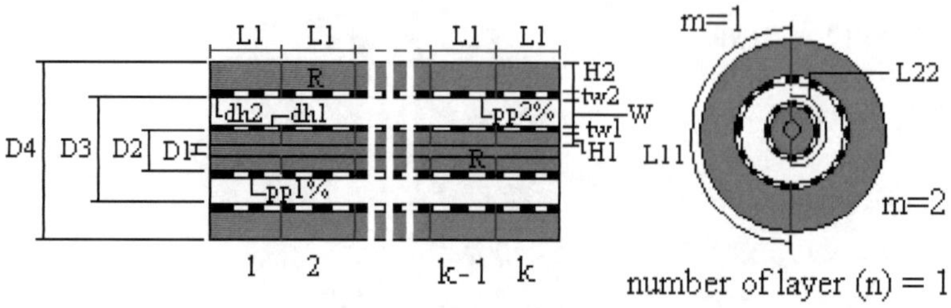

(A) Dimensions of muffler A

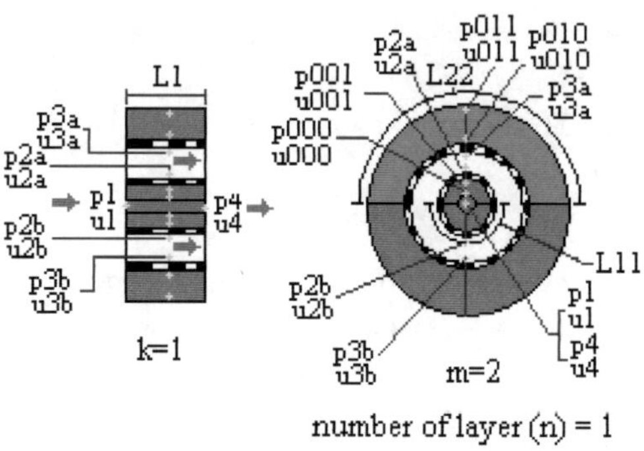

(B) Acoustical field of a segment of muffler A (section number: m=2, segment number: k = 1, layer number: n = 2)

Figure 3. The dimensions and acoustical field of muffler A.

3.2. Muffler B (a two-channel pod muffler)

The acoustical field of a segment of a cylindrical two-channel muffler (section number: m=2, segment number: k = 1, layer number: n = 2) is shown in Fig. 2(B) and Fig.4 where Z is the acoustic impedance, p is the acoustic pressure, and u is the particle velocity. As indicated in Fig. 4, the inner perforated plate is partitioned as $L_1{*}L_{1\text{-}11}$, and the outer perforated plate is partitioned as $L_1{*}L_{1\text{-}22}$. Based on the plane wave theory, the cutoff frequency (fc) is $c_0/2L_{max}$ where L_{max} is the maximum value of $(L_1,L_{1\text{-}11})$ and $(L_1,L_{1\text{-}22})$.

Similarly, the acoustical impedances of node 2-1a, node 2-1b, node 3-1a, and node 3-1b are

$$Z_{2\text{-}1a} = -j\cot[K_{fiber}(H_{1\text{-}1})]\cdot Z_{fiber} + \frac{\rho_o}{pp_{1\text{-}1}}\sqrt{8v\omega}\left(1+\frac{t_{w1\text{-}1}}{2d_{h1\text{-}1}}\right) + j\frac{\omega\rho_o}{pp_{1\text{-}1}}\left[\sqrt{\frac{8v}{\omega}}\left(1+\frac{t_{w1\text{-}1}}{2d_{h1\text{-}1}}\right)+t_{w1\text{-}1}+\delta_{1\text{-}1}\right] \tag{38a}$$

$$Z_{2\text{-}1b} = -j\cot[K_{fiber}(H_{1\text{-}2})]\cdot Z_{fiber} + \frac{\rho_o}{pp_{1\text{-}2}}\sqrt{8v\omega}\left(1+\frac{t_{w1\text{-}2}}{2d_{h1\text{-}2}}\right) + j\frac{\omega\rho_o}{pp_{1\text{-}2}}\left[\sqrt{\frac{8v}{\omega}}\left(1+\frac{t_{w1\text{-}2}}{2d_{h1\text{-}2}}\right)+t_{w1\text{-}2}+\delta_{1\text{-}2}\right] \tag{38b}$$

$$Z_{3\text{-}1a} = -j\cot[K_{fiber}(H_{1\text{-}1})]\cdot Z_{fiber} + \frac{\rho_o}{pp_{1\text{-}1}}\sqrt{8v\omega}\left(1+\frac{t_{w1\text{-}1}}{2d_{h1\text{-}1}}\right) + j\frac{\omega\rho_o}{pp_{1\text{-}1}}\left[\sqrt{\frac{8v}{\omega}}\left(1+\frac{t_{w1\text{-}1}}{2d_{h1\text{-}1}}\right)+t_{w1\text{-}1}+\delta_{1\text{-}1}\right] \tag{38c}$$

$$Z_{3\text{-}1b} = -j\cot[K_{fiber}(H_{1\text{-}2})]\cdot Z_{fiber} + \frac{\rho_o}{pp_{1\text{-}2}}\sqrt{8v\omega}\left(1+\frac{t_{w1\text{-}2}}{2d_{h1\text{-}2}}\right) + j\frac{\omega\rho_o}{pp_{1\text{-}2}}\left[\sqrt{\frac{8v}{\omega}}\left(1+\frac{t_{w1\text{-}2}}{2d_{h1\text{-}2}}\right)+t_{w1\text{-}2}+\delta_{1\text{-}2}\right] \tag{38d}$$

Also, the acoustical impedances of node 2-2a, node 2-2b, node 3-2a, and node 3-2b are

$$Z_{2\text{-}2a} = -j\cot[K_{fiber}(H_{2\text{-}1})]\cdot Z_{fiber} + \frac{\rho_o}{pp_{2\text{-}1}}\sqrt{8v v\omega}\left(1+\frac{t_{w2\text{-}1}}{2d_{h2\text{-}1}}\right) + j\frac{\omega\rho_o}{pp_{2\text{-}1}}\left[\sqrt{\frac{8v}{\omega}}\left(1+\frac{t_{w2\text{-}1}}{2d_{h2\text{-}1}}\right)+t_{w2\text{-}1}+\delta_{2\text{-}1}\right] \tag{39a}$$

$$Z_{2\text{-}2b} = -j\cot[K_{fiber}(H_{2\text{-}2})]\cdot Z_{fiber} + \frac{\rho_o}{pp_{2\text{-}2}}\sqrt{8v\omega}\left(1+\frac{t_{w2\text{-}2}}{2d_{h2\text{-}2}}\right) + j\frac{\omega\rho_o}{pp_{2\text{-}2}}\left[\sqrt{\frac{8v}{\omega}}\left(1+\frac{t_{w2\text{-}2}}{2d_{h2\text{-}2}}\right)+t_{w2\text{-}2}+\delta_{2\text{-}2}\right] \tag{39b}$$

$$Z_{3-2a} = -j\cot\left[K_{fiber}(H_{1-1})\right]\cdot Z_{fiber} + \frac{\rho_o}{pp_{1-1}}\sqrt{8\nu\nu\omega}\left(1+\frac{t_{w1-1}}{2d_{h1-1}}\right) + j\frac{\omega\rho_o}{pp_{1-1}}\left[\sqrt{\frac{8\nu}{\omega}}\left(1+\frac{t_{w1-1}}{2d_{h1-1}}\right) + t_{w1-1} + \delta_{1-1}\right] \tag{39c}$$

$$Z_{3-2b} = -j\cot\left[K_{fiber}(H_{2-2})\right]\cdot Z_{fiber} + \frac{\rho_o}{pp_{2-2}}\sqrt{8\nu\omega}\left(1+\frac{t_{w2-2}}{2d_{h2-2}}\right) + j\frac{\omega\rho_o}{pp_{2-2}}\left[\sqrt{\frac{8\nu}{\omega}}\left(1+\frac{t_{w2-2}}{2d_{h2-2}}\right) + t_{w2-2} + \delta_{2-2}\right] \tag{39d}$$

As indicated in Fig. 4, a sound wave propagates from the left side to the right side. According to the continuity equation and geometric data

$$p_1 = p_{2-1a} = p_{2-1b} = p_{3-1a} = p_{3-1b} = p_{2-2a} = p_{2-2b} = p_{3-2a} = p_{3-2b} = p_4; \tag{40a}$$

$$v_1 = v_{2-1a} + v_{2-1b} + v_{3-1a} + v_{3-1b} + v_{2-2a} + v_{2-2b} + v_{3-2a} + v_{3-2b} + v_4; \tag{40b}$$

$$S_1 = S_4 = \frac{\pi}{4}(D_3^2 - D_2^2)\cdot\frac{1}{M} = S_M; \tag{40c}$$

$$S_{2-1a} = S_{2-2a} = L_1 * L_{2-11a} * pp_{2-1a}; \tag{40d}$$

$$S_{2-1b} = S_{2-2b} = L_1 * L_{2-11b} * pp_{2-1b}; \tag{40e}$$

$$S_{3-1a} = S_{3-2a} = L_1 * L_{3-22a} * pp_{3-2a}; \tag{40f}$$

$$S_{3-1b} = S_{3-2b} = L_1 * L_{3-22b} * pp_{3-2b}; \tag{40g}$$

For the mass conservation theory

$$\rho_o S_M u_1 = \rho_o S_{2-1a}u_{2-1a} + \rho_o S_{2-1b}u_{2-1b} + \rho_o S_{3-1a}u_{3-1a} + \rho_o S_{3-1b}u_{3-1b} + \rho_o S_{2-2a}u_{2-2a} + \rho_o S_{2-2b}u_{2-2b} + \rho_o S_{3-2a}u_{3-2a} + \rho_o S_{3-2b}u_{3-2b} + \rho_o S_M u_4 \tag{41}$$

Based on the definition of acoustical impedance

$$Z_{2-1a} = \frac{p_{2-1a}}{u_{2-1a}};\ Z_{2-1b} = \frac{p_{2-1b}}{u_{2-1b}};\ Z_{3-1a} = \frac{p_{3-1a}}{u_{3-1a}};\ Z_{3-1b} = \frac{p_{3-1b}}{u_{3-1b}};\ Z_{2-2a} = \frac{p_{2-2a}}{u_{2-2a}};\ Z_{2-2b} = \frac{p_{2-2b}}{u_{2-2b}};\ Z_{3-1a} = \frac{p_{3-2a}}{u_{3-2a}};$$

$$Z_{3-2b} = \frac{p_{3-2b}}{u_{3-2b}};$$

$$u_{2-1a} = \frac{p_{2-1a}}{Z_{2-1a}};\ u_{2-1b} = \frac{p_{2-1b}}{Z_{2-1b}};\ u_{3-1a} = \frac{p_{3-1a}}{Z_{3-1a}};\ u_{3-1b} = \frac{p_{3-1b}}{Z_{3-1b}};\ u_{2-2a} = \frac{p_{2-2a}}{Z_{2-2a}};\ u_{2-2b} = \frac{p_{2-2b}}{Z_{2-2b}};$$

$$u_{3-2a} = \frac{p_{3-2a}}{Z_{3-2a}};\ u_{3-2b} = \frac{p_{3-2b}}{Z_{3-2b}}; \tag{42}$$

where $Z_{2a} = Z_{2b}; Z_{3a} = Z_{3b}$

Summing up the above formulas yields

$$\begin{cases} p_1 = p_4 \\ \rho_o c_o u_1 = \rho_o c_o \cdot u_4 + (\rho_o c_o \cdot 2\cdot\frac{S_{2-1a}}{S_M}\frac{1}{Z_{2-1a}} + \rho_o c_o \cdot 2\cdot\frac{S_{2-1b}}{S_M}\frac{1}{Z_{2-1b}} + \\ \rho_o c_o \cdot 2\cdot\frac{S_{3-1a}}{S_M}\frac{1}{Z_{3-1a}} + \rho_o c_o \cdot 2\cdot\frac{S_{3-1b}}{S_M}\frac{1}{Z_{3-1b}})p_4 \end{cases} \tag{43}$$

which can be expressed in a matrix form as

$$=>\begin{bmatrix} p_1 \\ \rho_o c_o u_1 \end{bmatrix} = \begin{bmatrix} 1 & 0 \\ \rho_o c_o\begin{pmatrix} 2\cdot\frac{S_{2-1a}}{S_M}\cdot\frac{1}{Z_{2-1a}} + 2\cdot\frac{S_{3-1a}}{S_M}\cdot\frac{1}{Z_{3-1a}} + \\ 2\frac{S_{2-1b}}{S_M}\cdot\frac{1}{Z_{2-1b}} + 2\frac{S_{3-1b}}{S_M}\cdot\frac{1}{Z_{3-1b}} \end{pmatrix} & 1 \end{bmatrix}\begin{bmatrix} p_4 \\ \rho_o c_o u_4 \end{bmatrix} \tag{44a}$$

Or in the form of

$$\begin{bmatrix} p_1 \\ \rho_o c_o u_1 \end{bmatrix} = \begin{bmatrix} T_{11} & T_{12} \\ T_{21} & T_{22} \end{bmatrix}\begin{bmatrix} p_4 \\ \rho_o c_o u_4 \end{bmatrix} \tag{44b}$$

The STL between nodes 1 and 4 is

$$STL = 10\log\left[\frac{|T_{11} + T_{12} + T_{21} + T_{22}|^2}{4}\right] \tag{45}$$

Furthermore, as indicated in Fig. 2(B), for a cylindrical two-channel muffler horizontally partitioned as k segments (with a section number m=M and a layer number n = 2), the matrix form from node 1 and node 2 to node k and node K+1 is

$$\begin{bmatrix} p_1^* \\ \rho_o c_o u_1^* \end{bmatrix} = \begin{bmatrix} 1 & 0 \\ \rho_o c_o \begin{pmatrix} M \cdot \frac{S_{2-1a}}{S_M} \frac{1}{Z_{2-1a}} + M \frac{S_{3-1a}}{S_M} \frac{1}{Z_{3-1a}} \\ + M \cdot \frac{S_{2-1b}}{S_M} \frac{1}{Z_{2-1b}} + M \cdot \frac{S_{3-1b}}{S_M} \frac{1}{Z_{3-1b}} \end{pmatrix} & 1 \end{bmatrix} \begin{bmatrix} p_2^* \\ \rho_o c_o u_2^* \end{bmatrix} \tag{46a}$$

$$\begin{bmatrix} p_2^* \\ \rho_o c_o u_2^* \end{bmatrix} = \begin{bmatrix} 1 & 0 \\ \rho_o c_o \begin{pmatrix} M \cdot \frac{S_{2-1a}}{S_M} \frac{1}{Z_{2-1a}} + M \frac{S_{3-1a}}{S_M} \frac{1}{Z_{3-1a}} \\ + M \cdot \frac{S_{2-1b}}{S_M} \frac{1}{Z_{2-1b}} + M \cdot \frac{S_{3-1b}}{S_M} \frac{1}{Z_{3-1b}} \end{pmatrix} & 1 \end{bmatrix} \begin{bmatrix} p_3^* \\ \rho_o c_o u_3^* \end{bmatrix} \tag{46b}$$

...

$$\begin{bmatrix} p_{K-1}^* \\ \rho_o c_o u_{K-1}^* \end{bmatrix} = \begin{bmatrix} 1 & 0 \\ \rho_o c_o \begin{pmatrix} M \cdot \frac{S_{2-1a}}{S_M} \frac{1}{Z_{2-1a}} + M \frac{S_{3-1a}}{S_M} \frac{1}{Z_{3-1a}} \\ + M \cdot \frac{S_{2-1b}}{S_M} \frac{1}{Z_{2-1b}} + M \cdot \frac{S_{3-1b}}{S_M} \frac{1}{Z_{3-1b}} \end{pmatrix} & 1 \end{bmatrix} \begin{bmatrix} p_K^* \\ \rho_o c_o u_K^* \end{bmatrix} \tag{46c}$$

$$\begin{bmatrix} p_K^* \\ \rho_o c_o u_K^* \end{bmatrix} = \begin{bmatrix} 1 & 0 \\ \rho_o c_o \begin{pmatrix} M \cdot \frac{S_{2-1a}}{S_M} \frac{1}{Z_{2-1a}} + M \frac{S_{3-1a}}{S_M} \frac{1}{Z_{3-1a}} \\ + M \cdot \frac{S_{2-1b}}{S_M} \frac{1}{Z_{2-1b}} + M \cdot \frac{S_{3-1b}}{S_M} \frac{1}{Z_{3-1b}} \end{pmatrix} & 1 \end{bmatrix} \begin{bmatrix} p_{K+1}^* \\ \rho_o c_o u_{K+1}^* \end{bmatrix} \tag{46d}$$

The system matrix between node 1 and node k+1 is

$$\begin{bmatrix} p_1^* \\ \rho_o c_o u_1^* \end{bmatrix} = \begin{bmatrix} T_{11} & T_{12} \\ T_{21} & T_{22} \end{bmatrix} \begin{bmatrix} p_{K+1}^* \\ \rho_o c_o u_{K+1}^* \end{bmatrix} \tag{47}$$

The STL between node 1 and node k+1 is

$$STL_2 = 10\log\left[\frac{|T_{11} + T_{12} + T_{21} + T_{22}|^2}{4}\right] \tag{48}$$

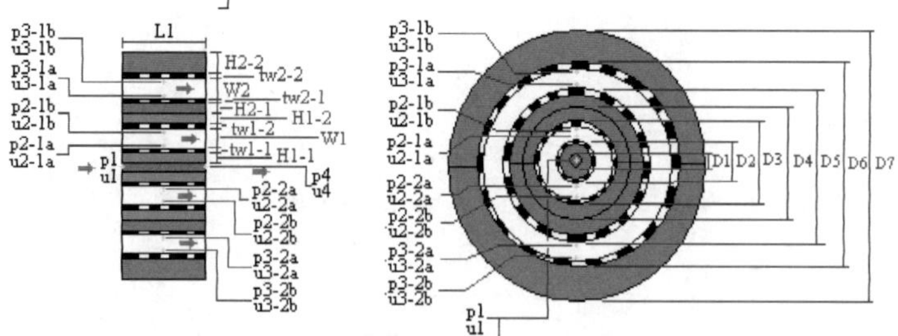

Figure 4. The acoustical number of a segment of muffler B (section number: m=2, segment number: k = 1, layer number: n = 2).

3.3. Muffler C (a three-channel pod muffler)

Similarly, as indicated in Fig. 2(C), for a cylindrical three-channel muffler horizontally partitioned as k segments (with a section number m=M and a layer number n = 3), the matrix form from node 1 and node 2 to node k and node K+1 is

$$\begin{bmatrix} p_1^* \\ \rho_o c_o u_1^* \end{bmatrix} = \begin{bmatrix} 1 & 0 \\ \rho_o c_o \left(\sum_{n=1}^{3} [M \cdot \frac{S_{(n+1)-1a}}{S_M} \frac{1}{Z_{(n+1)-1a}} + M \cdot \frac{S_{(n+1)-1b}}{S_M} \cdot \frac{1}{Z_{(n+1)-1b}}] \right) & 1 \end{bmatrix} \begin{bmatrix} p_2^* \\ \rho_o c_o u_2^* \end{bmatrix} \tag{49a}$$

$$\begin{bmatrix} p_2^* \\ \rho_o c_o u_2^* \end{bmatrix} = \begin{bmatrix} 1 & 0 \\ \rho_o c_o \left(\sum_{n=1}^{3} [M \cdot \frac{S_{(n+1)-1a}}{S_M} \frac{1}{Z_{(n+1)-1a}} + M \cdot \frac{S_{(n+1)-1b}}{S_M} \cdot \frac{1}{Z_{(n+1)-1b}}] \right) & 1 \end{bmatrix} \begin{bmatrix} p_3^* \\ \rho_o c_o u_3^* \end{bmatrix} \tag{49b}$$

...

$$\begin{bmatrix} p_{K-1}^* \\ \rho_o c_o u_{K-1}^* \end{bmatrix} = \begin{bmatrix} 1 & 0 \\ \rho_o c_o \left(\sum_{n=1}^{3} [M \cdot \frac{S_{(n+1)-1a}}{S_M} \frac{1}{Z_{(n+1)-1a}} + M \cdot \frac{S_{(n+1)-1b}}{S_M} \cdot \frac{1}{Z_{(n+1)-1b}}] \right) & 1 \end{bmatrix} \begin{bmatrix} p_K^* \\ \rho_o c_o u_K^* \end{bmatrix} \tag{49c}$$

$$\begin{bmatrix} p_K^* \\ \rho_o c_o u_K^* \end{bmatrix} = \begin{bmatrix} 1 & 0 \\ \rho_o c_o \left(\sum_{n=1}^{3} [M \cdot \frac{S_{(n+1)-1a}}{S_M} \frac{1}{Z_{(n+1)-1a}} + M \cdot \frac{S_{(n+1)-1b}}{S_M} \cdot \frac{1}{Z_{(n+1)-1b}}] \right) & 1 \end{bmatrix} \begin{bmatrix} p_{K+1}^* \\ \rho_o c_o u_{K+1}^* \end{bmatrix} \tag{49d}$$

$$= \begin{bmatrix} TCC1_{1,1} & TCC1_{1,2} \\ TCC1_{2,1} & TCC1_{2,2} \end{bmatrix} \begin{bmatrix} p_{K+1}^* \\ \rho_o c_o u_{K+1}^* \end{bmatrix}$$

The resulting system matrix between node 1 and node k+1 can be expressed as

$$\begin{bmatrix} p_1^* \\ \rho_o c_o u_1^* \end{bmatrix} = \begin{bmatrix} T_{11} & T_{12} \\ T_{21} & T_{22} \end{bmatrix} \begin{bmatrix} p_{K+1}^* \\ \rho_o c_o u_{K+1}^* \end{bmatrix} \tag{50}$$

The STL between node 1 and node k+1 is

$$STL_3 = 10 \log \left[\frac{|T_{11} + T_{12} + T_{21} + T_{22}|^2}{4} \right] \tag{51}$$

3.4 Overall sound power level

The silenced octave sound power level emitted from the k-th muffler's outlet is

$$SWL_{k-i} = SWLO(f_i) - STL_k(f_i) \tag{52}$$

where $SWLO(f_i)$ is the original *SWL* at the inlet of a muffler (or pipe outlet), and f_i is the relative octave band frequency; $STL_k(f_i)$ is the k-th muffler's *STL* with respect to the relative octave band frequency (f_i); SWL_{k-i} is the silenced *SWL* at the outlet of the k-th muffler with respect to the relative octave band frequency.

Finally, the overall SWL_T silenced by the k-channel pad muffler at the outlet is

$$SWL_{T-k} = 10 * \log\{\sum_{i=1}^{n} 10^{SWL_{k-i}/10}\}$$
$$= 10 * \log\left\{ 10^{[SWLO(f_1) - STL_k(f_1)]/10} + 10^{[SWLO(f_2) - STL_k(f_2)]/10} + 10^{[SWLO(f_3) - STL_k(f_3)]/10} + ... + 10^{[SWLO(f_n) - STL_k(f_n)]/10} \right\} \tag{53}$$

3.5. Objective function. STL maximization for a tone (f) noise

The objective function for maximizing the *STL* of muffler A at a pure tone (*f*) is

$$OBJ_{11} = STL_1(f, RT_1, RT_2, RT_3, RT_4, RT_5, RT_6, RT_7) \tag{54a}$$

RT_1=pp_1%=pp_2%; RT_2=dh_1=dh_2; RT_3=D_3/D_4; RT_4=D_2/D_3; RT_5=Df_1/H_1; RT_6=Df_2/H_2; RT_7=Ra (54b)

Similarly, the objective function for maximizing the *STL* of muffler B at a pure tone (*f*) is

$$OBJ_{12} = \left(Q, f, RT_1^*, RT_2^*, RT_3^*, RT_4^*, RT_5^*, RT_6^*, RT_7^*, RT_8^*, RT_9^*\right) \tag{55a}$$

RT_1*=pp_1%=pp_2%=pp_3%; RT_2*=dh_1=dh_2=dh_3; RT_3*= H_1;RT_4*= H_2; RT_5*=H_3; RT_6*=Df_1/H_1; RT_7*= Df_2/H_2; RT_8*= Df_3/H_3; RT_9*=Ra (55b)

Also, the objective function for maximizing the *STL* of muffler C at a pure tone (*f*) is

$$OBJ_{13} = \begin{pmatrix} Q, f, RT_1^{**}, RT_2^{**}, RT_3^{**}, RT_4^{**}, RT_5^{**}, RT_6^{**}, RT_7^{**}, RT_8^{**}, RT_9^{**}, \\ RT_{10}^{**}, RT_{11}^{**} \end{pmatrix} \tag{56a}$$

RT_1**=pp_1%=pp_2%=pp_3%=pp_4%; RT_2**=dh_1=dh_2=dh_3=dh_4; RT_3**= H_1; RT_4**=H_2; RT_5**= H_3; RT_6**= H_4; RT_7**= Df_1/H_1; RT_8**= Df_2/H_2; RT_9**= Df_3/H_3; RT_{10}**=Df_4/H_4; RT_{11}**=Ra (56b)

The related ranges of the parameters are shown in Table 1.

Table 1. The corresponding space constraints and the ranges of design parameters for the mufflers.

	Range of design parameters
Muffler A	Q=0.1 (m^3/s) ; L_1=0.6 (m); D_1=0.05 (m); RT_1:[5, 40] ; RT_2:[0.003, 0.015] ; RT_3:[0.5, 0.7] ; RT_4: [0. 5, 0.7] ; RT_5: [0.2, 0.99] ; RT_6: [0.2, 0.99] ; RT_7: [3000, 20000]
Muffler B	Q=0.1 (m^3/s) ; L_1=0.6 (m); D_1=0.05 (m); RT_1^*:[5, 40] ; RT_2^*:[0.003, 0.015] ; RT_3^*:[0.025, 0.075] ; RT_4^*: [0. 025, 0.075] ; RT_5^*: [0. 025, 0.075] ; RT_6^*: [0.2, 0.99] ; RT_7^*: [0.2, 0.99] ; RT_8^*: [0.2, 0.99] ; RT_9^*: [3000, 20000]
Muffler C	Q=0.1 (m^3/s) ; L_1=0.6 (m); D_1=0.05 (m); RT_1^{**}:[5, 40] ; RT_2^{**}:[0.003, 0.015] ; RT_3^{**}:[0.025, 0.05] ; RT_4^{**}: [0. 025, 0.05] ; RT_5^{**}: [0. 025, 0.05] ; RT_6^{**}: [0. 025, 0.05] ; RT_7^{**}: [0.2, 0.99] ; RT_8^{**}: [0.2, 0.99] ; RT_9^{**}: [0.2, 0.99] ; RT_{10}^{**}: [0. 2, 0.99] ; RT_{11}^{**}: [3000, 20000]

3.6. SWL minimization for a broadband noise

To minimize the overall SWL_T, the objective functions are

$$OBJ_{21} = SWL_{T-1}(f, RT_1, RT_2, RT_3, RT_4, RT_5, RT_6, RT_7) \tag{57a}$$

$$OBJ_{22} = SWL_{T-2}\left(f, RT_1^*, RT_2^*, RT_3^*, RT_4^*, RT_5^*, RT_6^*, RT_7^*, RT_8^*, RT_9^*\right) \tag{57b}$$

$$OBJ_{23} = SWL_{T-3}\begin{pmatrix} f, RT_1^{**}, RT_2^{**}, RT_3^{**}, RT_4^{**}, RT_5^{**}, RT_6^{**}, RT_7^{**}, RT_8^{**}, RT_9^{**}, \\ RT_{10}^{**}, RT_{11}^{**} \end{pmatrix} \tag{57c}$$

4. Case studies

A space-constrained muffler for a nitrogen venting system shown in Fig. 1 is connected with a high pressure nitrogen blow-down (with a mass flow rate(W) of 1000 kg/H) system for the purpose of eliminating noise. The upstream pressure (P_{oo}) and temperature (T_{oo}) of the safety valve are 20kg/cm^{2}G and 20 ^{o}C. The diameters of the valve throat and the downstream pipe are 1 inch each. The unsilenced sound power level (SWL) produced by the N_2 vents and blow-down is [14]

$$Lw_i = 10\log(P_{oo}) + 10\log(A_o) + 20\log(T_{oo}) + 20\log(\frac{1}{SG}) + 94 + CCC_i(fm) \tag{58}$$

where the peak frequency ($fm = \frac{0.2(Vj)}{dm}$) is a function of the jet's velocity (Vj) and the valve size (dm). The spectrum correction ($CCC_i(fm)$) is a function of the pipe's downstream diameter. A_o is the section area of the valve's throat. SG, the specific gravity of N_2, is 0.969. The maximal volume flow rate (Q) during venting is given as 0.1 (m^3/s). The calculated SWL at various octave frequencies is listed in Table 2.

To efficiently reduce the sound energy, a cylindrical multi-channel pod muffler is adopted. As shown in Fig. 1, the available space for a cylindrical muffler is 1.0 m in diameter. Three kinds of available lengths (1.5m, 2.5m, and 3.5 m) will be assessed during the muffler's shape optimization process. Moreover, to simplify the optimization, the thickness of the inner perforated plate (t=0.008(m)) is preset in advance. Before the minimization of the broadband noise is performed, a SA function check using the maximization of the STL at the targeted pure tone (1000 Hz) has been performed. The corresponding OBJ functions are summarized in Eqs.(54)~(57).

Table 2. Unsilenced SWL of a Nitrogen venting noise inside a duct outlet.

	Octave Band Frequencies (Hz)					overall
	125	250	500	1000	2000	
Original - dB	140.8	140.8	140.8	140.8	140.8	

IOP Conf. Series: Materials Science and Engineering **644** (2019) 012004 doi:10.1088/1757-899X/644/1/012004

Spectrum correction - CCC_i (fm=5664 (Hz))	-27	-19	-12	-7	-5	
A-weighted	-16	-9	-3	+0	+1	
SWLO – dB(A)	97.8	112.8	125.8	133.8	136.8	138.8

5. Influence of design parameters

In order to achieve an optimally shaped muffler, the selection of appropriate design parameters is essential. An investigation of the acoustical influence with respect to the design parameters in muffler A is exemplified. The resulting *STL*s with respect to design parameters pp% (porosity of the perforated plate), dh (diameter of the perforated hole), H (thickness of the sound absorbing ring), W (width of the air way), Df (thickness of the sound absorbing wool), R (flow resistivity of the sound absorbing wool), L1 (length of the sound absorbing rings), and tw (thickness of the perforated plate) are assessed and shown in Figs. 6 and 12. As indicated in Figs. 6 and 12, the *STL* is influenced by the design parameters RT_1~RT_7. Similarly, the acoustical influence with respect to the design parameters in muffler B and muffler C also indicate that the *STL* is influenced by the design parameters RT_1*~RT_9* and RT_1**~$RT_{11}**$. Therefore, the design parameters RT_1~RT_7 in muffler A, the design parameters RT_1*~$RT_{19}*$ in muffler B, and the design parameters RT_1**~$RT_{11}*$ in muffler C will be selected as the optimization parameters in section 6.

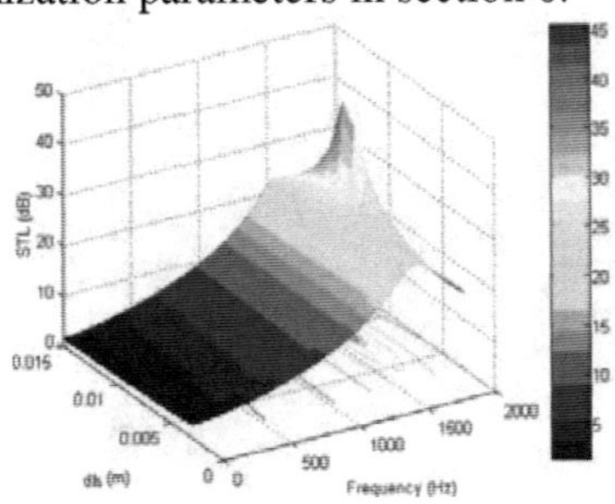

Figure 5. The resulting *STL* with respect to the design parameter - dh (diameter of a perforated hole).

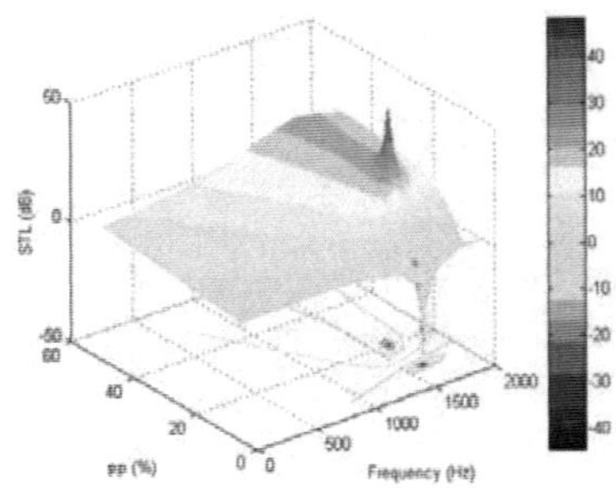

Figure 6. The resulting *STL* with respect to the design parameter - pp% (porosity of perforated plate).

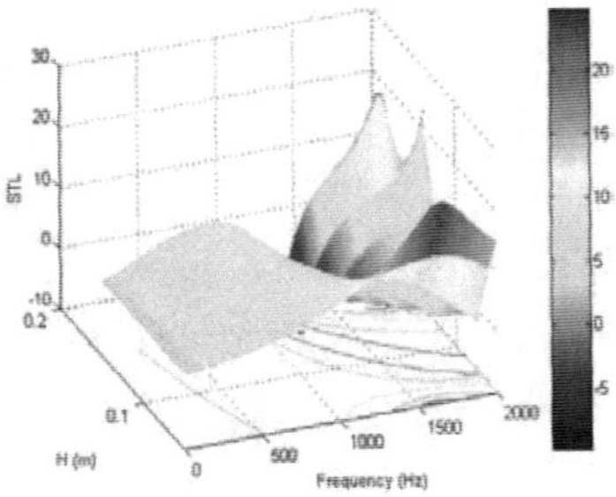

Figure 7. The resulting *STL* with respect to the design parameter - H (thickness of the sound absorbing ring).

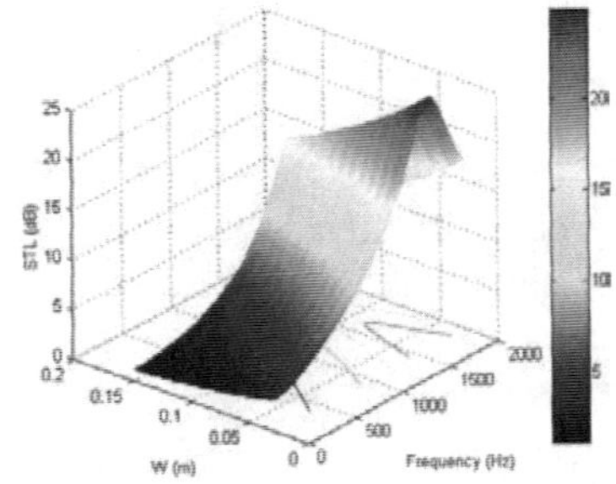

Figure 8. The resulting *STL* with respect to the design parameter - W (width of the air way).

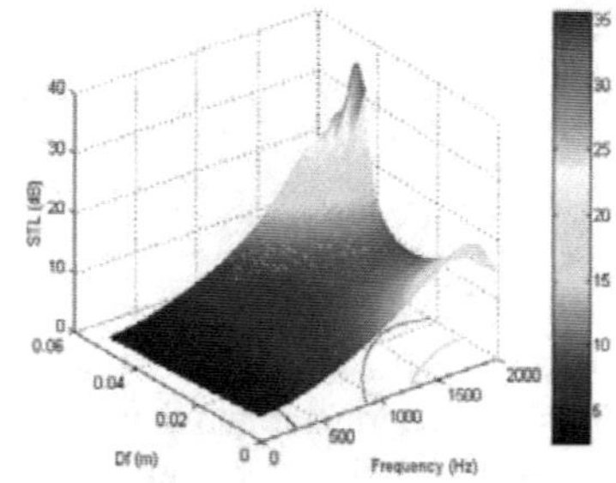

Figure 9. The resulting *STL* with respect to the design parameter - Df (thickness of the sound absorbing wool).

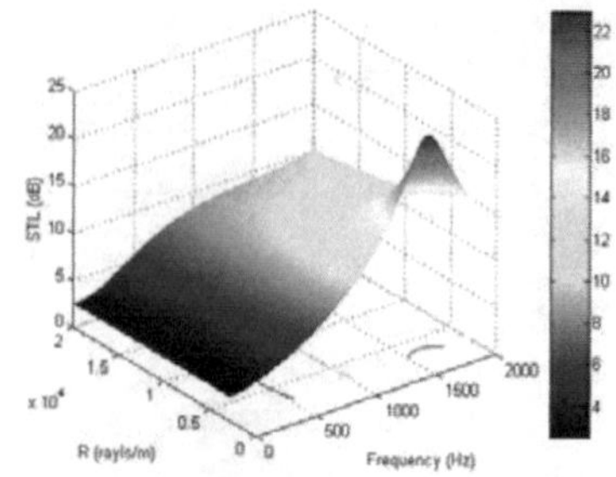

Figure 10. The resulting *STL* with respect to the design parameter - R (flow resistivity of the sound absorbing wool).

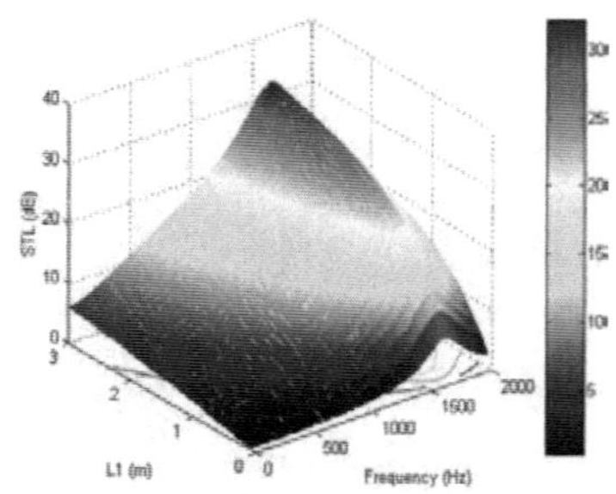

Figure 11. The resulting *STL* with respect to the design parameter – L1 (length of the sound absorbing rings).

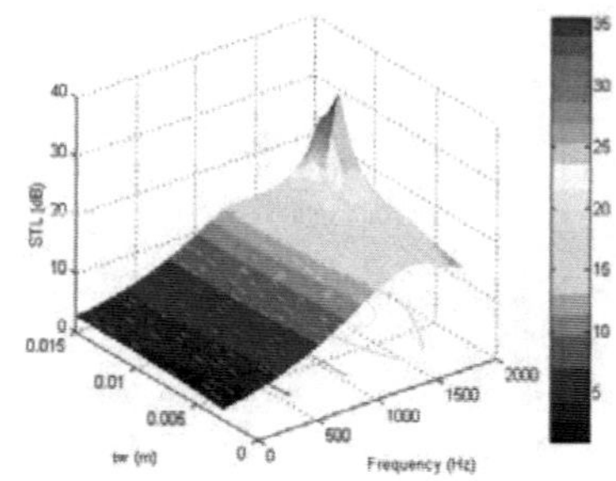

Figure 12. The resulting *STL* with respect to the design parameter – tw (thickness of the perforated plate).

6. Simulated annealing method

Because there is no need to choose starting data that is necessary for the classical gradient methods of *EPFM*, *IPFM* and *FDM* [15], the Simulated Annealing (*SA*), which is one of the best stochastic search method and does not need a good starting point for global searching, is convenient for the numerical assessment.

The simulated annealing (*SA*) was first introduced by Metropolis *et al.* [16] and further developed by Kirkpatrick *et al.* [17]. According to the philosophy of *SA*, the main issue is to bring the system from an arbitrary initial state to a state with minimum possible energy, which is the desired objective. Annealing is a slow cooling process that stabilizes a metal's temperature. Because of this, the particles remain close to the minimum energy state. In order to emulate the *SA*'s evolution, a new random solution (X') is chosen from the neighborhood of the current solution. If, possibly, there is a negative change in the objective function ($\Delta F \le 0$), the new solution will be recognized as the new current solution. If, conversely, there is not a negative change, the probability (pb(X')) of a transition to the new state will be calculated using the Boltzmann factor (pb(X') = $\exp(\Delta F / CT)$) where *C* and *T* are the Boltzmann constant and the current temperature.

$$pb(X') = \left(\begin{array}{c} 1, \Delta F \le 0 \\ \exp(\dfrac{-\Delta F}{CT}), \Delta f > 0 \end{array} \right. \tag{59a}$$

$$\Delta F = F(X') - F(X) \tag{59b}$$

IOP Conf. Series: Materials Science and Engineering **644** (2019) 012004 doi:10.1088/1757-899X/644/1/012004

For the purpose of escaping from the local optimum, *SA* also permits movement that results in inferior solutions (uphill movement). Hence, if the transition property (pb(X')) is greater than a random number of rand(0,1), the new inferior solution which results in a higher energy condition will be accepted; if not, it will be rejected. Each successful substitution of the new current solution will conduct to the decay of the current temperature as

$$T_{new} = kk * T_{old} \tag{60}$$

where *kk* is the cooling rate.

This process is repeated until the preset (*iter*) of the outer loop is reached [18, 19, 20, 21, 22, 23].

7. Results and discussion

7.1. Results

To achieve good optimization, two kinds of *SA* parameters including *kk* (cooling rate) and *iter* (maximum iteration) are varied step by step:

kk (0.91, 0.93, 0.95, 0.97, 0.99); *iter* (50, 100, 500, 1000, 5000, 10000).

Two results of optimization (one, pure tone noises used for *SA*'s accuracy check; and the other, a broadband noise occurring in a venting outlet) are described below.

7.2. Pure tone noise optimization

Before dealing with a broadband noise for muffler A~C, the *STL*'s maximization for muffler A with respect to a one-tone noise (1000Hz) is introduced for a reliability check using the *SA* method. By using Eq. (54), the maximization of the *STL* with respect to muffler A (a one-channel pod muffler) at the specified pure tone (1000Hz) and limited length of Lo=0.6 (m) was performed first. As indicated in Table 3, ten sets of *SA* parameters were tried during the muffler's optimization. Obviously, the optimal design data can be obtained with the tenth set of *SA* parameters at (*kk, iter*) = (0.99, 10000). Using the optimal design in a theoretical calculation, the optimal *STL* curves with respect to various *SA* parameters (*kk, iter*) are plotted and depicted in Figs. 13 and 14. As revealed in Figs. 13 and 14, the *STL* is precisely maximized at the desired frequency of 1000 Hz. Consequently, the *SA* optimizer is reliable in the optimization process. Using the same *SA* parameters (*kk, iter*) in mufflers A and C, the optimal *STL* curves obtained are shown in Table 4 and plotted in Fig. 15. As illustrated in Table 3, the *STL*s of mufflers A~C at 1000 Hz reach 12.3 dB, 6.3 dB, and 8.4 dB, respectively.

Table 3. Optimal *STL* for muffler A (equipped with two layers of sound absorbing rings) at various *SA* parameters (targeted tone of 1000 Hz and Lo=0.6 (m)).

item	SA Parameter	Control	Design parameters							STL (dB)
	kk	iter	RT1	RT2	RT3	RT4	RT5	RT6	RT7	
1	0.91	50	35.64	0.0135	0.675	0.675	0.8915	0.8915	17880	3.47
2	0.93	50	39.94	0.01498	0.6997	0.6997	0.9886	0.9886	19970	4.93
3	0.95	50	36.98	0.01397	0.6828	0.6828	0.9219	0.9219	18530	5.73
4	0.97	50	39.3	0.01476	0.696	0.696	0.9741	0.9741	19660	6.07
5	**0.99**	50	39.89	0.01496	0.6994	0.6994	0.9874	0.9874	19950	6.52
6	0.99	100	39.52	0.01483	0.6972	0.6972	0.9791	0.9791	19770	7.23
7	0.99	500	39.4	0.01479	0.6965	0.6965	0.9764	0.9764	19710	7.29
8	0.99	1000	39.39	0.01479	0.6965	0.6965	0.9762	0.9762	19700	7.40
9	0.99	5000	39.41	0.01480	0.6967	0.6967	0.9768	0.9768	19720	7.54
10	**0.99**	**10000**	35.8	0.01356	0.6760	0.6760	0.8953	0.8953	17960	12.25

Table 4. Optimal design data for three kinds of mufflers (mufflers A~C) (at a targeted tone of 1000 Hz and Lo=0.6 (m)) (kk=0.99, iter=10000).

	Design parameters											STL (dB)
Muffler A	RT1	RT2	RT3	RT4	RT5	RT6	RT7					12.25
	35.8	0.01356	0.6760	0.6760	0.8953	0.8953	17960					
Muffler B	RT1*	RT2*	RT3*	RT4*	RT5*	RT6*	RT7*	RT8*	RT9*			6.29
	39.25	0.01474	0.07392	0.07392	0.07392	0.9730	0.9730	0.9730	19630			
Muffler C	RT1**	RT2**	RT3**	RT4**	RT5**	RT6**	RT7**	RT8**	RT9**	RT10**	RT11**	8.41
	39.97	0.01499	0.04998	0.04998	0.04998	004998	0.9893	0.9893	0.9893	0.9893	19990	

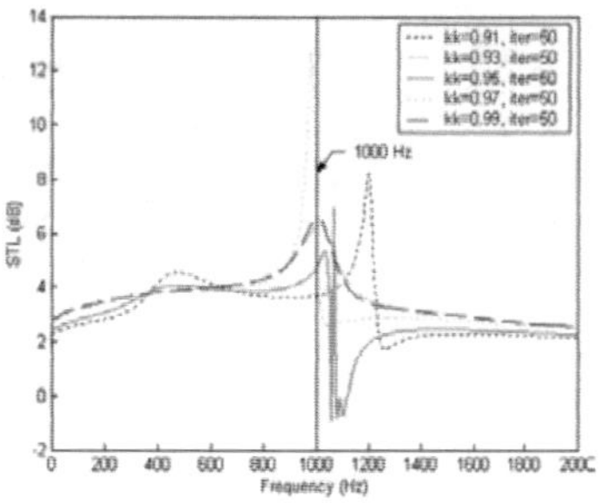

Figure 13. *STL* with respect to various *kk* [muffler A: target tone=1000 Hz].

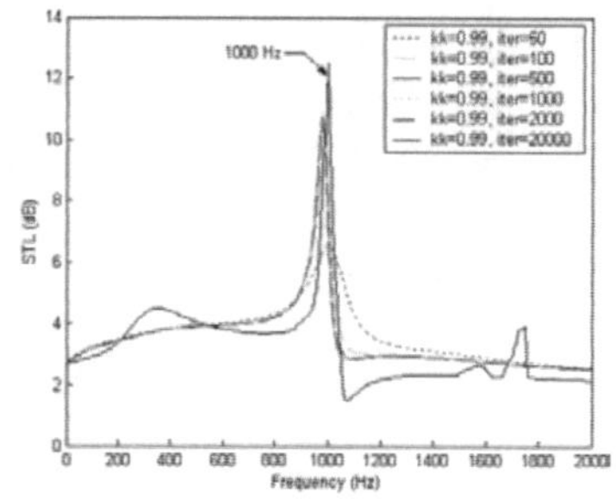

Figure 14. *STL* with respect to various *iter* [muffler A: target tone=1000 Hz].

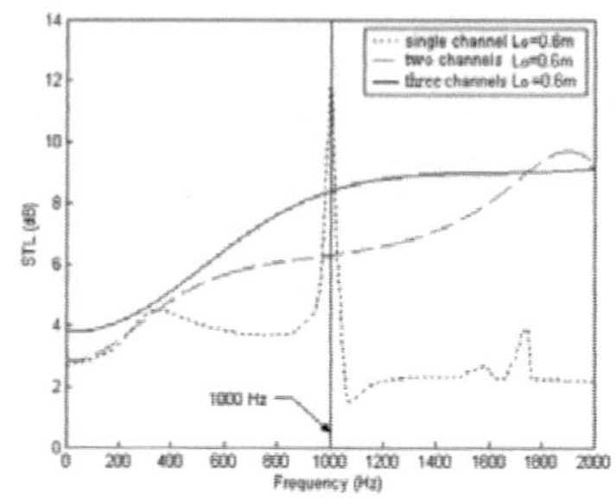

Figure 15. Comparison of the optimal *STL*s of three kinds of mufflers (mufflers A, B, and C) optimized at a targeted tone of 1000 Hz.

Adopting the same *SA* parameters of (*kk, iter*) = (0.99, 10000) and doing the minimization of the broadband noise at SWL_{T-1}, SWL_{T-2}, and SWL_{T-3} (mufflers A~C) at a limited length of Lo=1.5 (m), the resulting design parameters are obtained and shown in Table 5. Using the optimal design parameters in a theoretical calculation, the optimal *STL* curves with respect to various mufflers A~C are plotted and depicted in Fig. 16.

Similarly, using the same *SA* parameters of (*kk, iter*) = (0.99, 10000) and minimizing the broadband noise at SWL_{T-1}, SWL_{T-2}, and SWL_{T-3} (mufflers A~C) at a limited length of Lo=2.5 (m), the resulting design parameters are obtained and shown in Table 6. Using the optimal design parameters in a theoretical calculation, the optimal *STL* curves with respect to various mufflers A~C are plotted and depicted in Fig. 17.

Likewise, using the same *SA* parameters of (*kk, iter*) = (0.99, 10000) and minimizing the broadband noise at SWL_{T-1}, SWL_{T-2}, and SWL_{T-3} (mufflers A~C) at a limited length of Lo=3.5 (m), the resulting design parameters are obtained and shown in Table 7. Using the optimal design parameters in a theoretical calculation, the optimal *STL* curves with respect to various mufflers A~C are plotted and depicted in Fig. 18.

Table 5. Optimal design data for three kinds of mufflers (mufflers A~C) at Lo=1.5 (m) (broadband noise) (kk=0.96, iter=2000).

	Design parameters											SWL_T (dB)
Muffler A	RT1	RT2	RT3	RT4	RT5	RT6	RT7					127.3
	37.44	0.01412	0.6854	0.6854	0.9322	0.9322	18760					
Muffler B	RT1*	RT2*	RT3*	RT4*	RT5*	RT6*	RT7*	RT8*	RT9*			125.6
	39.32	0.01477	0.07403	0.07403	0.07403	0.9746	0.9746	0.9746	19670			
Muffler C	RT1**	RT2**	RT3**	RT4**	RT5**	RT6**	RT7**	RT8**	RT9**	RT10**	RT11**	124.3
	39.07	0.01468	0.04933	0.04933	0.04933	004933	0.9690	0.9690	0.9690	0.9690	19550	

Table 6. Optimal design data for three kinds of mufflers (mufflers A~C) at Lo=2.5 (m) (broadband noise) (kk=0.96, iter=2000).

	Design parameters											SWL_T (dB)
Muffler A	RT1	RT2	RT3	RT4	RT5	RT6	RT7					124.9
	37.42	0.01411	0.6852	0.6852	0.9317	0.9317	18750					
Muffler B	RT1*	RT2*	RT3*	RT4*	RT5*	RT6*	RT7*	RT8*	RT9*			122.1
	39.32	0.01477	0.07404	0.07404	0.07404	0.9748	0.9748	0.9748	19670			
Muffler C	RT1**	RT2**	RT3**	RT4**	RT5**	RT6**	RT7**	RT8**	RT9**	RT10**	RT11**	120.6
	39.14	0.01470	0.04938	0.04938	0.04938	004938	0.9705	0.9705	0.9705	0.9705	19580	

Table 7. Optimal design data for three kinds of mufflers (mufflers A~C) at Lo=3.5 (m) (broadband noise) (kk=0.96, iter=2000).

	Design parameters											SWL_T (dB)
Muffler A	RT1	RT2	RT3	RT4	RT5	RT6	RT7					121.9
	38.16	0.01437	0.6895	0.6895	0.9486	0.9486	19110					
Muffler B	RT1*	RT2*	RT3*	RT4*	RT5*	RT6*	RT7*	RT8*	RT9*			119.6
	39.32	0.01477	0.07404	0.07404	0.07404	0.9748	0.9748	0.9748	19670			
Muffler C	RT1**	RT2**	RT3**	RT4**	RT5**	RT6**	RT7**	RT8**	RT9**	RT10**	RT11**	118.1
	39.19	0.01472	0.04942	0.04942	0.04942	004942	0.9717	0.9717	0.9717	0.9717	19610	

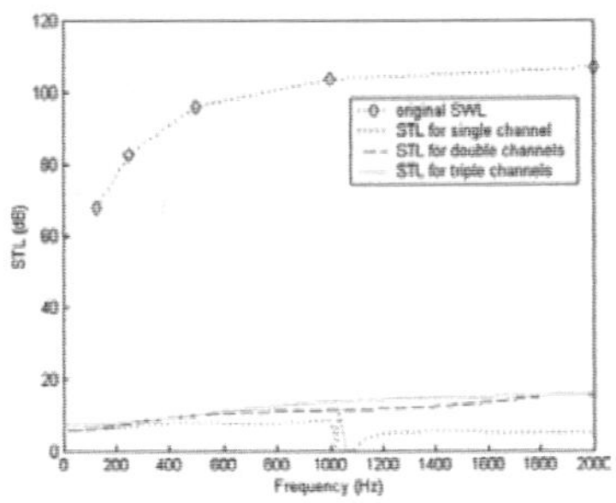

Figure 16. Comparison of the optimal *STLs* of three kinds of mufflers (mufflers A, B, and C) and the original *SWL* at Lo=1.5 (m) (broadband noise elimination).

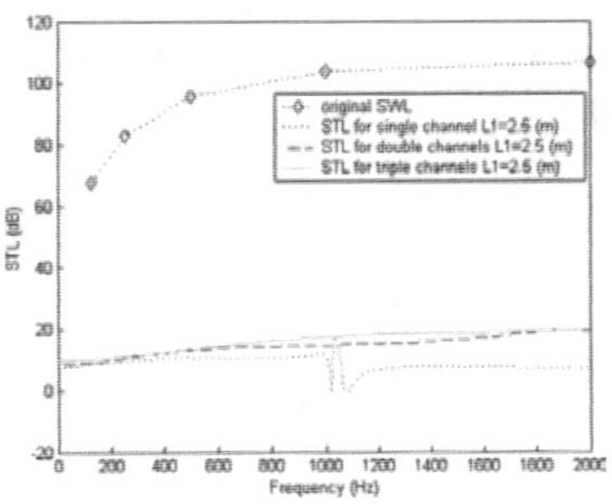

Figure 17. Comparison of the optimal *STLs* of three kinds of mufflers (mufflers A, B, and C) and the original *SWL* at Lo=2.5 (m) (broadband noise elimination).

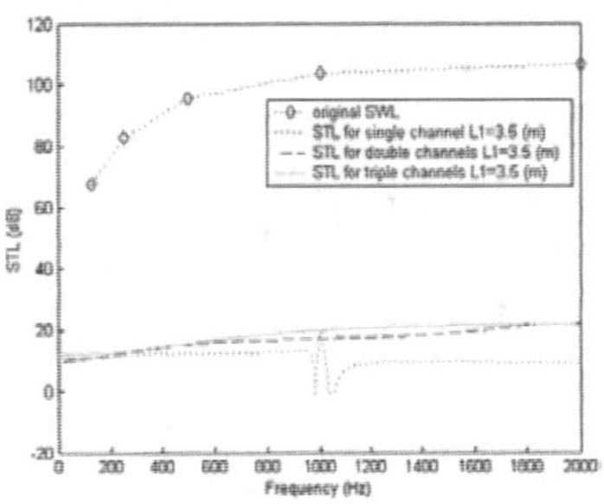

Figure 18. Comparison of the optimal *STLs* of three kinds of mufflers (mufflers A, B, and C) and the original *SWL* at Lo=3.5 (m) (broadband noise elimination).

7.4. Discussion

In order to achieve an optimally shaped muffler, the selection of design parameters is essential. An investigation of the design parameters' influence on muffler A's acoustical performance (*STL*) is introduced. The acoustical influence with respect to design parameters RT_1~RT_7 is shown in Figs. 6 and 12. Figs. 6 and 12 indicate that the *STL* is closely related to the design parameters RT_1~RT_7. Therefore, the design parameters RT_1~RT_7 in muffler A, the design parameters RT_1*~RT_9* in muffler B, and the design parameters RT_1**~RT_{11}** in muffler C are selected as the optimization parameters during the *SA* optimization.

To achieve a sufficient optimization, the selection of the appropriate *SA* parameter set is essential. As indicated in Table 3, the best *SA* set for muffler A at the targeted pure tone of 1000 Hz was found in

IOP Conf. Series: Materials Science and Engineering **644** (2019) 012004 doi:10.1088/1757-899X/644/1/012004

the eight set. The related *STL* curves with respect to various *SA* parameters are plotted in Figs. 13 and 14. Figs. 13 and 14 reveal that the predicted maximal value of the *STL* is precisely located at the desired frequency. Similarly, in dealing with pure tone noise (1000 Hz) in muffler B and muffler C, the profiles shown in Fig. 15 indicate that the maximum *STLs* of the mufflers are also located at the specified frequency.

In dealing with the broadband noise, three kinds of mufflers at three lengths (Lo =1.5(m), 2.5(m), and 3.5 (m)) are adopted in the optimal process. As illustrated in Table 5 and Fig. 16, the resultant sound power levels with respect to three kinds of mufflers at Lo=1.5 (m) have been reduced from 138.8 dB(A) to 127.3 dB(A), 125.6 dB(A), and 124.3 dB(A). Also, Table 6 and Fig. 17 indicate that the resultant sound power levels with respect to three kinds of mufflers at Lo=2.5 (m) have been reduced from 138.8 dB(A) to 124.9 dB(A), 122.1 dB(A), and 120.6 dB(A). Likewise, Table 7 and Fig. 18 reveal that the resultant sound power levels with respect to three kinds of mufflers at Lo=3.5 (m) have been reduced from 138.8 dB(A) to 121.9 dB(A), 119.6 dB(A), and 118.1 dB(A). The results mentioned above indicate that the acoustical performance of muffler C is superior to the other mufflers. This means that the overall sound transmission loss will increase when the number of the sound absorbing layer increases.

An investigation of the effect of length (Lo) on the acoustical performance (STL) is also explored and shown in Figs.19~21. As illustrated in Fig. 19, the STL of muffler A will increase when the length of Lo increases. In addition, as indicated in Fig. 20, the STL profile of muffler B will be promoted if the length of Lo increases. Consequently, Fig. 21 also reveals that the STL profile of muffler C will be promoted if the length of Lo increases.

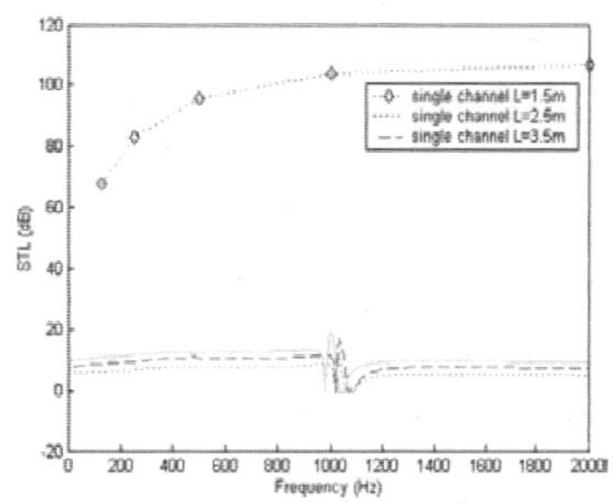

Figure 19. Comparison of the original *SWL* and the optimal *STLs* at various lengths (Lo=1.5 (m), 2.5 (m), and 3.5 (m)) (broadband noise elimination for muffler A).

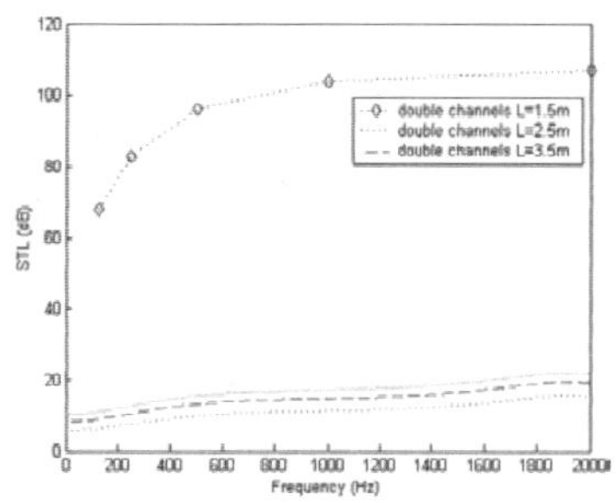

Figure 20. Comparison of the original *SWL* and the optimal *STLs* at various lengths (Lo=1.5 (m), 2.5 (m), and 3.5 (m)) (broadband noise elimination for muffler B).

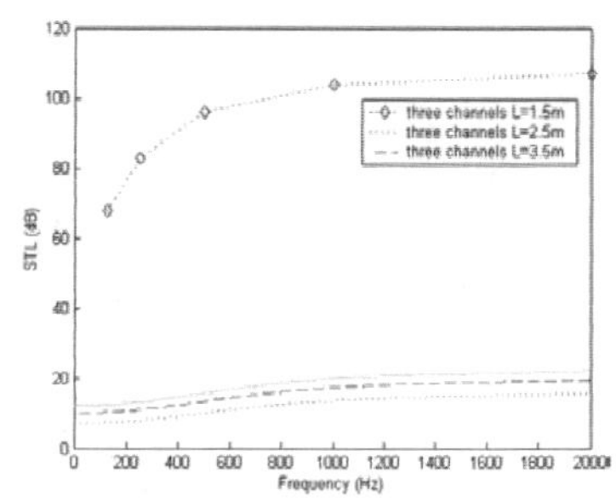

Figure 21. Comparison of the original *SWL* and the optimal *STLs* at various lengths (Lo=1.5 (m), 2.5 (m), and 3.5 (m)) (broadband noise elimination for muffler C).

8. Conclusion

It has been shown that pod mufflers hybridized with multiple layers of concentric sound-absorbing rings can be efficiently optimized within a limited space by using the acoustical lump analysis, a four-pole transfer matrix, and a *SA* optimizer. As indicated in Table 3, Figs. 13, and 14, two kinds of *SA* parameters (*kk* and *iter*) play essential roles in the solution's accuracy during *SA* optimization. Figs. 13, 14, and 15 indicate that the tuning ability established by adjusting design parameters in mufflers A~C is reliable. Moreover, the appropriate design parameters of three kinds of mufflers hybridized with multiple layers of concentric sound-absorbing rings (mufflers A~C) have been assessed.

Consequently, as indicated in Table 5, the resultant noise reductions with respect to mufflers A~C at Lo = 1.5 (m) are 11.5 dB, 13.2 dB, and 14.5 dB. Also, as indicated in Table 6, the resultant noise reductions with respect to mufflers A~C at Lo = 2.5 (m) are 13.9 dB, 16.7 dB, and 18.2 dB. Likewise, in Table 7, the resultant noise reductions with respect to mufflers A~C at Lo = 3.5 (m) are 16.9 dB, 19.2 dB, and 20.7 dB. Obviously, the muffler hybridized with more layers of concentric sound absorbing rings is superior to mufflers equipped with fewer layers of concentric sound absorbing rings. Also, the longer muffler(length (Lo)) is superior to shorter mufflers(length (Lo)).

Acknowledgements
The authors acknowledge the financial support of the Ministry of Science and Technology (MOST 104-2622-E-235-002 -CC3, ROC)

9. References

[1] P.M. Morse, Transmission of Sound inside Pipes, J. Acoust. Soc.Am., 11, 205-210 (1939).
[2] R.A. Scott, The Propagation of Sound between Walls of Porous Material, Proceedings of the Physical Society, 58, 358-368 (1946)
[3] S.H. Ko, Theoretical Analyses of Sound Attenuation in Acoustically Lined Flow Ducts Separated by Porous Splitters (Rectangular, Annular and Circular Ducts), Propagation of Sound between Walls of Porous Material, Journal of Sound and Vibration, 39, 471-487 (1975)
[4] Cummings, I.J. Chang, Sound Attenuation of a Finite Length Dissipative Flow Duct Silencer with Internal Mean Flow in the Absorbent, Journal of Sound and Vibration, 127, 1-17 (1988)
[5] K.S. Peat, A Transfer-matrix for an Absorption Silencer Element, Journal of Sound and Vibration, 146, 353-360 (1991)
[6] A.Selamet, I.J. Lee, Z.L. Ji, N.T. Huff, Acoustic Attenuation Performance of Perforated Concentric Absorbing Silencers, SAE Noise and Vibration Conference and Exposition, SAE Paper No. 2001-01-1435, Traverse City, MI, April 30-May 3, 2001
[7] A.Selamet, Lee I.J., N.T. Huff, Acoustic Attenuation of Hybrid Silencers, Journal of Sound and Vibration, 262, 509-527 (2003)
[8] M.L. Munjal, Analysis and Design of Pod Silencers, Journal of Sound and Vibration, 262, 497-507 (2003)
[9] M.L. Xu, A. Selamet, I.J. Lee, N.T. Huff, Sound Attenuation in Dissipative Expansion Chambers, Journal of Sound and Vibration, 272, 1125-1133 (2004)
[10] M.C. Chiu, Shape Optimization of One-chamber Perforated Mufflers Filled with Wool Using Simulated Annealing, Journal of Marine Science and Technology, 21(4), 380-390 (2013)
[11] M.C. Chiu, Numerical Assessment of Multi-Splitter Mufflers Using Neural Networks, the Boundary Element Method, and the Genetic Algorithm, Applied Mechanics and Materials, 58-60, 1049-1055 (2011)
[12] M.E. Delany, E.N. Bazley, Acoustical Properties of Fibrous Absorbent Materials, Applied Acoustics, 13, 105–116 (1969)
[13] L.L. Beranek, I.L Ver, Noise and Vibration Control Engineering, Wiley, New York, 232–243 (Chapter 8) (1992)
[14] B.G. Golden, Application of Noise Control in Vents and Blow Downs to Atmosphere, Proc. ASME Industrial Pollution Control Symposium, 137-145 (1984)
[15] Y.C. Chang, L.J. Yeh, M.C. Chiu, G.J. L ai, Shape Optimization on Constrained Single-layer Sound Absorber by Using GA Method and Mathematical Gradient Methods, Journal of Sound and Vibration, 286(4-5), 941-961 (2005)
[16] Metropolis, W. Rosenbluth, M.N. Rosenbluth, H.Teller, E. Teller, , Equation of Static Calculations by Fast Computing Machines, The Journal of Chemical Physics, 21, 1087-1092 (1953)
[17] S. Kirkpatrick, C.D. Gelatt, M.P. Vecchi, Optimization by Simulated Annealing, Science, 220, 671-680 (1983)
[18] M.C. Chiu, Y.C. Chang, Shape Optimization of Multi-chamber Cross-flow Mufflers by SA Optimization, Journal of Sound and Vibration, 312, 526- 550 (2008)
[19] M.C. Chiu, Shape Optimization of Double-chamber Side Mufflers with Extended Tube by Using Four-pole Matrix and Simulated Annealing Method, J. of Mechanics, 24, 31-43 (2008).
[20] Chiu, M. C., Numerical Assessment of Reverse-flow Mufflers Using a Simulated Annealing Method, The Canadian Society for Mechanical Engineering (CSME) Transactions, 34(1), 17-35 (2010)

IOP Conf. Series: Materials Science and Engineering **644** (2019) 012004 doi:10.1088/1757-899X/644/1/012004

[21] M.C. Chiu, Optimal Design of Multi-chamber Mufflers Hybridized with Perforated Intruding Inlets and Resonated Tube Using Simulated Annealing, ASME J. of Vibration and Acoustics, 132, 1-10 (2010)
[22] M. C. Chiu, Noise Elimination of a Multi-tone Broadband Noise Using Hybrid Helmholtz Mufflers by a Simulated Annealing Method, Archives of Acoustics, 37(4), 489-498 (2012) .
[23] M.C. Chiu, Numerical Assessment for a Broadband and Tuned Noise Using Hybrid Mufflers and a Simulated Annealing Method, Journal of Sound and Vibration, 332, 2923-2940 (2013)

The Design and Test of Elliptical-Circular Bicycle Sprockets

Ardi Lesmawanto, Shinn-Liang Chang, Sheng-Yuan Ceng, and Shin-Hao He

Department of Power Mechanical Engineering, National Formosa University, Huwei, Yunlin 632, Taiwan
E-mail: 1241891962@qq.com

Abstract. Most people have different strength and frequency of the movements between the left and the right body side. In riding a bicycle with circular sprocket, the torques produced by the right leg and the left leg are generally different. Usually, the dominant leg exerts larger force than the non-dominant leg. Based on these conditions, in this paper, the application of elliptical-circular bicycle sprockets with the right size and position on the leg can provide more advantages compared to the ordinary circular sprocket in cycling. The crank torque comparisons reveal that the application of the elliptical-circular sprocket can reduce the torque produced by the right leg, and increase the torque produced by the left leg. This proves that the elliptical-circular sprocket is useful for balancing the force produced between the left leg and the right leg. Then, from the leg joint comparisons, the elliptical-circular sprocket can reduce the torque in the hips, knees, and the ankles. This means that the use of elliptical-circular sprocket can reduce the legs fatigue in riding a bicycle.

1. Introduction

Most people have different strength and frequency of the movements between the left and the right body side [1]. Some studies have supported that the motion of humans are usually the right dominant of activities requiring mobilization and left dominant of activities requiring stabilization and postural strength [2-6].

Velotta et al. [7] reported that leg dominance seems to be a function of the type of activity a subject required to perform. When the task is manipulative in nature, most subjects use the right leg (most people are right-side preference), but when the task involves stabilization, most of the subjects in the study use left leg to perform the task. Spry et al. [3] reported that right leg dominant people tended to use the right leg for manipulative skills and the left leg for supportive or weight-bearing skills. The reverse was true for the left leg dominant people. Peters [8] defined the dominant leg as 'the leg used in order to manipulate an object or to lead out in movement'. Sadeghi et al. [9] reported during stabilization task, the opposite leg is used for dynamic counter-balance and since the right leg is better at manipulative activities, the left leg is used for stabilization. Duque et al. [10] investigated torque variation from pedal motion in cycling. Torques produced by the right foot and the left foot in pedalling are different. Different peak values indicate an asymmetry due to emphasizing an unequal pushing action during the downstroke. Faharani et al. [11] and Bini et al. [12] have similar result when investigating peak crank torque in pedalling as shown in Figure 1.

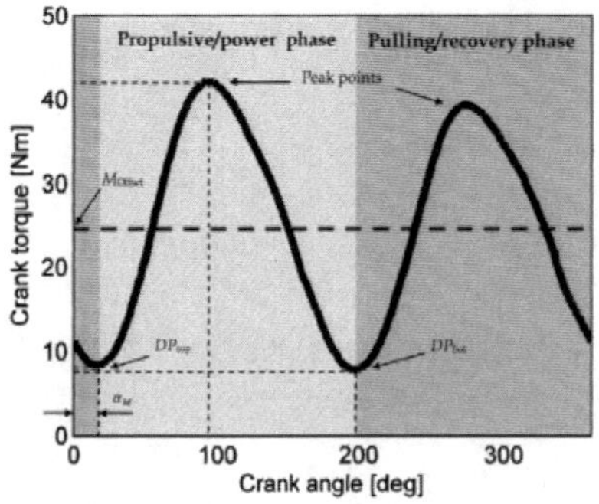

(a)

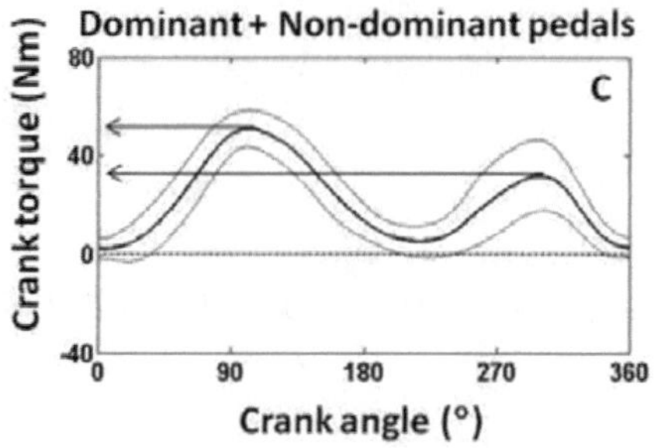

(b)

Figure 1. (a) The averaged crank torque on level terrain at preferred cadence [11], (b) Crank torque measured by dominant and non-dominant pedals [12].

In this paper, the radius of the ordinary circular sprocket (OC sprocket) were manipulated by adjusting the number of teeth and the radius when the non-dominant leg are pedalling in order to increase the level of leg torque. The sprocket produced in this study is a combination of ellipse and circle, namely elliptical-circular sprocket (EC sprocket).

2. Methods

2.1. Instrumentation

The measurement tool applied in this paper is a bicycle test platform that transmits the obtained data signals to the computer through variou s sensors and presents the required data via the software. The front and rear rotary encoders shown in Figures 2a are used to measure the rotational speed of the crank and the rear wheel, respectively. The bottom bracket and the smart training platform shown in Figures 2b and 2c, are equipped with pulleys and belts to drive the encoder to measure the rotation speed of the crank and the rear wheel. The strain gauges are placed on the front and back sides of the left and right cranks shown in Figure 2d and 2e. The wireless receiving module transmits the strain gauge signal generated by the cranks of both sides during pedalling to another wireless receiving module in the storage box connected to the computer.

In this paper, the bicycle load is created by the smart training platform produced by Wahoo Fitness and use a Bluetooth to connect the mobile phone to adjust the resistance as shown in Figure 2c. The simulated weight is 175 pounds (79.38 kg). The electromagnetic load applicator provides a stable electromagnetic load for the overall riding process, while dynamometer shown in Figure 2a measures the force generated by the training platform. The CCD camera is used to catch the illuminator device on the leg and on the pedal crank for the calculation of the angle formed.

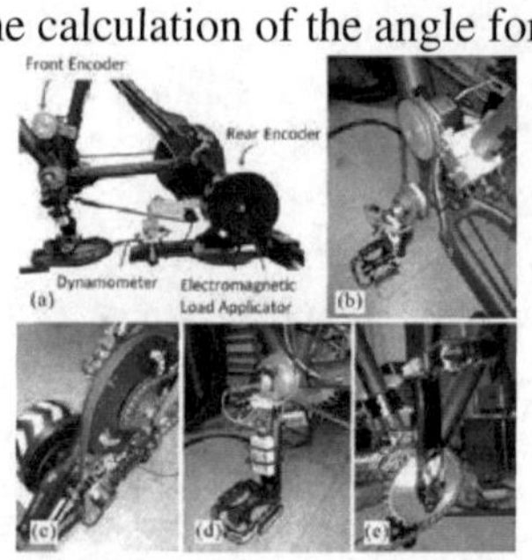

Figure 2. (a) Developed bicycle test platform, (b) The bottom bracket with belts to drive the front encoder, (c) the smart training platform produced by Wahoo Fitness, (d) strain gauge on the left crank, and (e) strain gauge on the right crank.

2.2. Novel sprocket design for averaging torque applied to the left and right

To overcome the unequal torque applied by the left and the right legs, the pitch curve of the sprocket driven by the left foot can be changed to be the ellipse, while the pitch curve for the right foot is not changed. The result will be an elliptical-circular sprocket (a combination of ellipse and circle). The sprocket then has two different shaped sections. The long axis is for the left foot forces section, thereby generating a large torque. The elliptical eccentricity needs to be limited, otherwise, the bicycle is likely caused a power drop in riding [13]. For the number of teeth, the number of teeth in the left and right section (ellipse and circle) must be even and the pitch must be equal. The combination of the sprocket is presented in table 1. In order to make sure the smooth in combination of two shapes, the radius of the short axis of the ellipse is equal to the radius of the circular section.

Table 1. Combination of left and right sprocket.

Type	Left foot section sprocket				Right foot section sprocket	
	Number of teeth	Eccentricity	Long axis (mm)	Short axis (mm)	Number of teeth	radius (mm)
EC 0.45-0 U	34	0.4502	72.5518	64.6806	32	64.6806
EC 0.59-0 U	36	0.5913	80.3301			

2.3. Sprocket generation

In this paper, the sprocket generation is based on the equations proposed by Chang [14] to make the sprocket. Firstly, the pitch curve must be calculated, and then the sprocket tooth profile shown in Figure 3 can be configured in the normal direction of ellipse. The circumference (S) of the pitch curve of ellipse is set to be an integral multiple of the chain pitch.

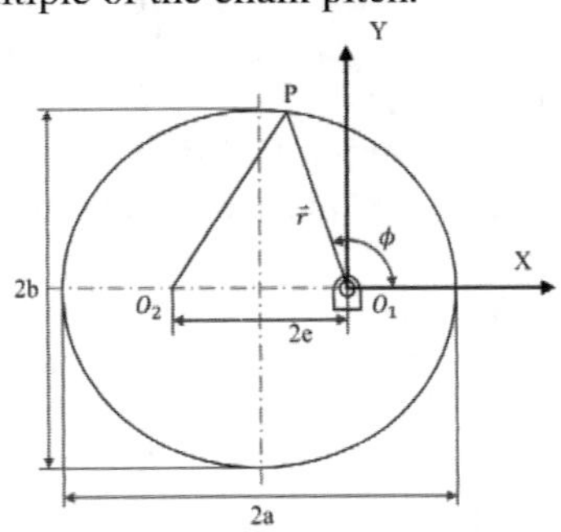

Figure 3. Pitch curve of the elliptical sprocket [14].

For the elliptical section of the sprocket, the distance from the focus to the point P on the ellipse is obtained by:

$$r = \frac{b^2}{a(1+\varepsilon\cos\phi)} \tag{1}$$

where ε is the eccentricity, a is the semi-major axis, b is the semi-minor axis, and ϕ is the angle of rotation. The integral formula of the circumference is obtained by:

$$S = \int_0^{2\pi} \frac{a(1-\varepsilon^2)\sqrt{\varepsilon^2+2\varepsilon\cos\phi+1}}{(1+\varepsilon\cos\phi)^2}\,d\phi \tag{2}$$

Further, the calculation can be solved by Simpson's 1/3 rule integral formula. Since the circumference of the pitch curve needs to be multiple integrals of the pitch, no incomplete tooth shape will occur when all the teeth profiles are placed. In this study, the bi-section method is used to obtain the position angle of each tooth. After obtaining the each position of tooth, the coordinate of each tooth can be obtained. The normal direction of sprocket on the ellipse can be established with the right-hand rule by:

$$\mathrm{n} = \frac{\varepsilon+\cos\phi}{\sqrt{\varepsilon^2+2\varepsilon\cos\phi+1}}\mathrm{i} + \frac{\sin\phi}{\sqrt{\varepsilon^2+2\varepsilon\cos\phi+1}} \tag{3}$$

then use the inverse trigonometric function to convert the normal direction to the angle:

$$\theta = \tan^{-1}\left(\frac{\sin\phi}{\varepsilon+\cos\phi}\right) \quad (4)$$

Substituting the eccentricity and the number of teeth required in the foot segment of Table 1 into the equation, then the elliptical-circular sprocket can be generated by using AutoCAD software. JIS U type tooth profile [15] is considered to be the teeth shape. The resulting elliptical sprocket curve is cut in half, then using the annular array method, the tooth can then be placed on the circular sprocket. The combined graphics of the sprocket are shown in Figure 4.

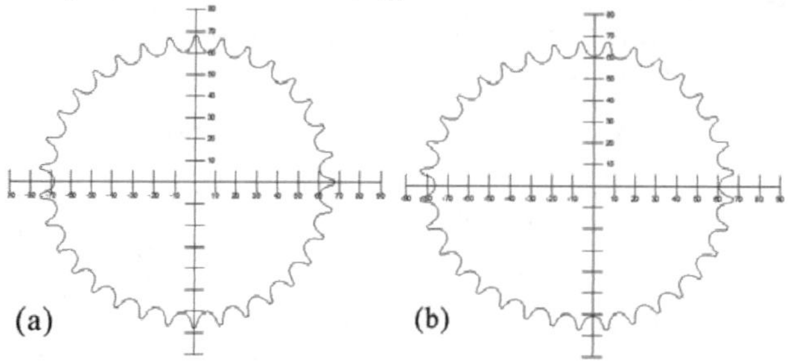

Figure 4(a). Elliptical-circular sprocket eccentricity 0.45-0U, and (b) eccentricity 0.59-0U.

3. Results and discussion

3.1. Crank speed

Figure 5 shows the comparison of the crank speed between the applications of the ordinary circular sprocket (OC sprocket), the elliptical-circular sprocket (EC sprocket) 0.45-0U, and 0.59-0U. From the figure, it reveals that the speed of the OC sprocket crank is stable at around 46-48 rpm. For EC 0.45-0U, the crank speed is greater than the OC sprocket. When the right leg is pedalling at 0°-180°, the crank speed is stable between 48-50 rpm. But when the left leg starts pedalling at 180°, the speed rises slightly then goes down until 270^0, reaches around 44 rpm. This is because the stroke is in the largest sprocket radius position. After that, the speed increases again until 360°. For EC 0.59-0U, the crank speed is higher than the others when the crank position is at 0-180°. It shows the same phenomenon with the EC 0.45-0U, the speed rises slightly then goes down until 270°, reaches around 41 rpm. This speed is the lowest compared to the other sprocket because it has the largest major axis.

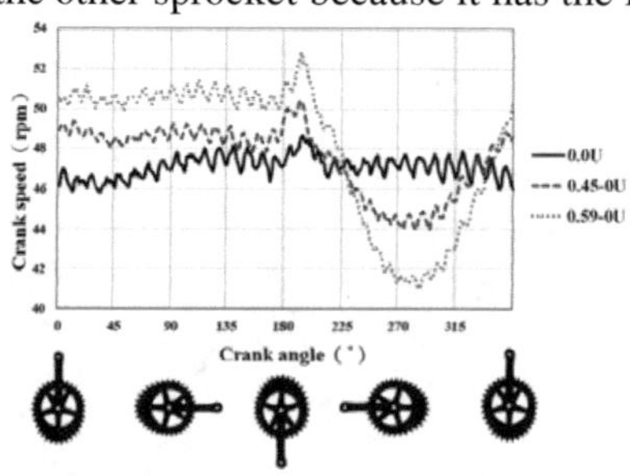

Figure 5. Crank speed comparison.

3.2. Crank torque

Figure 6 shows the comparison of the crank torque, which is similar to the results proposed by Bini and Hume [10]. In 0° – 180°, the force is applied by the right leg, while in 180° – 360° the force is applied by the left leg. The resulting torque difference between the right leg and the left leg is about 40%. The main reason is the strength of the left leg is less than the right leg. From the result of EC 0.59-0U, the right leg is pedalling at 0°-180°, the crank torque is smaller than the OC crank torque. When the left leg is pedalling at 180°-360°, EC 0.59-0U can produce more torque. From the comparisons, it reveals that the use of EC 0.59-0U and EC 0.45-0U is able to increase torque on the left leg and reduce the torque difference produced by the right leg and the left leg. This torque difference is around 27.7% using EC 0.59-0U, and around 31% using EC 0.45-0U.

IOP Conf. Series: Materials Science and Engineering **644** (2019) 012005 doi:10.1088/1757-899X/644/1/012005

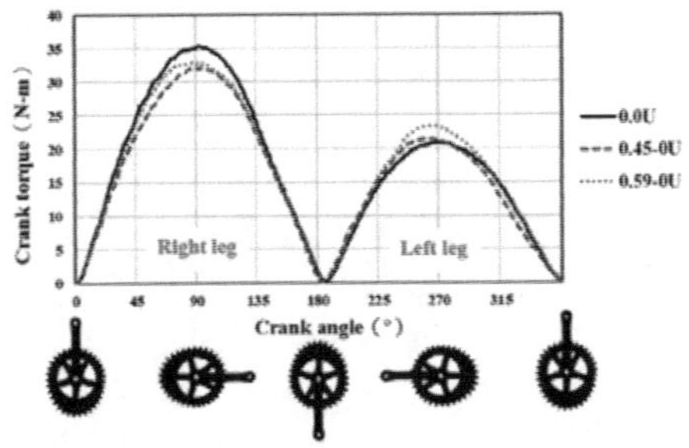

Figure 6. Crank torque comparison.

3.3. Leg joint torque

The leg-bicycle system can be modelled as a closed loop of five-bar linkage with the frame as the fixed link. There are consist of three joint human legs: hip joint, knee joint, and ankle joint, as shown in Figure 7.

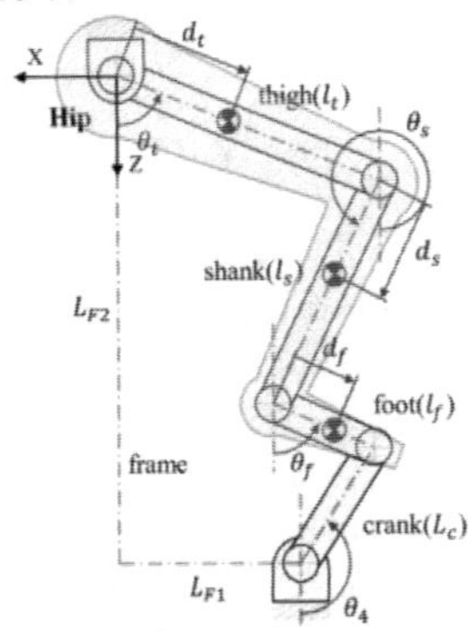

Figure 7. 5-bar linkage model of the leg-bicycle system [16].

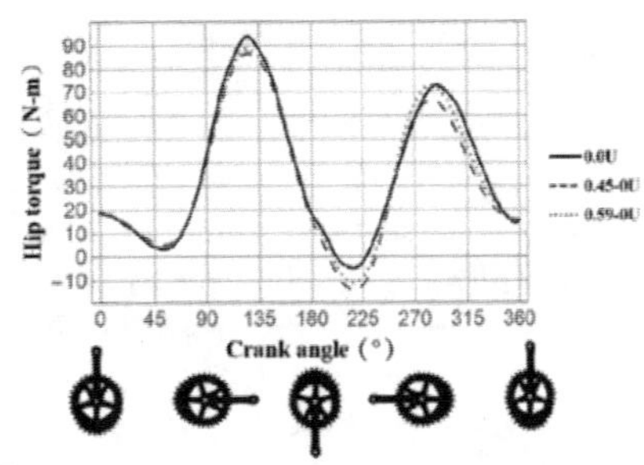

Figure 8. Hip joint torque comparison.

Figure 8 shows the hip joint torque compared with the application of OC sprocket, EC 0.45-0U sprocket, and EC 0.59-0U sprocket. The right hip joint generates peak torque when the crank angle position is 115°-135°. And the left hip joint generates peak torque when the crank angle position is 270°-300°. Even though in that position, the torque produced in the crank is not at the peak value. When using elliptical-circular sprocket (EC), especially EC 0.45-0U the peak torque produced by the hip joint can be the minimum.

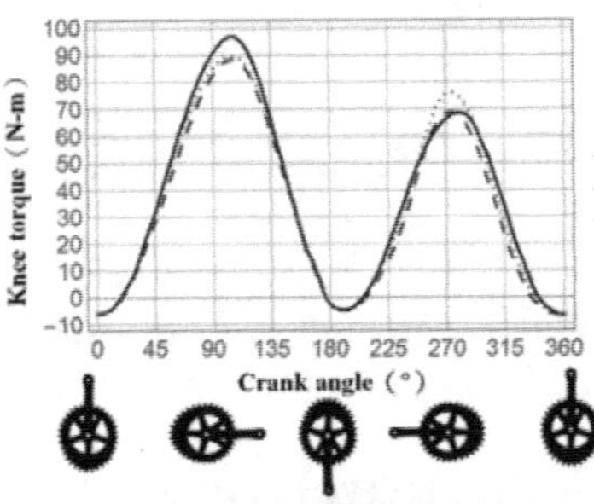

Figure 9. Knee joint torque comparison.

Figure 9 shows that the right knee joint generates peak torque compared with the application of OC sprocket at 90°-130°. And the left knee joint generates peak torque when the crank angle position at around 270°. Using EC 0.59-0U sprocket, the peak of the right knee joint can be reduced, but the torque value increases on the left knee when the crank position is at 270°, larger than the torque produced by the OC sprocket. The peak torque on the right knee joint can be reduced significantly by EC 0.45-0U.

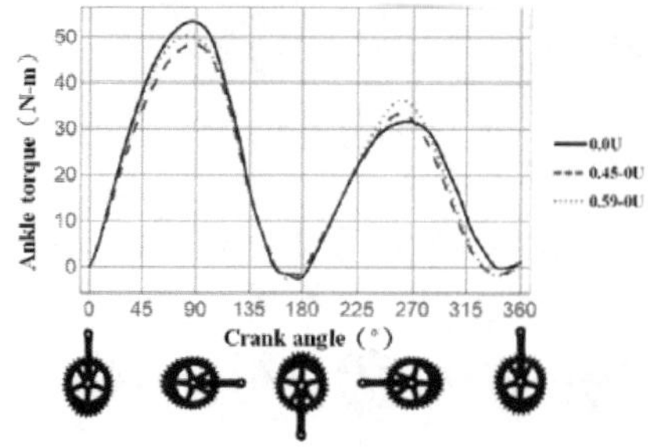

Figure 10. Ankle joint torque comparison.

The results shown on Figure 10 are similar to the knee joint torque comparison shown in Fig. 10. It appears that the peak value of torque is produced by the right ankle joint when the crank is in the position at around 90° using OC sprocket. This torque can be reduced by using EC sprocket. Using EC 0.45-0U, the torque on the right ankle joint can be reduce more than the application of EC 0.59-0U.

4. Conclusions

This paper shows how the elliptical-circular sprocket is designed to balance the torque produced by the dominant leg and the non-dominant leg in cycling, and compared with the ordinary circular sprocket. From the crank torque comparisons, it reveals that the right leg and the left leg produce different torques. The use of the elliptical-circular sprocket can reduce the torque produced by the right leg, and increase the torque produced by the left leg. This proves that the elliptical-circular sprocket is useful for balancing the force produced between the left leg and the right leg. In this case, the effect of elliptical-circular sprocket 0.59-0U is better than 0.45-0U.

From the leg joint comparisons, it can be seen that the use of the elliptical-circular sprocket can also reduce the torque in the hips, knees, and the ankles. The elliptical-circular sprocket 0.45-0U is better than the other sprocket in reducing torque in the leg joint. This means that the use of elliptical-circular sprocket can reduce the legs fatigue in riding a bicycle.

The author is grateful to the Ministry of Science and Technology of the R.O.C. for their partial support under grant MOST 106-2221-E-150-018-MY2 and to Mr Zheng Weiyuan, Vice General Manager of Chuan Wei Metal Products Co., Ltd. for assisting the production of the elliptical-circular sprocket.

5. References

[1] D. Cular, D. Miletic, and A. Miletic, Influence of Dominant and Non-Dominat Body Side, *Kinesiology*, 42, 184-193 (2010)

[2] V. Gentry, and C. Gabbard, Foot-Preference Behavior: A Developmental Perspective, *Journal of General Psychology*, **122(1),** 37-45 (1995)

[3] S. Spry, C. Zebas, and M. Visser, What is Leg Dominance, *Symposium of the International Society of Biomechanics in Sport* , 11 (1993)

[4] J.E. Whittington and P.N. Richards, The Stability of Children's Laterality Prevalences and Their Relationship to Measures of Performance, *Journal of Educational Psychology,* 57(1), 45-55 (1987)

[5] J.W. Chow and M.D. Tillman, Bilateral Strength and Activation Characteristics of Quadriceps in Experienced Soccer Player, *International Symposium on Biomechanics in Sports*, 23 (2005)

[6] N.V. Melick, B.M. Meddeler, T.J. Hoogeboom, M.W.G. Sanden, and R.E.H. Cingel, The Agreement between Self-Reported and Observed Performance, *Plos One Journals,* 12 (2017)

[7] J. Vellota, J. Weyer, A. Ramirez, J. Winstead, and R. Mahamonde, Relationship Between Leg Dominance Test and Type of Task, *Journal of Sport Sciences*, 11 (2011)

[8] M. Peters, Asymmetries in Foot Preference and Skill and Neuropsychological Assessment of Foot Movement, *Psychological Bulletin*, 103(2), 179-192 (1988)

[9] H. Sadeghi, P. Allard, F. Prince, and H. Labelle, Symmetry and Limb Dominance in Able-Boied Gait: A Review, *Gait and Posture*, 12, 34-45 (2000)

[10] Q. Duque, T. Dahmen, and D. Saupe, Estimation of Torque Variation from Pedal Motion in Cycling, *Journal of Computer Science in Sport*, 14(1) (2015)
[11] S.D. Faharani, W. Bertuccini, M.S. Andersen, M.D. Zee, and J. Rasmussen, Prediction of Crank Torque, *Structural and Multidiciplinary Optimization,* 51(1), 251-266 (2015)
[12] R.R. Bini and P.A. Hume, Bilateral Asymmetry Assessment in Cycling Using Commercial Instrumented Crank System and Instrumented Pedals, *Journal of Sport Physiology and Performance,* 9(5), 867-881 (2014)
[13] C. Sengyuan, *Bicycle Non-Circular Sprocket Design*, Master Thesis, National Formosa University (2017)
[14] S.L. Chang, *Design of Elliptical Involute Gears*, Master Thesis, National Tsing-Hua University (1992)
[15] JIS B-1802, Sprocket Tooth Form for Transmission Roller Chains, Japanese Industrial Standard (1989)
[16] M.L. Hull and M. Jorge, A Method for Biomechanical Analysis of Bicycle Pedalling, *Journal of Biomechanics*, 18(9), 631-644 (1985)

IOP Conf. Series: Materials Science and Engineering **644** (2019) 012006 doi:10.1088/1757-899X/644/1/012006

Security System Design in a Crop

Tung-Jung Chan[1],Min-Chie Chiu[1], Ho-Chih Cheng[2], Long-Jyi Yeh[3] , and Wei-Chong Haung[3]

1 Department of Electrical Engineering and Energy Technology, Chung Chou University of Science and Technology, Taiwan
2 Department of Intelligent Automation Engineering, Chung Chou University of Science and Technology, Taiwan, ROC
3 Department of Mechanical Engineering, Tatung University, Taiwan
E-mail:minchie.chiu@msa.hinet.net

Abstract. A traditional scarecrow does not scare birds; subsequently, crops will be demolished by birds during the growing season. Moreover, because farms are often invaded and crops stolen during harvest, the farmer and his dogs are forced to stay and protect the crops. However, this is not efficient for a large farm. Therefore, a remote crop monitoring security system is necessary. In order to maintain sufficient electrical power used by the system, green electrical energy generated by both a solar board and a wind generator are adopted. This system placed in each section of the farm includes five ultrasonic sensors, a microcontroller, a twinkling lighting, and a speaker unit. To ensure efficiency when using the crop security system, crops in huge areas will be divided into several small zones for monitoring. Each zone will have an individual crop security system. When a sensor detects an intrusion of birds, the alarm and light will be initiated. With this, any birds or other intruders will be scared and leave the field.

1. Introduction

Much research indicates that wildlife including birds, wild bears, wild boars, and primates are the primary villains of crop damage. Global Rice Science Partnership (GRiSP), in 2010, based on farmer surveys in 20 African countries, demonstrated that birds were the second most serious biotic constraint on African rice production after weeds. [1]. Klosterman et al. [2], in 2012, also identified blackbirds (Icteridae) as the significant cause of damage to both ripening corns and sunflowers. To reduce the crop damage, Linz et al. [3], in 2011, suggested using habitat management of roosting sites, plant desiccants to accelerate harvest time, and decoy crops. Kale et al. [4] assessed an interdisciplinary approach to develop an eco-friendly technique that would reduce the loss of both birds and crops. Their research also indicated that pesticide poisoning is another factor in bird extinction. They also found that the techniques like using sticks, scarecrows, reflecting ribbons, firecrackers, and catapult (Kavan) are not efficient in repelling birds, though these techniques are eco-friendly. In 2013, according to research from George and Linz [5], methyl anthranilate, which can be maintained for a long time due to high concentrations, was effective in repelling birds. However, it is expensive and causes damage to sunflowers. Additionally, huge amounts of manpower or huge airplanes are required when using methyl anthranilate. Demont [6], in 2013, indicated that extensive labour is required for controlling birds. Also, possible health and environmental hazards must be considered when using chemicals. Kale et al. [7], in 2014, found that techniques such as chemical repellent, nets, spike guards, shooting birds, and using fire crackers are either costly or lethal. Desoky [8], in 2015, reported that agricultural pests, such as birds and rodents, may cause significant damage to crops and reduce crop

IOP Conf. Series: Materials Science and Engineering **644** (2019) 012006 doi:10.1088/1757-899X/644/1/012006

production. In Taiwan, many birds (Pycnonotus sinensis and sparrow) cause serious crop damage. They are smart. Traditional scarecrows, noise, nets, and spike guards can not scare them. However, fire crackers are effective, but they are too expensive and time-consuming. In Taiwan, typical birds such as sparrows and light-vented bulbuls also destroy rice fields and orchards.

Crop damage caused by primates is one of the most wide spread and common examples of problems in areas that are close to a forest [9]. A study by Gebeyehu and Bekele [10] in 2009 also revealed that damage to ripened crops in the fields of the Zegie Peninsula (Ethiopia) by wild animals was equivalent to about 13,000 Ethiopian Birr per day. In Taiwan, Formosan Macaque (Macaca cyclopis) often destroys home gardens, fruit trees, and crops. The Formosan Macaque also raids a variety of crops that are adjacent to forests.

Hunting birds and monkeys is illegal and prohibited in Taiwan. Moreover, because of health and environmental considerations, chemical repellent (pesticide), which may seriously damage crops, is discouraged. Therefore, alternative methods used to repel animals without harm seems most reasonable. In addition, to overcome the drawbacks of a scarecrow, a crop monitoring security system presented here will be interactive during the detecting process. Likewise, to overcome the shortcomings of labor mentioned above, the crop security system is automatic.

2. Crop Monitoring security system

As indicated in Fig. 1, the crop monitoring system includes five ultrasonic sensors, a speaker unit (for alarming purpose), a twinkling lighting (for alarming purpose), a solar electricity generator, a wind-electricity generator, and a microcontroller (Arduino MeGa) shown in Fig. 2. Five ultrasonic sensors used to detect intruders such as birds, boars, bears, and monkeys, will be installed in the desired sections. As depicted in Fig. 3, considering the limited detecting distance of the ultrasonic sensor and the quality of monitoring/controlling, a large crop will be divided into several smaller sections. In addition, to ensure sufficient electrical power for the system, both a solar-electricity generator shown in Fig. 4 (a) [11] and a wind-electricity generator shown in Fig. 4(b) [12] are installed and connected to the system. Here, each part (ultrasonic sensors, lighting, and alarm) connected to the microcontroller plays essential roles in expelling intruders. The algorithm for intruder detecting is illustrated in Fig. 5. As indicated in Fig. 5, the ultrasonic sensors, online, will detect the specified crop zone when the crop security system is initiated. And, both the speaker unit and the twinkling lighting unit will also be continuously triggered for 1 minute when the intruding event happens. Both the sound and lighting alarm units will be shut OFF when the intruder disappears. Here, to efficiently scare the intruder, the speaker unit can loudly broadcast one kind of firecrackers or two kinds of gun shutting randomly. Also, a strong and strobe-like light will be emitted from the lighting unit.

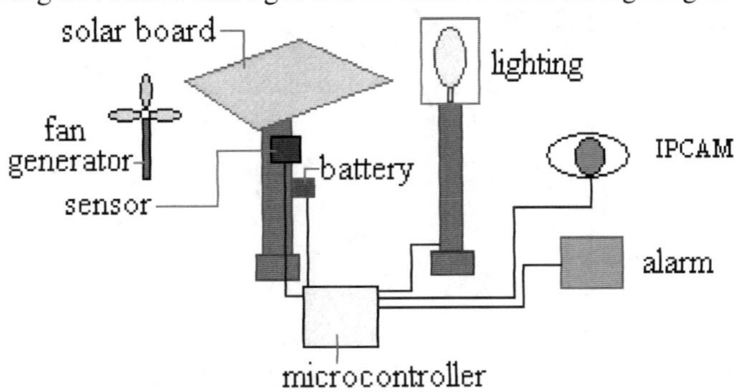

Figure 1. The crop security system.

IOP Conf. Series: Materials Science and Engineering **644** (2019) 012006 doi:10.1088/1757-899X/644/1/012006

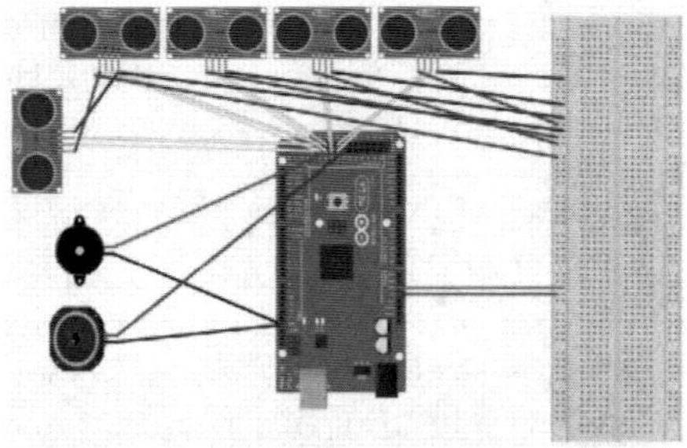

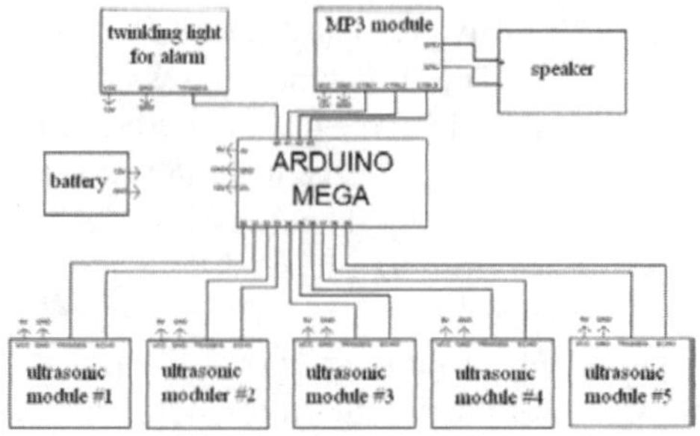

(a) practical wire connecting drawing (b) wire connection plot plan.

Figure 2. The diagram of the microcontroller (Arduino MeGa) and the connecting unit.

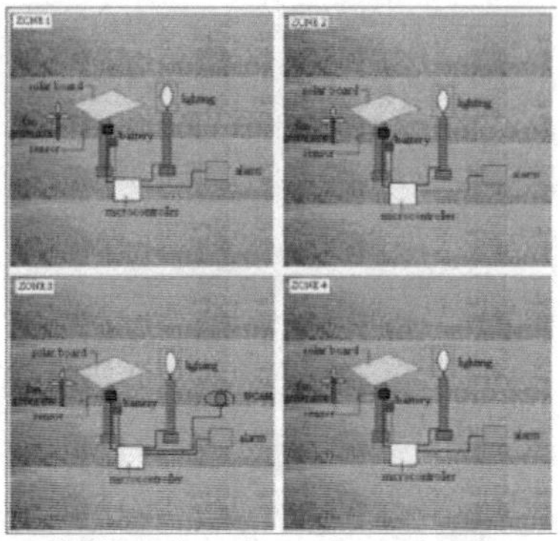

Figure 3. The planning of monitoring zones.

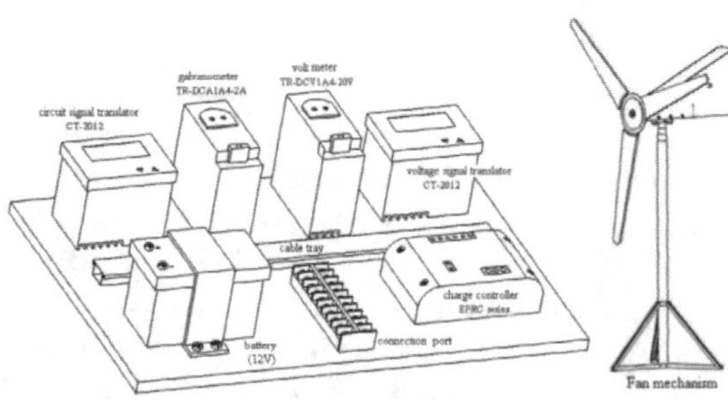

(a)The solar electricity generator (b) The wind electricity generator.

Figure 4. Green energy for the solar electricity generator and the wind electricity generator.

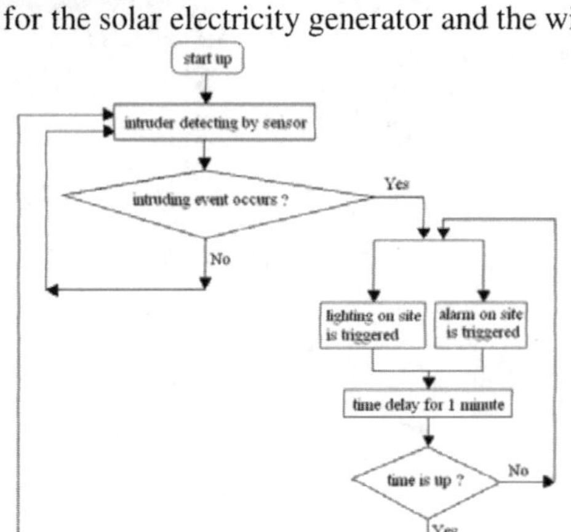

Figure 5. Algorithm for the intruder detecting and warning system.

3. Results and discussion

The crop monitoring system equipped with five ultrasonic sensors, a twinkling lighting unit, a speaker unit, a solar electricity generator, a wind electricity generator, and a microcontroller has been introduced. The alarm unit and lighting unit will be initiated via the microcontroller when the intruding message is sent by the ultrasonic sensors. A bird repelling system is shown in Fig. 6. The related interactive monitoring process for birds is shown in Fig. 7. As indicated in Fig. 7(a), a bird approaching the crop zone is detected by the ultrasonic sensor. Subsequently, as shown in Fig. 7(b) and Fig. 7(c), both the alarm unit and the lighting unit are initiated. Crop security also can be used in preventing crop loss from humans. The interactive monitoring process for a human being is also shown in Fig. 8. As indicated in Fig. 8(a), the intruder approaches the crop at night. The intruder is detected by the ultrasonic sensor in Fig. 8(b). Fig. 8(c) and Fig. 8(d) are subsequently depicted that both the alarm unit and the lighting unit are initiated. As illustrated in Fig. 8(e), the intruder is scared and runs away from the crop zone. As a result, the intruder disappears and the detecting system of the crop security system will continue.

Consequently, the prototype of the crop monitoring system has been introduced and shown in Fig. 9.

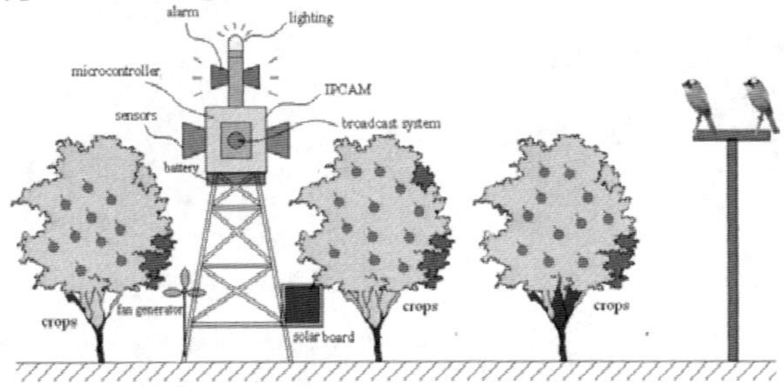

Figure 6. A bird repelling system.

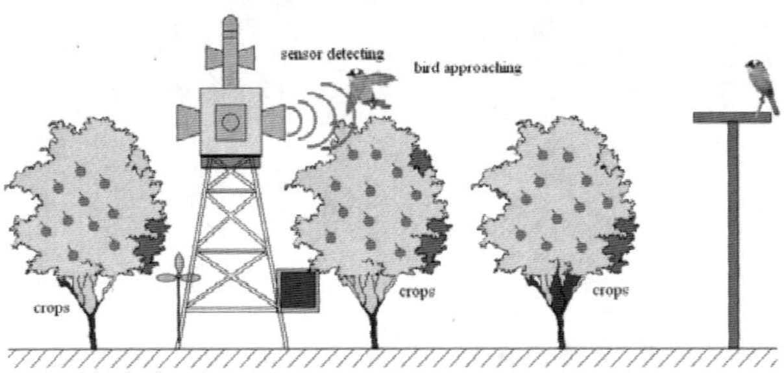

(a) Bird is approaching the crop zone and is detected by the ultrasonic sensor

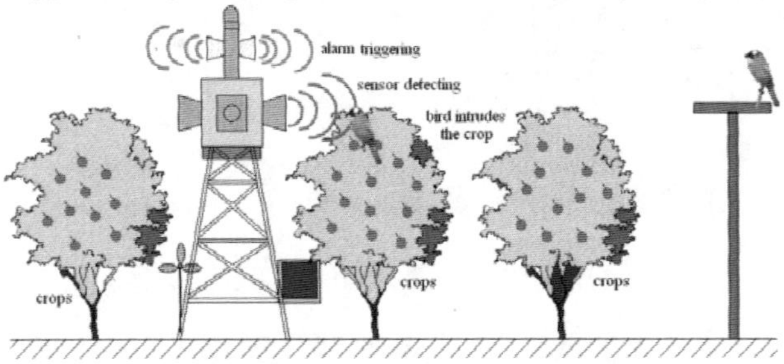

(b) The alarm unit is triggered

IOP Conf. Series: Materials Science and Engineering **644** (2019) 012006 doi:10.1088/1757-899X/644/1/012006

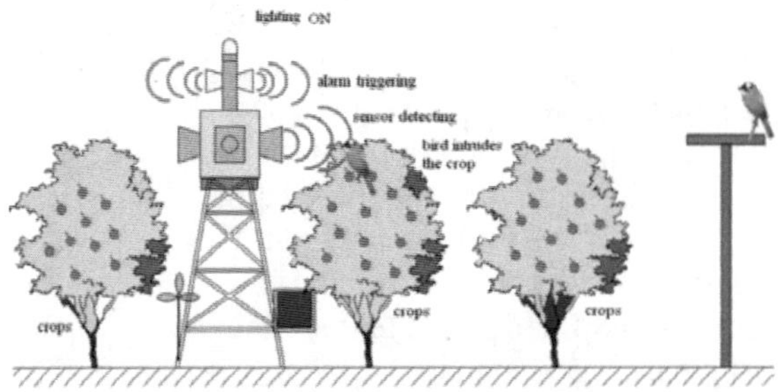

(c) The lighting unit is triggered

Figure 7. The interactive monitoring and warning.

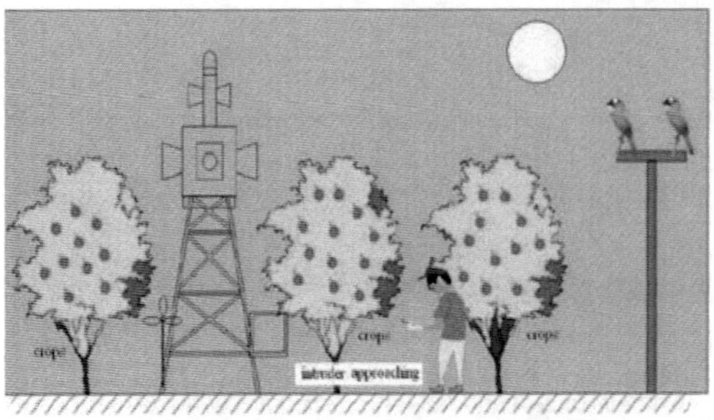

(a) An intruder is approaching the crop

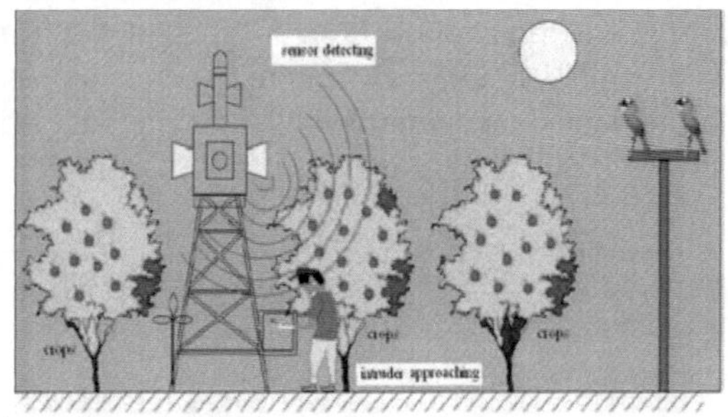

(b) An intruder is detected by the ultrasonic sensor

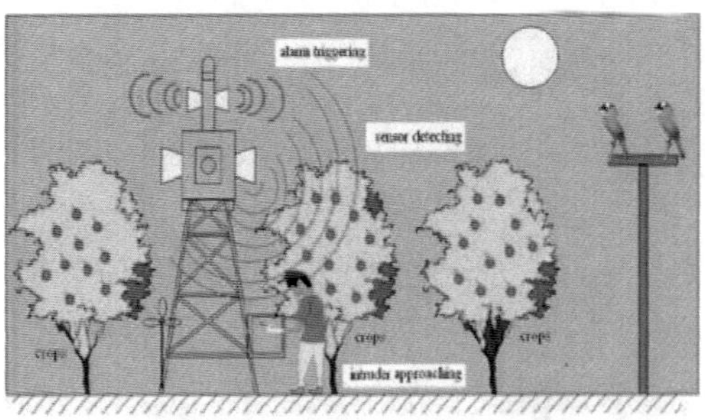

(c) Alarm unit is triggered

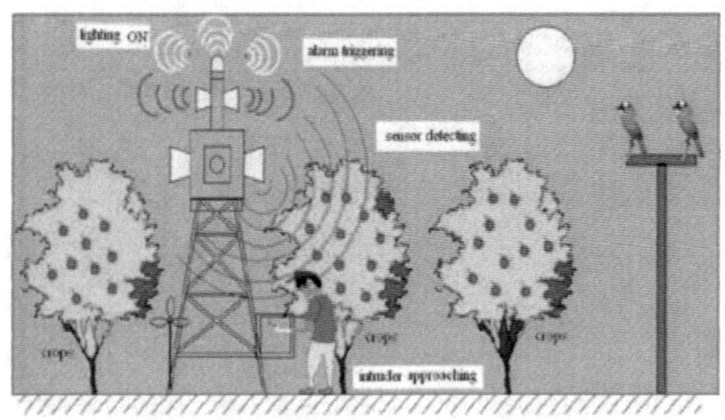

(d) Lighting unit is triggered.

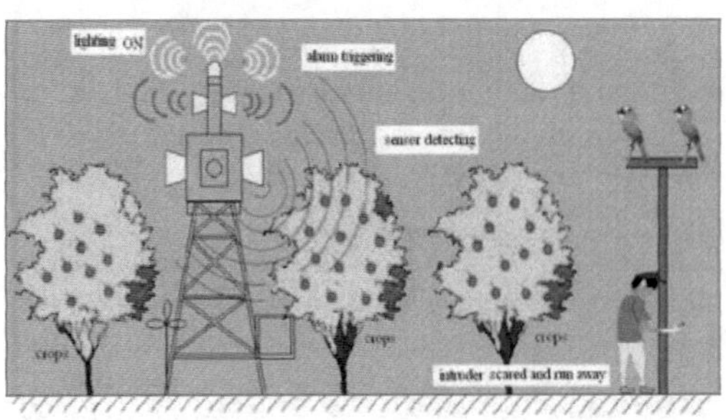

(d) Intruder is scared and runs away

Figure 8. The interactive monitoring and warning for humans.

IOP Conf. Series: Materials Science and Engineering **644** (2019) 012006 doi:10.1088/1757-899X/644/1/012006

(a)Ultrasonic wave sensors (b)Lighting device (c) electrical control panel

Figure 9. The prototype of the crop security system.

4. Conclusion

It has been shown that a crop security system can automatically perform intruder detecting, and warning. The microcontroller connects to the ultrasonic sensors and alarm. Both solar electricity and wind electricity connected to the microcontroller will provide electrical power to the crop security system. Results reveal that the interactive system used in repelling birds in a crop is easy and efficient. Consequently, the automatic crop detecting and monitoring has been established and demonstrated.

5. References

[1] IRRI, AfricaRice and CIAT, Global rice science partnership (GRiSP), International Rice Research Institute, Manila, Philippines; Africa Rice Center, Cotonou, Benin; and International Center for Tropical Agriculture, Cali, Colombia (2010)
[2] M.E. Klosterman, G.M. Linz, AA. Slowik, W. J. Bleier, Proceedings of the 14th WDM Conference, pp.119-123 (2012)
[3] G.M. Linz, H. . Homan, S.J. Werner, H. M. Hagy, W. J. Bleier, BioScience 61, 12, (2011)
[4] M. Kale, B. Balfors, U. Mörtberg, P. Bhattacharya, S. Chakane, , J. Agri. Tech. 8, 1 (2012)
[5] G.M. Linz, H.J. Homan, Demonstration of Avian Control Bird Repellent (a.i., Methyl Anthranilate) for Managing Blackbird Damage to Ripening Sunflower, U.S. Department of Agriculture, Wildlife Services, National Wildlife Research Center (2013)
[6] Y. de Mey, M. Demont, M., Realizing Africa's Rice Promise pp.241-249 (2013)
[7] M.A. Kale, N. Dudhe, R. Kasambe, P. Bhattacharya, I. J. Biodiversity 947683, (2014)
[8] A. Desoky, S.S. El-Aleem, Academic Research J. Agri. Sci. and Research 3, 1 (2015)
[9] C.M. Hill, Uganda. Environmental Conservation. 26 (1998)
[10] G. Gebeyehu, A. Bekele, Ethiopian J. Sci. 32, 2 (2009)
[11] H.C. Cheng, M.C. Chiu, Applied Mechanics and Materials. 336-338 (2013)
[12] H.C. Cheng, M.C. Chiu, J. Information & Optimization Sci. 34, 6 (2013)

IOP Conf. Series: Materials Science and Engineering **644** (2019) 012007 doi:10.1088/1757-899X/644/1/012007

A Digital Analysis Approach for Estimating Decarburization Layer Depth of Carbon Steel via a Portable Device

Wesley Huang[1,2], Chia-Sui Wang[1], Yih-Feng Chang[1], and Chia-Mao Yeh[3]

1 Chia Nan University of Pharmacy & Science, No. 60, Sec 1, Erren Road, Rende District, Tainan City, 717, Taiwan

2 First Optoech Co., Ltd., No. 1, University Rd., Yan Chao Dist., Kaohsiung, 824, Taiwan

3 National Sun Yat-sen University, No. 70, Lienhai Rd., Kaohsiung City, Taiwan
E-mail: box1025@mail.cnu.edu.tw

Abstract. Decarburization layer depth has great influence on the quality of metal forming process. For determining the depth value, the inspector progresses the measurement by manual nearly, which is subjective and inconsistency. This paper presents a digital analysis approach for measuring the decarburization depth automatically. Meanwhile, this analysis can be carried out by a portable device. For the digital analysis approach, the morphology image process is developed by comparing the variation of gray gradient of image from edge to core region. While the histogram of grayscale gradient falling down sharply, the distance from edge to the point was defined as the ferrite decarburization depth. Meanwhile, if the histogram keeps decreasing and stopping the decrease at one point, then the distance from edge to that point was defined as the totally decarburization depth. For the portable device, the analysis software is settled on a cloud-based calculating system. The analyst can capture the metallographic image shown in the monitor by smart glasses. Then, the analysis result is shown in smart glasses directly. All analysis and calculation process are progressed on a cloud-computing system. Finally, a digital metallographic analysis approach for estimating decarburization depth of carbon steel specimen by a portable device can be carried out.

1. Introduction

Decarburization phenomenon refers to that the carbon on steel surfaces reacts with the deoxygenation in heat treatment furnaces due to heat treatment and causes reduction in carbon content on steel surfaces as well as hardness. Decarburization depth refers to the distance from the place where the steel starts to be decarburized to the surface. There are two major approaches to manually measure the decarburization layer depth. One is hardness test, referring to using a micro Vickers hardness meter to gradually knock into the core along the surface, so as to record the hardness changes. It is the beginning of decarburization when the difference between the measured hardness and the core hardness is more than Hv30. The distance from the point to the surface measured with a microscope is defined as the decarburization layer depth, denoted as DHT. Another is to measure with optical microscopes. In this way, an inspector selects the fixed magnification (normally 100 to 200 times) for observation in the view field and determines the start point of decarburization, to measure the distance from the start point to the surface to be the position of the decarburization layer, denoted as DMT [1]. There are two approaches in measuring decarburization layer depth with metallographic images, including manual measurement and digital measurement, and this study used the latter to discriminate automatic measurement.

IOP Conf. Series: Materials Science and Engineering **644** (2019) 012007 doi:10.1088/1757-899X/644/1/012007

About material image inspection, the digital metallographic inspection is mostly applied to failure analysis [2-4]; and some applications are in abrasion calculation and spacemen life forecasting [5, 6]. Papa and Albuquerque applied digital image processing to analysis on grey casting iron and nodular casting iron, to distinguish the types of grey casting iron and to globular graphite statistics [7, 8].
Conerning the grain analysis, Hanning and Engelberg tried to compensate the missing grain boundary after corrosion of 304 Austenite stainless steel by the method of image comparison [9]. Vieira and Paciornik [10] calculated the area proportion of pearlite by the method of uncertainty evaluation, but the research only issued the calculation results and the uncertainty of two different materials.
This study carried out an automatic digital algorithm, and used the gray gradient in combination with the way of determining the threshold value through hardness re-correction, to increase the accuracy of automatic calculation results of the decarburization layer depth. In application, the image capture function of smart glasses can be applied to take the image and send back to the cloud to estimate the decarburization layer depth, so as to achieve the inspection efficiency.

2. Digital image analysis

Combining with hardness correction, the main inspection approach applied in this study was optical measurement for automatic decision making. This approach is relatively simple but objective in inspection process. The main steps are as below: (1) measuring hardness, and deciding position and depth used in hardness method and decarburization method; (2) based on this decarburization layer depth, correcting the threshold value measured by optical method, so that the measurement is more consistent with the current status of hardness variation; (3) conducting automatic calculation with the corrected composite software to calculate the optical decarburization layer depth.
The image processing procedure mainly consists of the following steps: (1) gray scaling, (2) sharpening, (3) binarization, (4) analytic statistics and (5) decision making.
Gray scaling refers to, assuming that the color temperature has no effect on optical input, making color images grayscale reference to existing grayscale parameters to avoid calculation errors caused by different colors [11]. Sharpening function is to increase the grayscale gradient of all pixels in grayscale images, so as to improve the accuracy of the following binarization [12]. Binarization is to convert a grayscale image into a black-and-white image based on a threshold value. Assuming that, for an image, the width and high pixels are w and h, separately. The grayscale is G_i when the pixel is at the position i, and it is defined black ($G_i = 0$) and white ($G_i = 255$) while $0 \leq G_i \leq 255$. When the binarization threshold value is λ, the pixel B_i after binarization is

$$B_i = \begin{cases} 0, if\ G_i \leq \lambda \\ 255, if\ G_i > \lambda \end{cases}; \tag{1}$$

where the quantity of i is $w \times h$. Afterword, a binarization pixel statistics is carried out after binarization, and the analysis method is mainly based on the binarization pixel distribution in the specimen. The major way is to divide the specimen image into ten equal parts (from Area [1] to Area [10]) from the core to the surface, and to calculate the binarized pixel mean values in all areas from $\bar{G}[1]$ to $\bar{G}[10]$. In these ten binarized pixel mean values, the first threshold value is determined the location with the maximum gradient of pixel counts. The interval of $\bar{G}[i]$ $(1 \leq i \leq 10)$ is divided into ten equal parts, so as to calculate the grayscale mean values from $\bar{\bar{G}}[1]\ to\ \bar{\bar{G}}[10]$. Likewise, the second threshold value is determined the location with the maximum binarized pixel gradient. In decision making, to take $\bar{\bar{G}}[j]$ $(1 \leq j \leq 10)$ of the second stage and the median position (x, y) of its interval can be determined as the DMT. As the algorithm issued above, the DMT measurement is the distance from this position to the surface, i.e., the decarburization layer depth. The second threshold value can be regulated through manually correction by hardness method, which is measured by means of optical microscopes. In this study, the threshold value was determined through manually hardness correction method firstly, and then the digital optical measurement is progressed only in actual inspection.

3. Metallographic inspection service system of a portable device

Regarding the cloud metallographic inspection service system used in this study, combining with a portable device and the cloud computing platform. The digital inspection mentioned above was used to establish a cloud-based metallographic inspection platform, and the relevant patent were obtained. In this study, the structures of cloud-computing platform shown in Fig. 1 mainly consist of four modules, including the portable device module, the cloud inspection module, the account management module and the history query module, respectively

3.1. Portable device module

Connected to the network of the cloud metallographic inspection server, the portable device module used in this study is a pair of smart glasses with an application software (APP). The pair of smart glasses used in this study consists of a pair of portable inner projection glasses and a separate auxiliary controller. The controller runs an APP to use the camera on the smart glasses to capture the images of the decarburization layer displayed on the observation screen of the microscope, and is connected to the cloud inspection server through a module linked by WiFi wireless signals, so as to achieve cloud network connection and data transmission. A display module is used to display the photos obtained by the image capture module or the image transceiver module, and an image transceiver module is used to send the photos obtained by the image capture module or a mobile device operation system is used to receive the images from the cloud inspection server. The auxiliary controller contains an input module for the users to control to input data or instructions, and the display module displays the data or instructions.

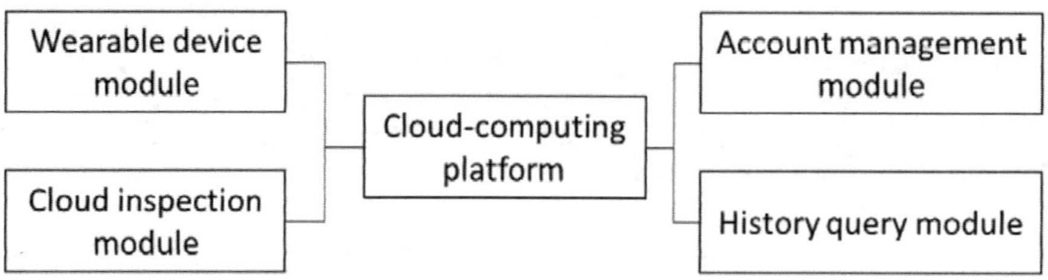

Figure 1. Architecutre of cloud-computing platform.

3.2. Cloud inspection module

The cloud inspection module is the core of this study, after logging in, corresponding type of digital metallographic inspection can be selected or other digital problems can be analyzed by using the manual inspection. The users can select inspection types, upload images, server, transmit data to the digital computer, select parameters, call corresponding algorithms and calculate the results of digital decision making.

The results of digital decision will be firstly sent to the reviewers to confirm rationality of the inspection contents. After the rationality of the inspection contents is confirmed, the reports will be sent to the inspectors. In the above way, the original inspection duration of 4 to 5 days for the post metallographic inspection analysis can be shortened to 1 hour, and the inspection time is shortened by above 99%.

3.3. Account management module

For the cloud fastening inspection service system, the user account management can be performed to confirm valid accounts as well as the grade and categories of users, in addition to general user information.

3.4. History query module

The system in this study can be used to inquire the history of inspection results. The major function is that, after inspection, the users can directly inquire on the cloud inspection service system without saving the inspection results on their own. In this method, not only the users' trouble of data saving and maintenance at local ends can be eliminated, but also the historical inspection data can be accessed directly and rapidly, so that the metallographic inspection efficiency can be improved.

4. Experimental description

In this study, the optical microscope used was Olympus BX51M, the metallographic camera was Sentech STC-MC152USB, the image capture pixel was 1360 × 1024, RGB was used for image capture, and images were sampled at an optical magnification of 500 times. The captured images were output to a 22" 16:9 screen for display. A pair of Epson smart glasses BT-300 was used for capture, a 5-megapixel CMOS camera was used, the CPU was Intel-based, and the operation system was Android V5.0.

This study decided the decarburization layer depth after spheroidization anneal of AISI 1022 steel. The chemical components composed are: C: 0.18-0.23%; S: 0.1%; Mn: 0.7-1.0%; P: 0.03%; S: 0.035%. The conditions of spheroidization anneal are heating to 700 ℃, cooling the furnace temperature to 500℃ after holding the temperature for 10 hours, and taking out to cool in air.

5. Results and discussion

From the hardness test, the core hardness of the specimen was Hv 127. The distance from the position of (Hv-30) to the wire surface was 0.024 mm, which was the total decarburization depth for verification of this study object. By microscopic method, digital metallographic images were captured through smart glasses, and then the calculation was made after returning to the cloud analysis system. Fig. 2 shows the images and analysis results used for inspection in this study. Fig. 2(a) shows the metallographic images taken digitally before inspection. Fig. 2(b) shows the inspection result which is 0.021 mm analyzed by microscopic method based on the analysis mode in this study with a personal computer. Fig. 2(c) shows the inspection images taken by smart glasses and transmitted to the cloud, and the inspection result is 0.028 mm shown in Fig. 2(d). The errors compared with hardness detection and general cloud analysis are 0.003 and 0.007 mm, respectively, which are able to be acceptable in the metal manufacturing industries.

The difference between personal computer and smart glasses is mainly caused by two conditions. Firstly, the light condition is the main factor very much in demand. The inhomogeneous grey-scale distribution is carried out by environment light with smart glasses. Meanwhile, the reflection of monitor also causes the image distortion, as shown in Fig. 2(c) and 2(d), what we can see is the lower part of image exhibiting relative white than origin image [Fig. 2(a)].

Secondly, the magnification also performs divergence with different distance between smart glasses and monitor. As shown in Fig. 3, the magnification is calculated as: original specimen for 1 time, objective lens for 50 times, eyepiece for 10 times, reduce lens for 0.5 time, capturing camera for 20 times, controlled monitor for 1 time. And the magnifications can be estimated as $1 \times 50 \times 10 \times 0.5 \times 20 \times 1 = 500$ times.

When the smart glasses is obtained in calculation, through the resolution can be adjusted with the same height of captured image of camera on smart glasses and shown image in monitor, the measuring difference is also generated for the difficult to tuning posture of the head mount device. This measuring difference also causes little influence on image recognition.

When the analysis of different magnifications was proceeded as shown in Fig. 4 which can be found that same two conditions happened. Fig. 4(a) and 4(b) show the metallographic images observed under 100 times of magnification, whose decarburization depths are 0.079 and 0.091 mm, respectively. Fig. 4(b) performs a larger image than Fig. 4(a) apparently, though two images are the same one. The inspector could not keep a perfect posture and make the difference. Fig. 4(c) and 4(d) show metallographic images observed under 200 times of magnification, whose decarburization depths are 0.024 and 0.030 mm, respectively. The distance difference between two images is 0.006 mm, which is made from different light distribution.

IOP Conf. Series: Materials Science and Engineering **644** (2019) 012007 doi:10.1088/1757-899X/644/1/012007

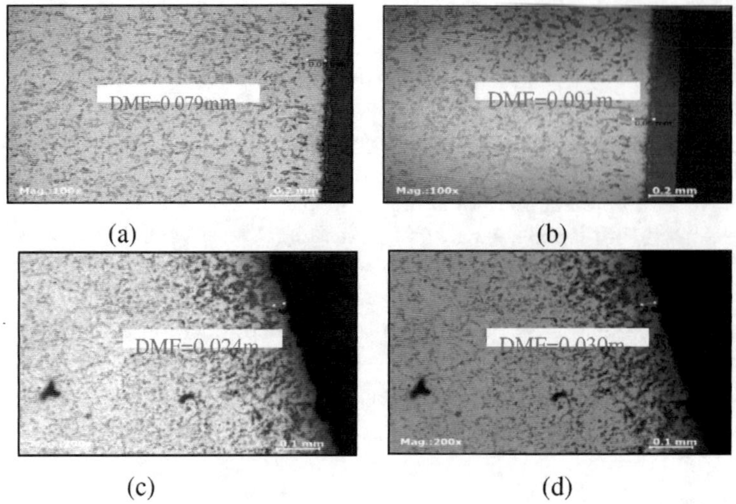

Figure 2.(a) shows the original metallographic captured by microscope camera; (b) shows the analysis result with original metallographic; (c) shows the metallographic captured by the portable device; (d) shows the analysis result with metallographic captured by the portable device.

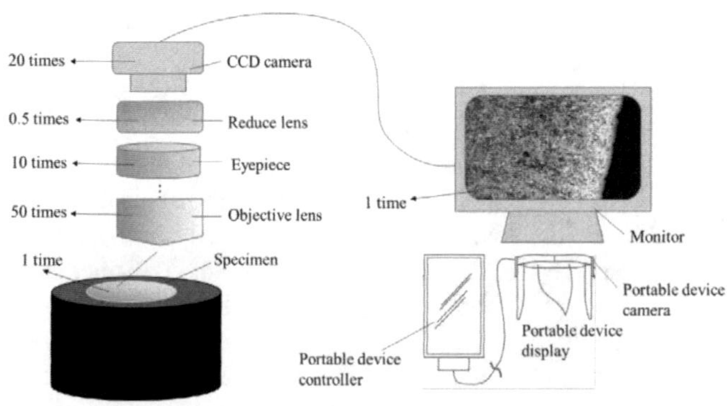

Figure 3. Schemia of magnification calculation.

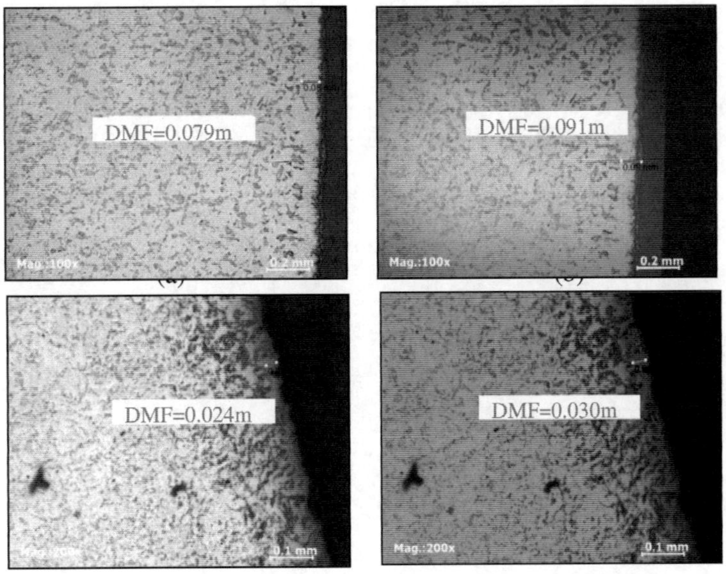

(c) (d)

Figure 4. (a) shows the analysis result with original metallographic under 100 times of magnification; (b) shows the analysis result with metallographic captured by the portable device under 100 times; (c) shows the analysis result with original metallographic under 200 times; (d) shows the analysis result with metallographic captured by the portable device under 200 times.

6. Conclusion

This study applied portable devices to develop a cloud computing platform with portable cloud metallographic inspection services for measuring the decarburization layer depth. Combining with the database system of mySQL, this system used the present PHP web language to call the local applications to digitally decide and calculate the uploaded metallographic images, and then the decision results were projected onto the inner projection lens by using the glasses and were transmitted to the users, to complete the service. The conclusion is shown as below:

1. The study results can be used to actually conduct digital cloud decision for metallographic structures with images of decarburization layer depth.
2. The mainly difference is caused by two condition. One is the light factors, which includes environment light and reflection of monitor. The other is change of magnification with distance between smart glasses and monitor.
3. The errors compared with hardness detection and general cloud analysis are 0.003 and 0.007 mm, which can be accepted in industrial use.

7. References

[1] Standard Test Methods for Estimating the Depth of Decarburization of SteelSpecimens. *ASTM E1077-01* (2001)
[2] B. Pawłowski, P.Bała, J. Krawczyk, M. Stępień, T. Śleboda, Eng. Failure Anal. 82, 533 (2017)
[3] C.A. Duarte, E. Espejo, J.C. Martinez, Eng. Failure Anal. 79, 704 (2017)
[4] A. Lanzutti, A. Gagliardi, A. Raffaelli, M. Simonato, M. Magnan, F. Andreatta, L. Fedrizzi, Eng. Failure Anal. 79, 634 (2017)
[5] T. Mikołajczyk, K. Nowicki, A. Kłodowski, D.Y. Pimenov, Mech. Sys. Signal Process 88, 100 (2017)
[6] T. Mikołajczyk, K. Nowicki, A. Kłodowski, D.Yu. Pimenov, Mech. Sys. Signal Process. 104, 503 (2018)
[7] J.P. Papa, R. Y.M. Nakamura, V. H. C. Albuquerque, A. X. Falcão; J. M. R. S. Tavares, Expert Sys. Appl. 40, 590 (2013)

[8] V.H.C. Albuquerque, A.R. Alexandria, P.C. Cortez, J. M. R.S. Tavares, NDT&E Int. 42, 644 (2009)
[9] F. Hanning, D.L. Engelberg, Mater. Charact. 94, 111 (2014)
[10] P.R.M. Vieira, S. Paciornik, Mater. Charact. 47, 219 (2001)
[11] M. Grundland, N. A. Dodgson, *Color to Grayscale Conversion* (UCAM-CL-TR-649, University of Cambridge, 2005)
[12] R.C. Gonzalez, R.E. Woods, *Digital Image Processing* (Prentice Hall, Third edition, NJ, USA, 2008)

IOP Conf. Series: Materials Science and Engineering **644** (2019) 012008 doi:10.1088/1757-899X/644/1/012008

An Energy Consumption Evaluation of Non-TIM Strategy in IEEE 802.11ah

Tain-Lieng Kao[1], Huan-Chun Wang[2], Chang-Han Lu[2], and Ting-Hsu Cheng[2]

1 Department of Communication Engineering, I-Shou University, 84001 Kaohsiung, Taiwan
2 Department of Electronics Engineering, National Taiwan University of Science and Technology, 10607 Taipei, Taiwan
E-mail: tlkao@isu.edu.tw

Abstract. IEEE 80211ah is a low speed and long distance communication protocol which has two modes of operation: one is that stations in RAW need to listen Beacon for data transfer, and the other is that stations in Non-TIM TWT mode need not to synchronous with AP by listening Beacon. 802.11ah with Non-TIM TWT strategy is very suitable for many IoT applications with low speed and low energy consumption. However, there is no further discussion of energy consumption for stations under Non-TIM TWT mode in 802.11ah networks. Therefore, this paper proposes an 802.11ah Non-TIM Implicit TWT simulation which is built with the 802.11 Legacy simulator added to the transmit traffic distribution, Target Wake Time mechanism and Channel Model. This simulation analysis will evaluate energy consumption performance under different IoT applications (agricultural monitoring, smart metering, and industrial automation).

1. Introduction

Recently, IoT technology has been widely used in many applications which require remote monitoring, such as: agricultural monitoring, smart metering, and industrial automation. These applications remotely collect data through embedded sensors and exchange data by wireless communication transceivers. People get information about things through the wireless network, and then manage and monitor them.

Because the sensing nodes are widely deployed, that is, they are usually far away from the data collector of the wireless network. Sensing data transmitted by the node is usually very short, for example, the temperature data usually needs only one byte or less. The period during which a node transmits data is usually not very frequent, for example, sending temperature data once in 15 minutes. Therefore, how to choose the appropriate wireless network technology in the application to make the sensing node run for the longest time with the least power is an important issue in the current Internet of Things.

At present, long-distance wireless network technologies such as 802.11ah and LoRa both have the ability to make the sensing node consume the least amount of power and exchange data stably in a long-term operation. Both technologies can adapt to different IoT applications in different modes. When comparing LoRa and 802.11ah in low speed applications, 802.11ah under Non-TIM TWT mode saves listening to Beacon time and uses more efficient CSMA/CA compared to LoRa in Class B mode [1, 2, 3, 4].

There are two modes of operation in 802.11ah: Non-TIM TWT and TIM RAW. Since TIM RAW requires more power for each node to listen Beacon for synchronization of data transfer, it is not suitable for applications which need to save power when transferring a tiny data.
In the 802.11ah Non-TIM TWT mode, AP and stations come up with an agreement on the TWT Service Period in the association procedure. According to the TWT SP, stations know when to transmit data, and then enter the sleep mode after the data transmission is completed, thus avoiding a large number of stations competing for channels at the same time and saving more power.
Stations send an Association Request with TWT IE and Power Capability IE to AP in the association procedure. AP determines whether to connect with a station according to the Power Capability IE. AP then informs stations with the Association Response to operate TWT using Implicit or Explicit. There are two types in IEEE 802.11ah TWT, namely Implicit TWT and Explicit TWT. Implicit TWT is a periodic use of TWT SP. Explicit TWT is a non-periodic use of TWT SP. The control message such as ACK packet, TWT Information packet or normal packet with TWT information is transmitted through AP to inform stations of the next TWT time. This paper will explore how 802.11ah Non-TIM mode uses the Implicit TWT strategy with grouping functions.
The second section introduces energy analysis of the station in 802.11ah. The third section describes the establishment of 802.11ah TWT simulation environment. The fourth section provides the simulation analysis of energy consumption of the system under different IoT applications. Finally, the fifth section gives some conclusions.

2. Energy consumption analysis of 802.11ah stations

Since the station's energy consumption for transmission is an important factor in IoT applications. In this section, the theoretical equations for energy consumption of 802.11ah station are explained. It contains how the traffic is generated, the calculation of the transmission power for channel and the energy consumption of the stations in the transmitting, receiving, idle and sleeping states.

2.1. Periodic uniform traffic model

A large number of stations are often deployed in IoT applications. Each station usually transmits data once in a long period of time. In order to avoid stations unable to transmit data due to competitive collisions in timeslot, the probability of collision is greatly reduced by using the allocated transmission traffic. Therefore, the average power consumption of the station is reduced to achieve power saving. An IoT station uses uniform traffic model as the traffic flow, which increases the average time to generate packets [5, 6]. The station does not need to enter the idle state to continue to stay in sleeping state when no transmission traffic is generated. Therefore, the station can sleep longer and save more power than the traditional 802.11.

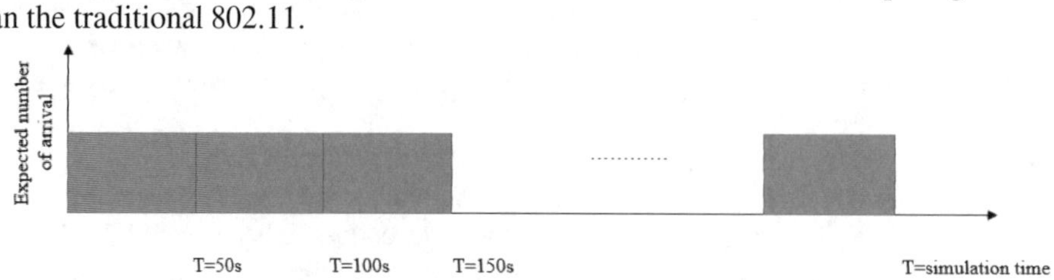

Figure 1. Periodic uniform traffic model.

The periodic uniform traffic model used in this paper is shown in Figure 1. Traffic is periodically generated and packets are distributed using uniform traffic model in each cycle.

2.2. Transmission Power for Channel

Rayleigh Block Fading Channel is a channel model used in IEEE 802.11ah [7]. According to different SNRs, the packet error rate is as shown in the equation (1):

$$\mathrm{PER} = 1 - e^{\frac{-r_{th}}{r}} \quad (1)$$

where r is the average SNR value at the receiver and r_{th} is the SNR threshold.

In general, the greater the given transmit power, the smaller the packet error rate will be. However, how to meet the system's PER with the minimum of transmit power is a tradeoff for IoT applications. In order to let the station transmit longer distances in 802.11ah, this paper uses BPSK modulation and 0.65Mbps Data Rate to establish channel model. Table 1 uses the PER of 0.1 as the parameter condition of the channel model to calculate how much of the transmit power is needed at different distances.

Table 1. Simulation parameters of transmit power.

Transmit Distance(m)	100	200	300	400	500	600	700	800	900	1000
Transmit Power(mw)	0.12	0.4	3.2	5	10	20	40	63	100	158

2.3. Energy consumption in the receiving state

In this paper, the station only receives the ACK packet. The energy consumption of the station in the receiving state is as shown in the equation (2):

$$Rx_{ACK} = Rx\_power \times (\frac{ACK_{length}}{r}) \quad (2)$$

Where Rx_{ACK} is the energy consumed to receive an ACK packet, Rx_power is the energy consumed by the receiver per second, ACK_{length} is the length of the ACK packet, *r* is the system data rate.

2.4. Energy consumption in the transmitting state

Based on different distance from AP, the transmitter of the station uses different transmit power. The energy consumption of the station in the transmitting state is as shown in equation (3):

$$Tx_{packet} = Tx\_power_{distance} \times (\frac{Packet_{length}}{r}) \quad (3)$$

Where Tx_{packet} is the energy consumed to transmit a packet, $Tx\_power_{distance}$ is the energy consumed by the transmitter according to different distance from AP per second, and $Packet_{length}$ is the length of the packet, *r* is the system data rate.

2.5. Energy consumption in the leeping and idle state

The energy consumption of the station in the sleeping and idle state is as shown in equation (4):

$$P = T_{sleep} \times P_{sleep} + T_{idle} \times P_{idle} \quad (5)$$

Where P is the total energy consumption, T_{sleep} and T_{idle} is the time of station's sleeping state and idle state, and P_{sleep} and P_{idle} is the energy consumed of the station in sleeping and idle state.

3. Simulation design of 802.11ah TWT

3.1. System description

The simulation architecture of 802.11ah TWT is shown in Figure 2. The simulation system was built based on the 802.11 legacy simulator with additional modules, include such as Channel Model, Traffic Model and TWT strategy [8, 9, 10].

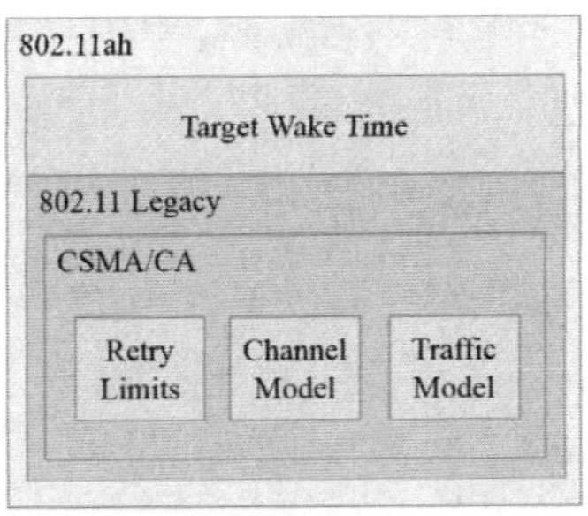

Figure 2. Architecture of 802.11ah TWT simulation.

IOP Conf. Series: Materials Science and Engineering **644** (2019) 012008 doi:10.1088/1757-899X/644/1/012008

First, the 802.11 legacy simulator adds the Retry Limits module to allow the system to perform Packet Error Rate operations. Traffic Model module allows the system to generate a periodic uniform traffic. Target Wake Time strategy implements the allocation of TWT SP and TWT grouping functions in 802.11ah. Channel Model module performs simulation of the transmit power based on the transmission distance of stations in 802.11ah networks. The parameters of the 802.11ah simulation are shown in Table 2 [11].

3.2. Application scenarios

TWT SP periods will affect the amount of power consumption. A long TWT SP period may enable more stations to have opportunities for transmission, but also make more stations remain in idle state. Such stations are very wasteful of power. Conversely, if the TWT SP period is too short, the station will have no time to transmit data at all. Therefore, this paper uses Packet Drop Rate as an index to find a corresponding TWT SP periods to make the system work well and save power. Finally, the simulator implements the TWT SP grouping function and adjusts the TWT SP periods to observe changes in Power Consumption, Packet Drop Rate, and Packet Delay Time.

The following three common applications are used to evaluate the energy consumption performance of 802.11ah Non-TIM TWT simulation.

(1) Agricultural monitoring: This application has to know whether it needs irrigation for a period of time in order to control the humidity of the environment. This application deploys 3,500 stations in a radius of 1000 meters. Each station generates a packet with 100 bytes every 120 seconds.

(2) Smart metering: This application can monitor the power consumption, gas volume and water consumption of all users in large communities. This application uses 1500 sensor stations in a radius of 500 meters, sending a packet with 100 bytes every 50 seconds.

(3) Industrial automation: For example, it is used in the refrigerating room to get the temperature to ensure that things such as fruits and vegetables in the refrigerating compartment can be kept fresh. This application uses 500 stations in a radius of 250 meters, producing a packet of 100 bytes in every 180 seconds.

Table 2. Simulation parameters of system.

Parameter	Value
Data Rate	0.65Mbps
Packet Length	100Bytes
Retry Limits	6
DIFS	264us
SIFS	160us
Slot	52us
ACK	24 Bytes
MAC Header	12 Bytes
PHY Header	6 Bytes
Packet Error Rate	0.1
Idle Power	1.5mw
Sleep Power	10uw
Receive Power	135mw

3.3. Simulation Results

This section will discuss the simulation results of three different applications with the TWT Grouping function. Based on the different transmission distances and TWT lengths, the power consumption, Packet Drop Rate and Packet Delay Time of the three applications are evaluated.

Agricultural monitoring is one of the Internet of Things applications, and the number of stations that can be provided is also the largest of the three applications in this simulation. Figure 3 shows the changes in power consumption when the TWT SP length and the location of the station are adjusted in the case of TWT Grouping function using the Periodic Uniform Traffic Model under agricultural monitoring scenario. Figure 4 displays the changes in Packet Drop Rate and Packet Delay Time when the TWT SP length is adjusted. The simulation of agricultural monitoring shows that the shorter the TWT time, the more power consumption decreases. The Packet Delay Time decreases, and Packet Drop Rate increases. However, the changes are quite small.

Smart metering simulation provides a scenario for more dense stations. Figure 5 and Figure 6 show the simulation results of smart metering.

Industrial automation scenario deploys 500 stations in a radius of 250 meters. Since the data period transmitted by each station is 180 seconds, the length of the TWT SP also becomes longer. Figure 7 and Figure 8 display the simulation results of industrial automation.

3.4. *Agricultural monitoring*

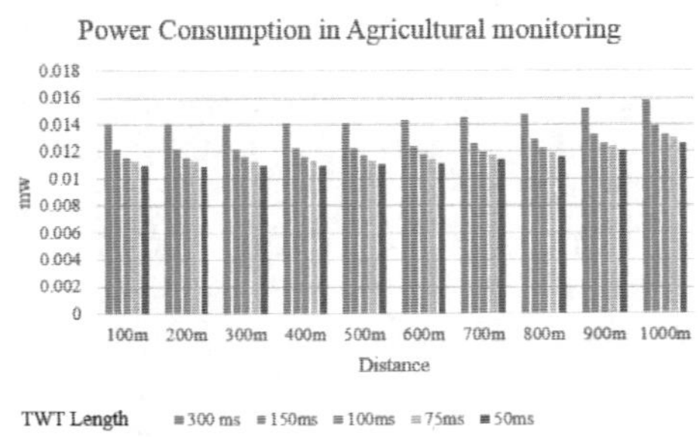

Figure3. Power consumption in agricultural monitoring.

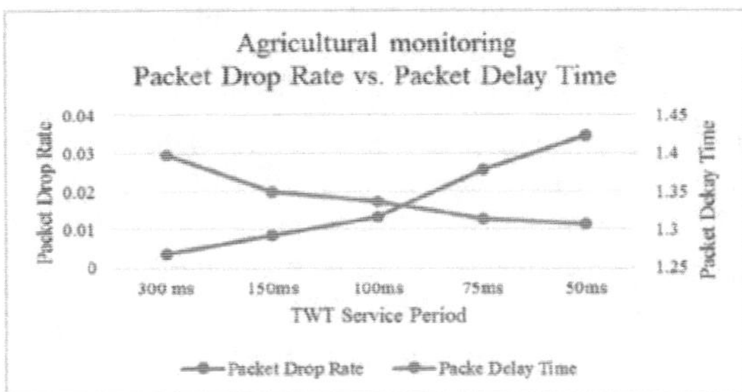

Figure 4. PDR vs. PDT in agricultural monitoring.

3.5. *Smart metering*

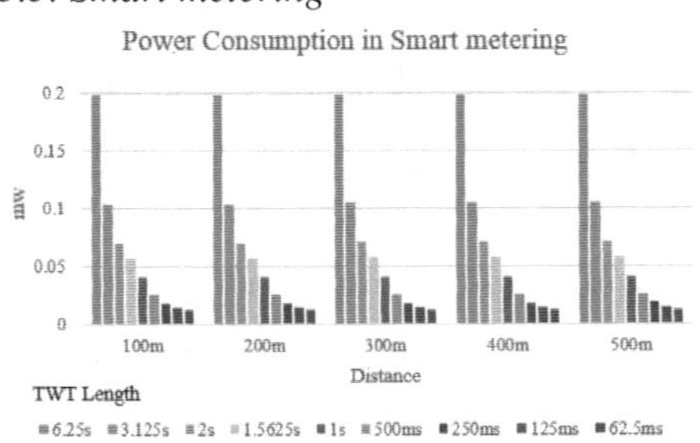

Figure 5. Power consumption in smart metering.

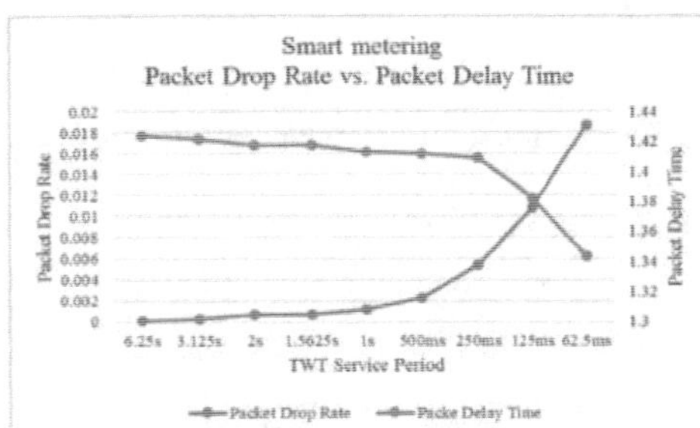

Figure 6. PDR vs. PDT in smart metering.

3.6. Industrial automation

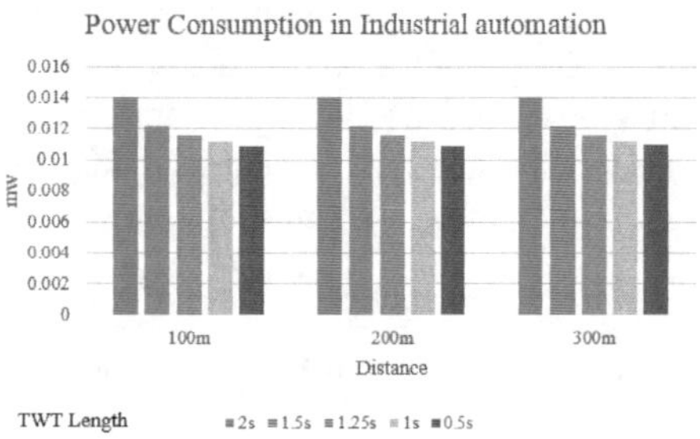

Figure 7. Power consumption in industrial automation.

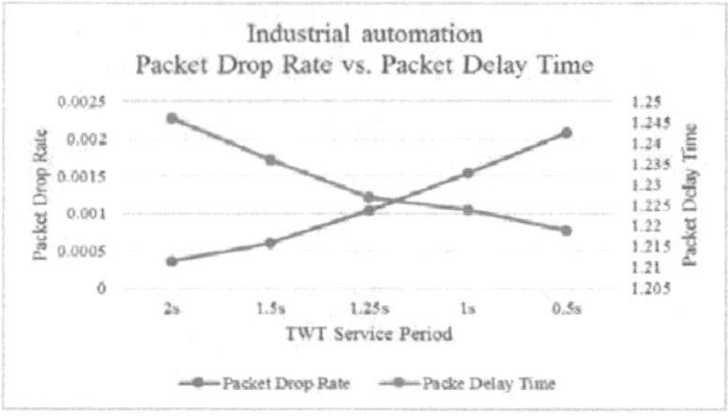

Figure 8. PDR vs. PDT in industrial automation.

4. Discussion of results

(1)Since different applications use different TWT SP periods, it can be observed that smart metering consumes more power than other applications. The TWT SP period directly affects the system power consumption.

(2)In short distances, the power consumption of the station will vary according to distance, but the difference is not obvious. In long distances, the power consumption of station is greatly increased under the application of agricultural monitoring. Therefore, in the case of using TWT grouping, the power consumption of the station in the idle state is the main cause of power consumption in short distances.

(3) The smaller the TWT SP is, the more power is saved. Packet Delay Time will continue to drop, and the Packet Drop Rate will slowly increase but not much affect the system. When the TWT SP reaches the minimum value, each TWT SP will only have one station to transmit, so there is no collision problem and Packet Drop Rate will suddenly drop to zero.

(4) When the system adopts TWT grouping and each station uses a TWT SP, this usage is similar to RAW, but TWT grouping differs from RAW in that it does not need to listen to Beacon. TWT grouping has better power saving effect than RAW.

5. Conclusions

This paper simulates three applications in 802.11ah with Non-TIM Implicit TWT, and analyzes the relationship between the energy consumed by the three applications, Packet Drop Rate and Packet Delay Time. It also discusses the impact of using different TWT periods on energy consumption in three applications, and provides a reference for deploying 802.11ah IoT systems.

6. References

[1] LoRaWAN™ 1.1 Specification
[2] Wireless LAN medium access control and physical layer specification-Sub 1 Ghz license exempt operation (802.11ah Specification)
[3] B. Domazetovic, E. Kocan, 25th Telecommunication Forum, Belgrade, Serbia, 1-4 (2017)
[4] E. Khorov, A. Lyakhov, A. Krotov, & A. Guschin, Computer Communications, 58, 53-69 (2015)
[5] M. Laner, P. Svoboda, N. Nikaein, M. Rupp, ISWCS 2013 The Tenth International Symposium on Wireless Communication Systems, Ilmenau, Germany, 1-5 (2013)
[6] G. C. Madueño, C. Stefanovic and P. Popovski, IEEE Wireless Communications Letters, 3(4), 429-432 (2014)
[7] P. Serrano, A.G. Saaverdra, G. Bianchi, IEEE/ACM Transactions on networking, 23(4), 1243-1256 (2015)
[8] P. Chatzimisios, A.C. Boucouvalas, V. Vitsas, GLOBECOM '03. IEEE Global Telecommunications Conference, San Francisco, USA, 2, 950-954(2003)

[9] P. Raptis, V. Vitsas, K. Paparrizos, P. Chatzimisios, A.C. Boucouvalas, 6th IEEE international Symposium on a World of Wireless Mobile and Multimedia Networks, Taormina-Giardini Naxos, Italy, 299-304 (2005)
[10] A. Bel, T. Adame, B. Bellalta, IEEE Wireless Communications, 13, 9(2014).
[11] X. Wang, J. Yin, D.P. Agrawal, *Wireless Communications & Mobile Computing*, 5(1), 113-122(2005)

IOP Conf. Series: Materials Science and Engineering **644** (2019) 012009 doi:10.1088/1757-899X/644/1/012009

Design and Research of Permanent Magnet Brushless DC Direct Drive Hub Motor

Zhong-Shu Liu

The Key Laboratory for Automotive Electronics and Electric Drive of Fujian Province /School of Information Science and Engineering, Fujian University of Technology, Minhou, Fuzhou City, 350118, China
E-mail: 1241891962@qq.com

Abstract. The permanent magnet brushless DC motor has many advantages such as simple structure, high power density, small volume, good mechanical properties and high efficiency, so it is very suitable for electric vehicle. However, the traditional brushless DC motor has the disadvantages of flux weakening speed regulating difficulty and torque ripple. According to the characteristic of permanent magnet motor, so a kind of special structure brushless DC hub motor with outer rotor was designed in this paper. The structure of interlaced arrangement of permanent magnet and non-salient pole of the outer rotor greatly increased the average inductance of the stator winding, so as the range of flux weakening speed regulating increased greatly as well. At the same time, large chute in stator core and fractional stator winding reduced torque ripple of the motor. The simulation model was established by using Ansys Maxwell software and the magnetic field was calculated and analyzed. It turned out that the magnetic field was rational and the optimum designed motor had less flux leakage. Simulation and test results showed that stator performance indicators of the prototype could meet the requirements of electric vehicle fully. So it has great value of generalization in practical application.

1. Introduction

Due to the increasingly serious energy shortage and environmental pollution in currently, electric vehicles have gradually become the focus of the global automotive industry, the direct drive hub motor which is directly connected to the wheel drives the car without engine, gear shift and transmission shaft simplify the transmission system and chassis structure, improve the transmission efficiency. It should be pointed out that as the hub motor of the electric vehicle, its performance is directly related to the driving performance of the vehicle [1]. Among all kinds of motor vehicles, permanent magnet brushless DC motor (BLDCM) has become a particularly suitable motor for electric vehicles due to its high efficiency, high power density, good reliability and other advantages. Compared with the traditional DC motor, permanent magnet brushless DC motor has no brush and commutator, so it makes its structure very simple. Meanwhile, its rotor is excited by permanent magnets, so there has no rotor resistance loss, and its efficiency is higher. The temperature rising of permanent magnet brushless DC motor mainly comes from the stator winding, it is easy to cool down. Finally, it has small electromagnetic time constant and good dynamic performance. However, although the permanent magnet brushless DC motor has so many advantages, but it has the disadvantages of flux weakening speed regulating difficulty and torque ripple. Therefore, in view of the above problems, this paper designed a six phase structure of the permanent magnet brushless DC motor so as to expand the speed range, reduce the ripple torque and increase the output torque [2]. The optimum designed permanent magnet brushless DC motor fully meets the requirements of electric vehicle.

2. Structure and Optimum Design of Direct Drive Hub Motor

2.1. Motor structure

The structure diagram of the direct drive hub motor designed in this paper is shown in figure 1. It is an outer rotor brushless DC motor (BLDCM) with external rotor and characteristics of low speed, large torque and high speed constant power. Meanwhile, this structure does not need deceleration mechanism, which can not only make the driving system structure simple, but also further improve the efficiency.

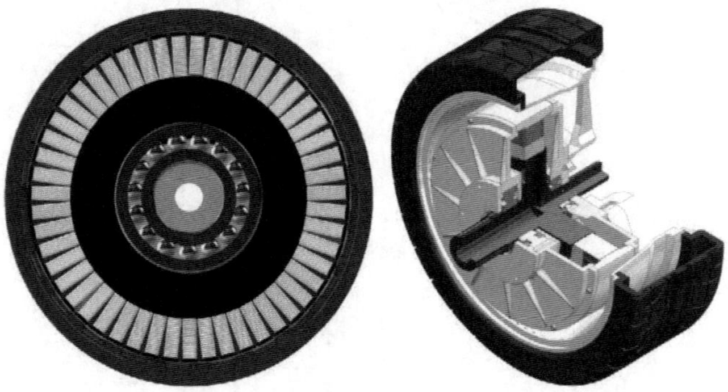

Figure 1. Structure of outer rotor BLDCM.

2.2. Motor design

The design of BLDCM mainly includes stator design and rotor design. Specifically, the main size of the motor, namely the stator outer diameter Da and the effective length of the iron core L_{eff}, should be determined first [3], and then other sizes of the motor could be determined. The steps of motor design are shown in figure 2.

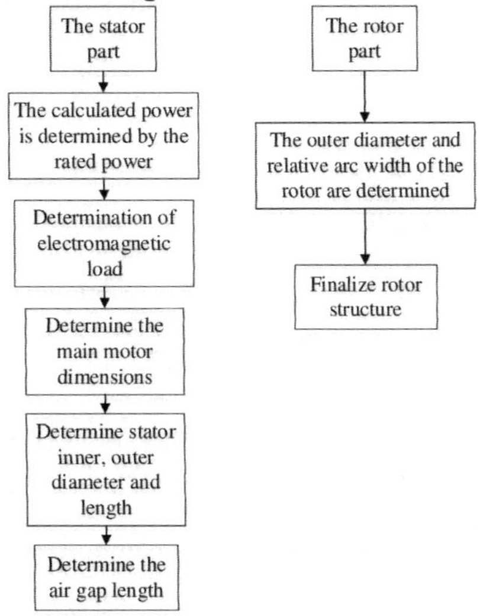

Figure 2. Blok diagram of main dimensions determination of BLDCM.

2.3. Design of stator winding

Whether the permanent magnet brushless DC motor (BLDCM) can produce the desired opposite potential waveform depends on the air gap magnetic field of the permanent magnet and the spatial distribution of stator winding. The design of stator winding should meet the following basic requirements: first, stator winding is energized to produce the same number of poles as rotor magnetic poles [4]. Second, the processing process of winding is simple. Finally, it can eliminate the high

IOP Conf. Series: Materials Science and Engineering **644** (2019) 012009 doi:10.1088/1757-899X/644/1/012009

harmonic in the phase potential and reduce torque ripple as well as cogging torque. Based on the above requirements, the stator winding of the prototype adopts fractional slot concentrated double-layer short pitch winding. Because the fractional slot winding group of coils in the same phase is distributed under different polar s and phases, the slots under each phase is staggered from each other, which reduces the cogging torque and improves the sinusoidal degree of the back electric motive force waveform. The prototype designed in this paper has 46 poles and 57 stator slots. The spatial distribution diagram of its stator winding is shown in figure 3.

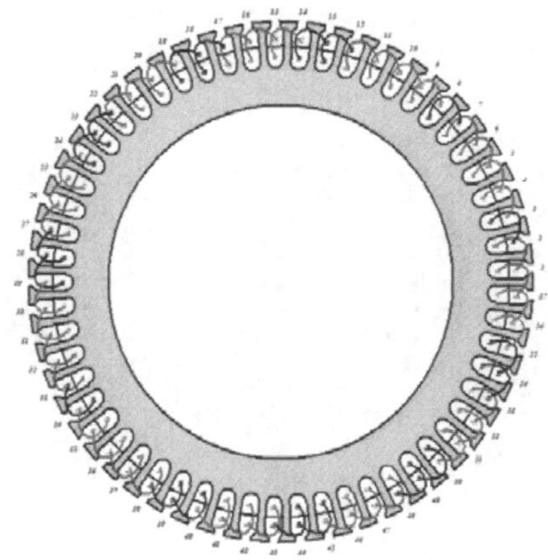

Figure 3. Stator winding spatial distribution diagram.

2.4. Design of stator core

The prototype designed in this paper adopted the semi-closed pear shaped slot as its stator slot. And the stator iron core adopted chute design, so as to significantly reduce the tooth harmonic potential and the influence of cogging torque by the criss-cross of the phase of the stator conductor tooth induced harmonic electric motive force [5]. By using Ansys Maxwell, we can analyze the influence of different chute width on the cogging torque. It can be concluded that when the width of the stator core chute is an integer multiple of the stator chute slot pitch, the cogging torque of the motor reaches the minimum, and the motor output torque ripple is the minimum. Everything taken into consideration, the width of stator core chute used in the prototype is 2 times the stator slot pitch.

2.5. Design of rotor

The design of rotor refers to two parts, including the design of permanent magnet and the rotor structure. Because the outer rotor is directly applied in the electric car wheel of hub motor, so the shock, vibration and temperature varies greatly [6]. Through the analysis and comparison of the performance of all kinds of permanent magnetic materials, Nd-Fe-B permanent magnet material was adopted because of its high coercive force, high magnetic energy product, linear demagnetization characteristic and convenient processing. Firstly, an appropriate nominal working point of the air gap flux density is determined, then the air gap magnetic fall could be found in the table, according to the air gap flux density to calculate gas gap magnetic potential and the stator yoke of flux density, finally the stator yoke magnetic potential drop could be calculated. Thus the total magnetic potential could be calculated totally. Then the thickness and the width of the magnet steel could be determined [7].

The rectangular magnetic steel designed above is uniformly adhered to the surface of the outer rotor core, and the rotor structure with staggered permanent magnet and rotor hidden pole makes the average inductance of stator windings increase, thus greatly increasing the range of weak magnetic growth.

2.6. Optimum design results

Based on the basic theory and method of motor design above, the electromagnetic structure design of hub motor was optimized. Its rated voltage is 216V, 6 phases, rated power is 6.8kw and fractional slot concentrated double-layer winding, Y connection. The structure parameters of the hub motor are shown in table 1.

Table 1. Hub motor structure parameters.

Parameter	Value (mm)	Parameter	Value (mm)
Outer diameter of the stator	379	Outer diameter of the rotor	420
Inner diameter of the stator	120	Inner diameter of the rotor	381
The length of the core	50	The air gap length	1
The thickness of the magnet steel	3.5	The width of the magnet steel	26

3. Finite element magnetic field analysis

3.1. Grid subdivision

Two-dimensional finite element analysis of magnetic field was carried out on the prototype, the finite element method is a numerical calculation method based on discretization, using Maxwell 2D software to grid subdivision of motor, finite element discretization grid subdivision is the most critical step [8]. Figure 4 is the grid subdivision of 6.8 kW hub motor. So through different grid density subdivision we can save computing resources and ensure the accuracy of the calculation.

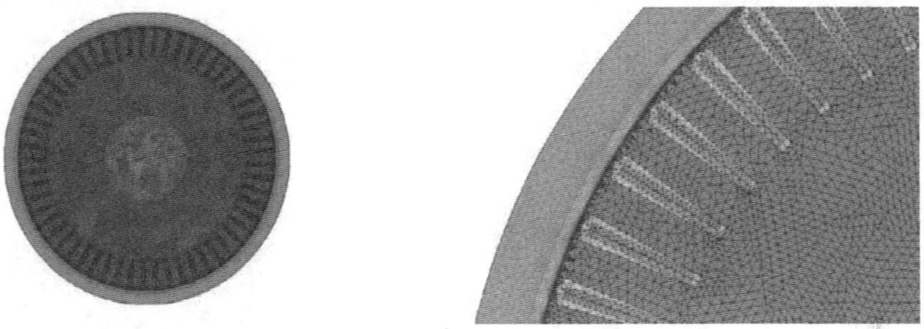

Figure 4. Grid subdivision of BLDCM.

3.2. Flux density distribution

Maxwell can be used to analyze the flux density distribution diagram of the motor under no-load and load, as shown in fig.5 (a) and fig.5 (b).

It can be seen from the figure 5 that the saturation degree in load greatly increase compared with no-load. Meanwhile, from the diagram, it can be concluded that the flux density of the motor is mainly between 1.22T and 1.58T, and its average is about 1.46T, which meet the design requirements fully [9]. The flux density of the stator tooth has reached saturation, and the flux density and the saturation value of stator yoke and the rotor yoke are very different. So under the condition of ensuring the mechanical strength of motor, the yoke of motor can be narrowed down appropriately so as to improve material utilization.

IOP Conf. Series: Materials Science and Engineering **644** (2019) 012009 doi:10.1088/1757-899X/644/1/012009

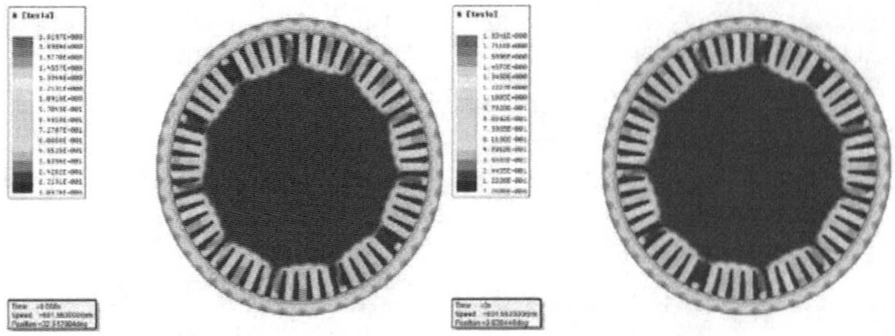

(a) No-load magnetic density (b) Load magnetic density
Figure 5. Flux density distribution diagram of BLDCM.

3.3. Control of six-phase permanent magnet BLDCM

The prototype adopts two separate three-phase windings with 30° electric angle between each other. Both windings adopt Y type joint and each with independent midpoint [10]. The six- phase winding is driven by double three-phase PWM wave. During every 30° electric angle, only a set of bridge changes phase, another set operates normally. The motor control circuit is shown in figure 6.

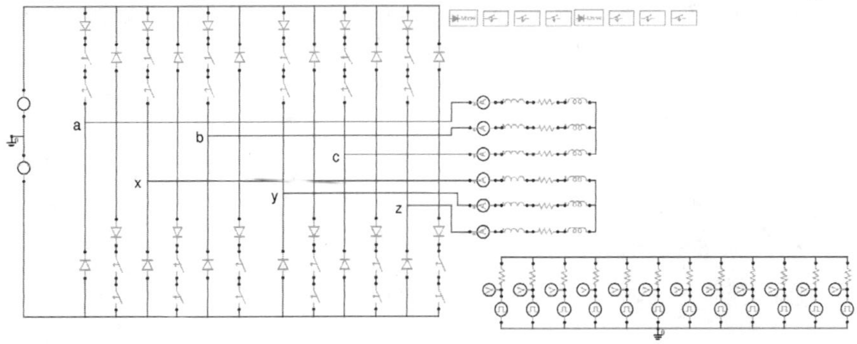

Figure 6. double Y winding circuit diagram of six-phase BLDCM.

The back electric motive force (EMF) waveform of the second set winding of six-phase permanent magnet BLDCM leg 30° electric angle than the first set of winding. The inductance of six-phase BLDCM is the same as three-phase BLDCM, and the resistance of six-phase is the double of three-phase, the current of six-phase is half of three-phase. Because the process of phase change is very short, so the back EMF can be thought of constant [11]. During non-commutation period, the current of shut off phase is equal to zero, so it can reduce the torque ripple. The analysis shows that the torque ripple of six-phase BLDCM can be reduced to half of that of three-phase.

4. Simulation results

The current characteristic, electromagnetic torque characteristic and iron loss characteristic of the permanent magnet BLDCM can be analyzed by using Maxwell simulation software, as shown in figure7, figure8 and figure9 respectively [12].

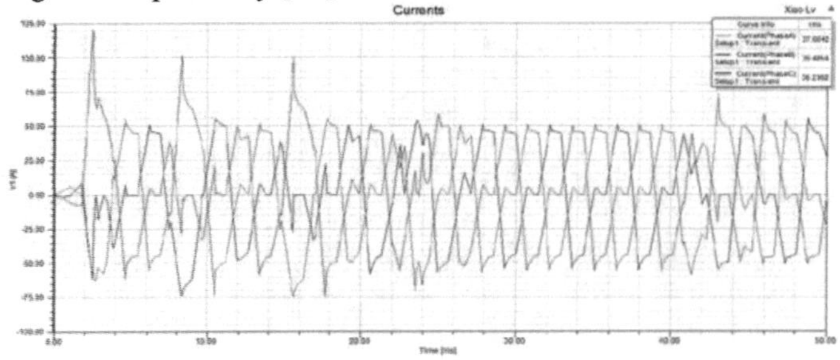

Figure 7. Current characteristic of BLDCM.

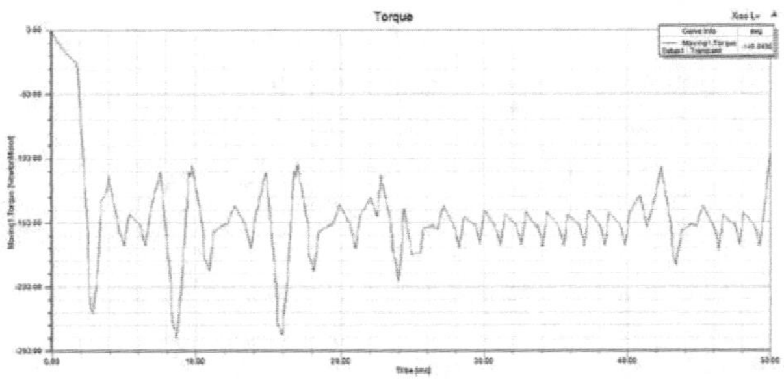

Figure 8. Electromagnetic torque characteristic of BLDCM.

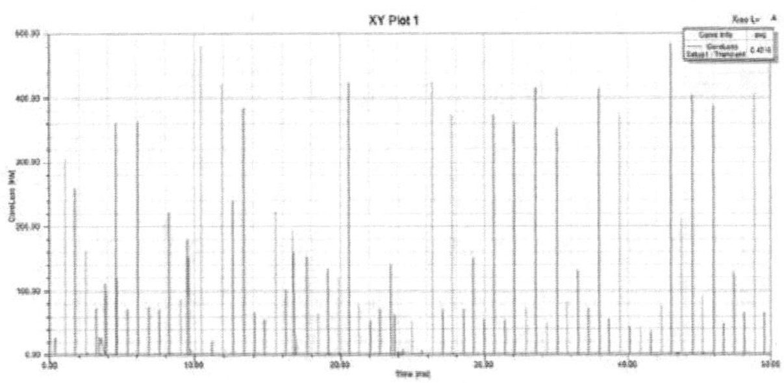

Figure 9. Iron loss characteristic of BLDCM.

Based on the above parameter characteristic diagram, when the speed equals 601.5r/min, we can get the efficiency of hub motor as follows:

$$\eta=\frac{P_{em}}{P_{em}+p_{cu}+p_{Fe}+p_{s}+p_{f}} =\frac{9246}{9246+310.6+421+377+471.3}\times 100\% = 85.6\% \quad (1)$$

In above, P_{em} is electromagnetic power; p_{cu} is copper loss, p_{Fe} is iron loss, p_s is stray loss, p_f is wind friction loss.

5. Conclusion

Hub motor drive system is a new drive system with obvious competitive advantages. It has become an important development direction of electric vehicles. In this paper, a direct drive six-phase permanent magnet brushless DC hub motor for electric vehicles is designed and analyzed. The stator adopts fractional slot concentrated winding and large chute structure to reduce the cogging torque, reduce torque ripple and expand the range of flux weakening speed regulating. Maxwell Ansys finite element analysis software was used to analyze the magnetic field and optimize the parameter design. The simulation results showed that the prototype had the characteristics of high power density and low torque ripple, which fully meet the design requirements.

Acknowledgments My work was supported by the Natural Science Foundation of Fujian Province of China (Grant No. 2017J01667) and the Science and Technology Program Foundation of Fuzhou Science and Technology Bureau. (Grant No. 2016-G-57)

6. References

[1] N. Lin, Transaction of China Electrotechnical Society, 31, 19 (2016)
[2] M.H. Hung, Q. Chen, Y. Chen, Journal of Network Intelligence, 3,1 (2018)
[3] H.H. Wen, Control Engineering of China, 21, 142 (2014)
[4] J.X. Shen, Transactions of China Electrotechnical Society, 28, 2 (2013)
[5] N. Chen, Y. Zhang, W.H. Gui et al. Control Theory & Application, 30, 6 (2013)

[6] H. Zhou, L. Chen, G.H. Liu et al. Electric Machines and Control, 18,9 (2014)
[7] J.P. Wen, B.G. Cao, Small & Special Electrical Machines, 43, 3 (2015)
[8] X.J. Shen, Q. Zhao, P.T. Xu, Journal of Electric Power, 29, 1 (2014)
[9] X. Qiu, W.X. Huang, F.F. Bu, Transaction of China Electrotechnical Society, 30, 22 (2015)
[10] S.G. Li, R. An, Z.F. Xu, D. Meng, Electric Power Automation, 35, 2 (2015)
[11] B. Zhang, Z,H. Jao, L.C. Huo, Small & Special Electrical Machines, 37, 2 (2009)
[12] B.H. Mei, D., X.G. Zhang, Small & Special Electrical Machines, 46, 10 (2018)

IOP Conf. Series: Materials Science and Engineering **644** (2019) 012010 doi:10.1088/1757-899X/644/1/012010

A High Reliability 3D Object Tracking Method for Robot Teaching Application

Yan-Chun Chen, Wei-Chan Weng, and Shi-Wei Lin

Metal Industries Research & Development Centre in Taiwan
E-mail: strongerlin@mail.mirdc.org.tw

Abstract. 3D Object tracking is the task of capturing the 3D position and pose of an object from each time-series image frame. As we known, 3D sensing technique can be realized by stereo vision, structured light, and ToF (time-of-flight) camera. All of them can capture the point cloud data for describing the depth information in a workspace. In past research, the reliability in 3D object tracking was a big problem for real industrial application, therefore, we address a different way to enhance the tracking accuracy and stabilize the tracing path for raising the reliability. In order to build a 3D tracking model and the workspace environment, we adopted an RGB-D camera which is the Intel® RealSense™ D400 Series depth modules to collect the cloud point data and RGB values. The built 3D tracking model should contains the information which includes points, normal and texture for producing many 2D object images with different perspectives. Then the produced images were fed to a SSD (single-shot detector) neural network to learn the object's features for 2D tracking. In dynamic tracking process, the image frames were through the sematic image segmentation by DeepLabV3+ for only extracting the object information without hands and background. Thus, the reserved data only included object's cloud point data and texture information in workspace. Then we use the iterative closest point (ICP) algorithm and the RGB intensity correlation method to confirm the object's position and posture in workspace. The result shows that our method has a better performance than SSD method for tracking a self-predefined object.

1. Introduction

The 3D object pose estimation and tracking are widely implemented in computer vision and machine vision. Computer vision is processed nature problem, such as video surveillance system, virtual reality and augmented reality. The video surveillance system is designed to capture sequence frames using the camera, then we analysis the behavior of humans, car, and 5animals. Analytical approaches in 2D image were often adopted to track specific object technologies for notification or warning purposes in many papers. Medical care or terrorism and attack prevention is a popular application in this method, however, the object's posture in 2D image is difficult to detect in a slight change. Therefore, 3D object pose estimation techniques are necessary to be developed for tracking and estimating objects' pose in realistic street conditions [1]. 3D pose estimation was also applied on virtual reality and augmented reality which emphasize the human interaction in a virtual environment that can be extended to the applications of education, games and medical rehabilitation [2-5]. Machine vision is used for analysing product defects and guiding robotic arm to execute a pick & place mission in industrial applications. In the past, machine vision always delivered a single point data to robot arm's controller to execute a movement program. Nowadays, the trend is that computer vision integrated with robotic arm will make them friendlier for operators, and object tracking and pose estimating [6] will be the key.

Well image pre-processing can obtain good results for detecting an object's trajectory and posture. Object detection method can be divided into four types [7], including point detectors, segmentation, background modelling and machine learning. The first type is point detectors which means interest point detectors, such as harris interest point detector [8], SIFT detector [9] and moment matrix computed with uniform Gaussian scale-space [10].The second type is traditional classic numerical method for segmentation such as active contours [10], graph cut [11],etc. In this methods, image's color-scale, texture, position by using statistical methods to achieve the purpose for segmentation. Recently, machine learning was also adopted to solve the segmentation problems [12]. In the third type, the background modelling was used for separating dynamic objects and static background. Generally, the target object which we need to detect will move continuously in natural environment, like human, animals, cars, and so on. Background modelling is often used for pedestrian detection in the surveillance field. But there is a disadvantages in this method, the object will be regarded as same as background when it moves very slowly. The classic background modelling methods includes mixture of gaussians model [13], dynamic texture background [14]. The last type is machine learning technique. Machine learning can be divided into supervised classifiers and unsupervised classifiers. Supervised classifiers must use the pre-marked data for analysing, such as support vector machine (SVM) [15] and Neural network [16-17].Unsupervised classifiers analysis are based on the characteristics of the original data by using special statistical methods [18]. At present, machine learning can be applied on different field problems with the newest development of neural network models.

Our research dealt with the 3D tracking and pose estimating by RGBD image for a specific object which we predefined. The 3D rotation and translation can be obtained through depth information. Th We adopted two kind of techniques in this research, including feature-base and model-base[19]. Feature-base is used for tracking by the object's texture and feature description through Statistical methods, such as kalman filter [20] and particle filter [21]. Model-base is used for considering the edge and geometry of object and also matches the features description. In object tracking, template matching [7] for considering the 6D pose estimation and the most famous method is the ICP(iterative closest point) method [22]. Although there are many estimation methods which was mentioned, the performances are still not good as ICP.

This paper proposes a reliable analysis procedure to solve 3D object tracking and pose estimating by using a single RGBD camera. In Section 2, we introduce different method for capturing the depth image and 3D image process method in this research. In Section 3, the result shows that our method has a better performance than SSD method for tracking a self-predefined object.

2. Methodology

Our approach is to process the xyz-rgb data from an RGBD camera. We remove background data and only retain dynamic data by a threshold value from the depth information. Then we will discuss how to create a complete object and region of interest (ROI) by image segmentation method. Finally, we use a 3D template model to match the dynamic image for to tracking and pose estimating by ICP algorithm.

2.1. Background removal from depth information

In robotic vision, to select a right hardware can simplify many problems and get better results in different operation environments. Usually, an RGBD camera can catch the depth information for processing and calculating objects' trajectory, position relationship, and 3D environment construct. Depth information can be obtained by different techniques for computing the distance on each pixel from cameras. The techniques include [24] Stereo vision, Time-of-Flight (ToF) and Structured light. Stereo vision is to calculate the disparity between the camera and image according to the edge information. Moreover, this system must be calibrated before using, the captured images should be consider the sufficient texture features and ambient light source. They always add a light source with some image pattern to enhance edge and surface information.

IOP Conf. Series: Materials Science and Engineering **644** (2019) 012010 doi:10.1088/1757-899X/644/1/012010

ToF means time of flight. The camera adopt active sensors that determine the depth value by measuring the time obtained by emitted IR light with high frequency to the object and back to the sensors. The depth information obtained by ToF is more accurate than Stereo vision. However, the noise generated during reflective process must be improved by conventional filters.
A structure light 3D image device composed of pattern projector, and a camera. The depth values are calculated from the reflective pattern imaged from objects and correlates it against the original pattern on a plane. The disadvantage of this technology has a short working distance, but it has the highest resolution of depth information than others.
In this paper, we must consider the resolution and accuracy of depth value, therefore the structure light based RGBD camera was used for obtain the xyzrgb data. In order to obtain the cloud point data of the dynamic object from the camera, the depth information extract out by (1).

$$f(x,y) = \{I_{RGB} | z \geq T_d\} \quad (1)$$

In (1), where $f(x,y)$ is foreground image. The foreground image was extracted by setting a threshold T_d was set for each position of the depth information z.

2.2. Object segmentation

We used deep learning for image segmentation that must be considered the spatial resolution problem. These problems can be improved by pooling and stride. Moreover, the category classification error I is caused by object pose change. It can be solved by atrous convolution and Encoder-Decoder neutral network architectures. In traditional classification structure, the feature map of the last layer size is 1/32 than original image size after pooling. This size is not good for pixel by pixel classification results. The atrous convolution method can process larger range of resolution of an image, and it does not need to change the input and output channels by this kernel. This method adopt various convolution kernels with insert holes to create different feature maps for replacing the last layers of the network. Defined a 1D signal Equation as (2):

$$y[i] = \sum_{l=1}^{L} x[i + r \cdot l]\, w[l] \quad (2)$$

Assume that a one-dimensional signal is input $x[i]$, dilated rate parameter *r*, output $y[i]$,and l of filter size. For instance, we use this convolution method on ResNet's network structure as shown below:

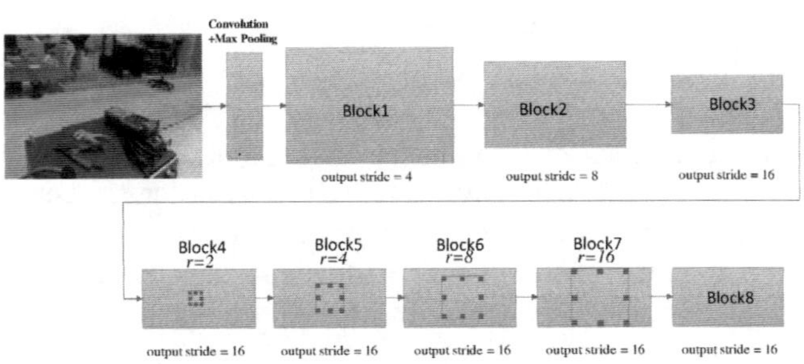

Figure 1. Consider atrous convolution in ResNet to increase the reception field.

In the original architecture, the last layer of the network is block4. The blocks 5 to 8 are copied from the block 4 to get more multi-scale cascade context (Fig1). Finally, this segmentation approach is used for dealing with multi-scale problems by the spatial pyramid pooling method with atrous convolution. That is shown below:

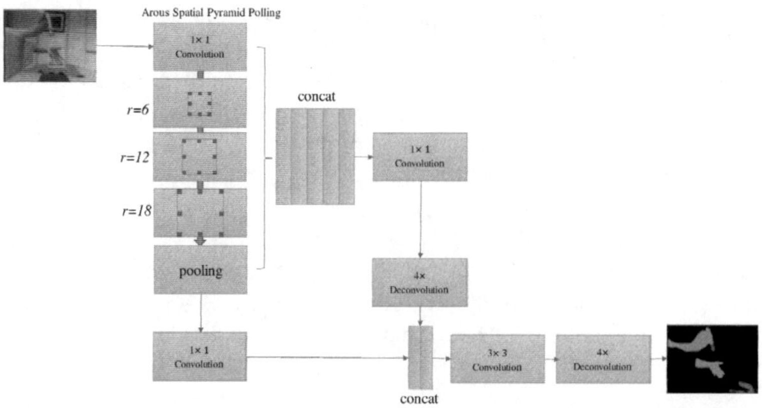

Figure 2. Deep learning architecture for image segmentation.

In Fig. 2, the Arous Spatial Pyramid Pooling (ASPP) is an architecture with different kernel size by various dilated rate. Besides, the architecture also includes an 1×1 convolution and an 1×1 pooling. The image was reconstructed by using two 4x deconvolution mask and added the low-level information.

2.3. Tracking and pose estimating

For tracking a d pose estimating a dynamic object, we converted 2D object image to 3D point cloud. It was considered the camera's internal parameters after the image segmentation of our method. The internal parameters format as below (3):

$$C = \begin{bmatrix} f_x & 0 & c_x \\ 0 & f_y & c_y \\ 0 & 0 & 0 \end{bmatrix} \tag{3}$$

In (3), the f_x and f_y refer to the camera's x and y focal length. The c_x and c_y refers to camera's aperture.

$$x = (u - c_x) \cdot \frac{z}{f_x} \tag{4}$$

$$y = (v - c_y) \cdot \frac{z}{f_y} \tag{5}$$

$$z = d/s \tag{6}$$

In Eqs. (4), (5) and (6), the *x, y,* and *z* are coordinates of 3D point cloud. The conversion calculation is through the coordinates of *u, v,* and *d* on the image with camera internal parameters. The 3D point cloud is the coordinate set of space merge color information, such as RGB values. We improve processing efficiency by using three-dimensional coordinate in our approach. Suppose the two sets are *P* and *Q* that used euclidean distance calculate the most similar points in the two sets.

$$\vec{p_i} \in P, \vec{q_i} \in Q$$
$$d(\vec{p_i}, \vec{q_i}) = \left\| \vec{p_i} - \vec{q_i} \right\| \tag{7}$$

Afterward, we consider translation and rotation changes for two sets of similar points and establish a translation matrix *T* and a rotation matrix *R*. The two matrix use the least squares iterative variation to find a set of the most similar *R* and *T* matrix. According (7), it establish the objective function of the least square access as equation (8).

$$E = \sum_{i=1}^{N} \left| \left(R\vec{p_i} + T\right) - \vec{q_i} \right|^2 \tag{8}$$

IOP Conf. Series: Materials Science and Engineering **644** (2019) 012010 doi:10.1088/1757-899X/644/1/012010

3. Experiments

In this paper, the tracking method was implemented by using TensorFlow 1.09 and cuDNN 9 on python and the device is PC with Intel Core i5 CPU and NVIDIA Titan GPU card. Our primary aim is to achieve tracking the posture of hand tools in a video. The image with the size of 1280×720 pixels in the video analyzed can be divided into three categories including equipment-base, console-base, and operator-base.

The experiment is to compare the detection accuracy between the proposed method and the single shot multi-box detector (SSD). The training data from the video contains 600 frames including two types of hand tools. There are two different strategy, proposed method and SSD method for comparison in this paper:

1. Strategy by using SSD model:
   i. Use MobileNetV2 to separate features.
   ii. Train the SSD model.
   iii. We got the best result after iterating 2500 epoch.
2. Proposed strategy:
   i. Separate object into front and background part by using depth image, and remove background image—equipment.
   ii. Use PASCAL VOC 2012 database for training segmentation model to separate operator and hand tool in the front part of the image. After 2000 iterations we got the best result.
   iii. Finally, cloud points of the hand tool are left in the image only. We use ICP function in the MVTec's software product HALCON to track the posture of a hand tool.

Following two figures are the results after SSD method and out method.

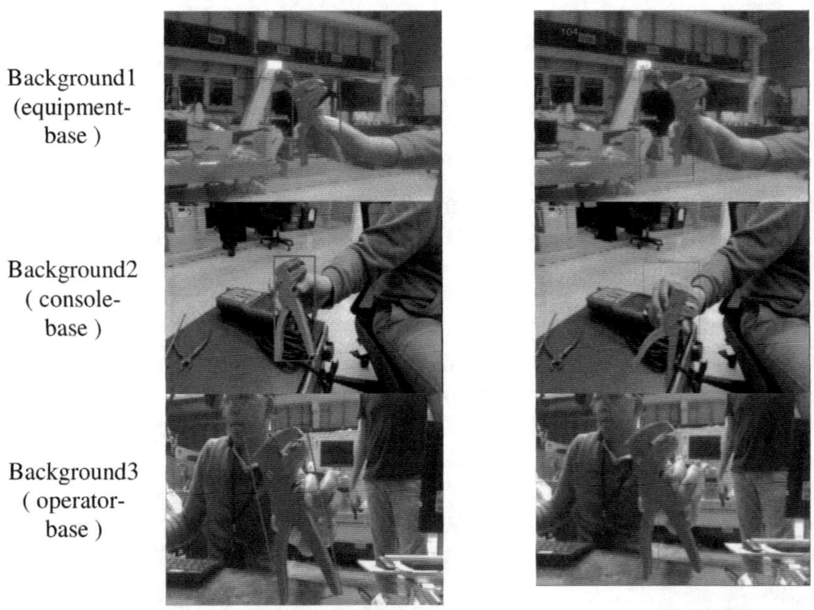

(a)Our Method (b)SSD

Figure 3. The illustrations show that SSD and proposed methods to detect object in different backgrounds.

IOP Conf. Series: Materials Science and Engineering **644** (2019) 012010 doi:10.1088/1757-899X/644/1/012010

The Fig. 3 Shows the analyzed frames in different backgrounds which are equipment-base, console-base and operator-base. It shows that our method has robust results for tracking and pose estimating in different backgrounds. The score of detection rate in SSD and proposed method are listed in Table 1, and the scores prove that our method is better than the SSD method.

Table 1. Each space is (correctly framed /total frame number) in two method.

Background Type \ Method	Our Method	SSD Method
Background1 (equipment-based)	0.81(300/369)	0.135(50/369)
Background 2 (console-based)	0.75(250/330)	0.181(60/330)
Background 3 (operator-based)	0.92(210/226)	0.442(100/226)

(a)vise object

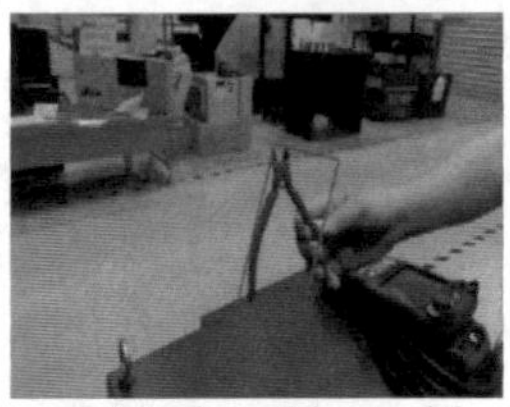
(b) vise object with different poses

(c) Multiple objects

(d) pose estimation

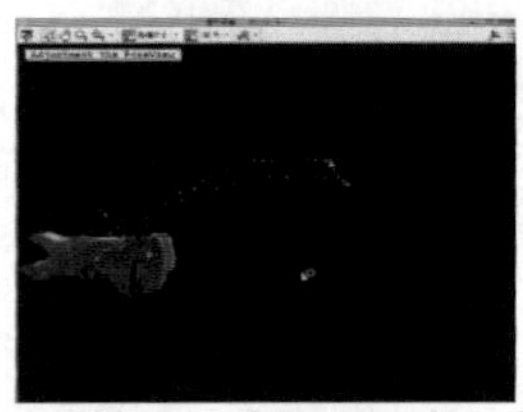
(e) tracking and pose estimation

Figure 4. (a)(b) The illustration show deteion results of metal hand tools with different posture.(c) Multiple objects can be detected by the proposed method.(d)(e) The object and posture were tracked and estimated by ICP in this paper.

As shown in Fig 4(a), 4(b), 4(c), our method performs well while changing to a metal tool or including multi-objects in a frame. In Figure 4(d) and 4(e), the hand tool's posture and the moving path are shown by using HALCON's ICP function.

4. Conclusion

The strategy combined both using 3D depth information and 2D object segmentation can improve the detection accuracy. Especially, our segmentation model without need retrain to achieve robust results in complex background or scale changes. Our method has a better performance than SSD while in the different environments, changing hand tools' posture and multi-objects detection in a frame. However, this method cannot do well with highly reflective or slender objects. We will find another way to improve these problems in the future. For example, the high-reflection objects will be suppressed by image enhancement method.

5. References

[1] M. Andriluka, S. Roth, B. Schiele, Monocular 3D pose estimation and Tracking by Detection, CVPR(2010)

[2] T. Darrell, B. Moghaddam, A.P. Pentland, Active face tracking and pose estimation in an

interactive room, CVPR(1996)
[3] F. Ababsa, M. Mallem, Robust camera pose estimation using 2D fiducials tracking for real-time augmented reality systems,VRCAI(2004)
[4] I. Rabbi, S. Ullah, A Survey on augmented reality challenges and tracking. Acta Graphica, 24, 29-46(2013)
[5] K. Pauwels, L. Rubio, J.D´ıaz, E. Ros, Real-time model-based rigid object pose estimation and tracking combining dense and sparse visual cues, CVPR (2013)
[6] T. Drummond, R Cipolla, Real-time visual tracking of complex structures, IEEE Transactions On Pattern Analysis And Machine Intelligence, 24,7(2002)
[7] A.Yilmaz, O. Javed, M. Shah, Object Tracking: A Survey. ACM Computing Surveys, **38**(4),1-45(2006)
[8] C. Harris, M. Stephens, A combined corner and edge detector.147-151(Alvey Vision Conference,1988)
[9] G. Lowe, Distinctive image features from scale-invariant keypoints. Int.J. Comput. 60(2), 91-110(2004)
[10] M. Kass, A. Witkin, D. Terzopoulos, Snakes: Active Contour Models, **1**(4), 321-331(1988)
[11] I. Xu, N. Ahuja, Object contour tracking using graph cuts based active contours. ICIP. 277-280(2002)
[12] L.C. Chen, G. Papandreou, I. Kokkinos, K. Murphy, A.L. Yuille, Deep lab: Semantic image segmentation with deep convolutional nets, atrous convolution, and fully connected CRFs. arXiv:160600915(2016)
[13] C. Stauffer, W. Grimson, Learning patterns of activity using real time tracking. Transactions On Pattern Analysis And Machine Intelligence. 22(8),747–767(2000)
[14] A. Monnet, A. Mittal, N. Paragios, V. Ramesh. Background modeling and subtraction of dynamic scenes, ICCV, 13–16(2003)
[15] C. Papageorgiou, M. Oren, and T. Poggio, A general framework for object detection, ICCV(1998)
[16] A. Rowley, S. Baluja, T .Kanade. Neural network-based face detection. IEEE Trans. Patt. Analy.Mach. Intell. 20(1), 23-38(1998)
[17] W. Liu, D. Anguelov, D. Erhan, C. Szegedy, and S. Reed, SSD: Single shot multibox detector. ECCV(2016)
[18] D. Comaniciu, P. Meer, Mean shift:A robust approach toward feature space analysis. IEEE Trans. Patt. Analy. Mach. Intell. 24(5), 603-619(2002)
[19] B. Seo, H. Park, J. Park, S. Hinterstoisser, and S. Ilic. Optimal local searching for fast and robust textureless 3d object tracking in highly cluttered backgrounds. IEEE Transactions on Visualization and Computer Graphics, 20(1), 99-110(2014)
[20] T.J. Broida, R. Chellappa, Estimation of object motion parameters from noisy images.IEEE PAMI, 90-99(1986).
[21] C. Choi, H. I. Christensen, Robust 3D visual tracking using particle filtering on the special euclide an group: A combined approach of keypoint and edge features. Int. J. Robot.Res., 31(4):498–519(2012)
[22] A. Tagliasacchi, M. Schroeder, A. Tkach, S. Bouaziz, M. Botsch, M. Pauly, Robust articulated-icp for real-time hand tracking, Technical report (2015)
[23] L. Maddalena, A. Petrosino. Background subtraction for moving object detection in rgbd data: A survey. Journal of Imaging, 4(5), 71 (2018)

Design of Single-Ended Forward Switching Power Supply Based on TOP246Y Chip

Chia-Wei Huang, Yi-Yi Huang, and Wei-Tai Hsu

Department of Electrical Engineering, Zhaoqing University, 526061, China
E-mail: wthsu1@gmail.com

Abstract. This paper is designed a single-ended forward switching power supply controlled by TOP246Y chip. The chip is high efficiency, small, low in cost, and can avoid damage caused by overvoltage of the switch tube caused by the feedback. Less peripheral components are advantageous for miniaturization of the forward power adapter. The output DC voltage by applying the PWM circuit is stable by using the Multisim simulation. The load can be stabilized from 50Ω~50KΩ. The output DC voltage is 11.85V (about 12V), the ripple coefficient is kept at 1.4%, and the voltage regulation rate is about 0.068%. The load regulation rate β is about 0.025%, which verifies that the power supply has excellent dynamic response characteristics and strong load capacity.

1. Switching Power Supply Working Principle

The TOP246Y chip is suitable for high efficiency, small size, low cost, fully sealed switching power supply modules or power adapters [1]. The basic structure of the switching power supply is shown in Figure 1. It consists of 220V AC input voltage, primary-side rectification, lowpass filter, electronic switch, high-frequency transformer, secondary-side rectification, and a smoothing filter. The output voltage is controlled by a comparator-controlled by a reference voltage, an error amplifier, and feedback to control the duty cycle of the pulse width modulation (PWM) to achieve an adjusted output [2].

This circuit is a forward-excited circuit. In work, the power source of the transformer core working on the hysteresis loop side can realize the one-way change of the magnetic flux. If the electronic switch is turned on, it enters the transformer through rectification and filtering. The core coupling is used to transfer the electrical energy of the primary coil to the secondary winding, and after secondary rectification into the output. If the electronic switch is turned off, the energy storage inductor can form a discharge loop with the diode to power the load. The excess inductor flows back to the power supply through the magnetic reset of winding to protect the chip [3].

IOP Conf. Series: Materials Science and Engineering **644** (2019) 012011 doi:10.1088/1757-899X/644/1/012011

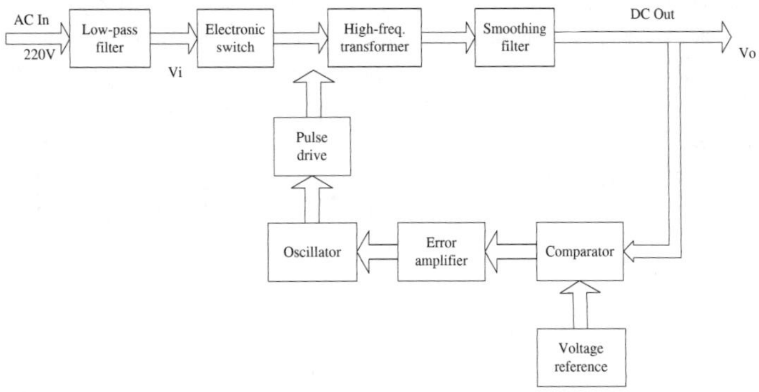

Figure 1. The basic structure of switching power supply.

2. Magnetic circuit design

2.1. Forward transformer design

The primary function of the transformer is to electrically isolate the output and output, increase or decrease the amplitude of the AC voltage after PWM, and protect the chip through the magnetic reset of winding [4].

2.2. Winding turns of the transformer

The number of winding turns can be determined by the following formula:

$$N_1 = \frac{V_{in}}{4fB_{\max}A_c} \times 10^8 \tag{1}$$

where A_c: active cross-sectional area of the core in cm^2;

V_{in}: input working voltage in V;

B_{max}: maximum working flux density in G(wh/cm^2).

The voltage drop of the output rectifier cannot be ignored. The number of turns in the secondary winding of the transformer is determined by the following equation:

$$N_2 = \frac{1.1(V_{out} + V_f)}{N_1 V_{in(\min)} DC_{\max}} \tag{2}$$

where V_f: forward voltage drop of output rectifier;

DC_{max}: maximum duty cycle;

$V_{in(min)}$: minimum input voltage.

This formula is used to calculate the expected minimum input-voltage value of the secondary winding turns. (The regulator will lose its regulation if the input voltage is lower than this value) [5]

Since the core can only be around an integer number of turns, the number of turns calculated by the formula is generally not an integer, so the integer of an approximate value is usually taken. Therefore, this causes an error in the output voltage. It is required to calculate the range of the error, and then to compare the output voltage of each turn before and after the rounding. If the error is too large, it is necessary to replace the bridge rectifier with a higher or lower forward voltage drop. If the output voltage cannot meet the requirements, it is considered adding a coil to the primary output winding, or whether the secondary side needs isolation, and whether the type of autotransformer on the secondary side is used [6][4].

2.3. Winding method with low leakage inductance

The leakage inductance is that the magnetic lines of force generated by the coil cannot both generate leakage inductance through the secondary coil. The existence of leakage inductance,

; when the switching device is turned off. It makes back electromotive force, which is easy to the over-voltage breakdown of the switching equipment. The leakage inductance can also form an oscillating circuit with the distributed capacitance in the loop and the distributed capacitance of the transformer coil, so that the circuit oscillates and radiates electromagnetic energy outward, causing electromagnetic interference [7].

The main factors affecting the leakage inductance of the winding are winding turns, winding width, and the insulation thickness of the winding. The wider the winding, the smaller the leakage inductance and the leakage inductance is also affected by the coupling effect [7].

3. Circuit design

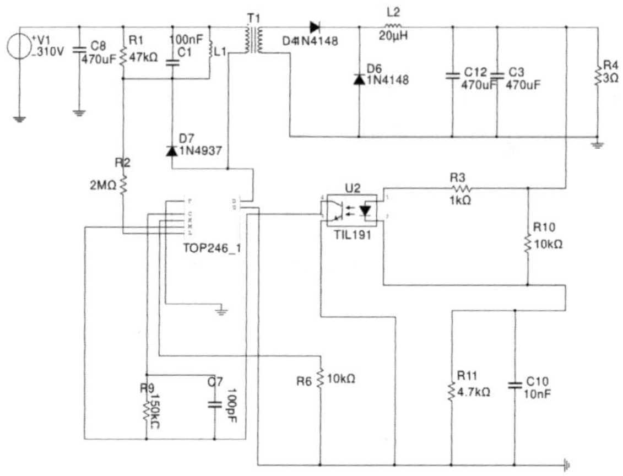

Figure 2. The circuit diagram of single-ended forward switching power supply.

3.1. Design of the input part

A resistor, a capacitor, a Zener diode, etc. are used to form a start-up circuit, so that the transformer is rectified and filtered using a diode, a capacitor, etc., to achieve a control power supply current limiting. The auxiliary winding control circuit of the transformer achieves low voltage start-up, and when the driving current is large, power consumption is reduced [11].

3.2. Design of control part

The TOP246Y chip is used as the core of the control part. The chip contains an error amplifier and a MOSFET. The alternating magnetic field is formed by the on and off of the MOSFET, and the secondary coil of the transformer generates a corresponding induced electromotive force [11].

3.3. Reset circuit

Using the residual current device (RCD) resets circuit [9] when the switch is disconnected, the transformer induces a tremendous electromotive force. If there is no reset circuit, the magnetizing inductance oscillates with the junction capacitance of the MOSFET, and the oscillating capacitance is caused the breakdown for the MOSFET [11].

3.4. Feedback part design

The TL191 electrically isolates the input and output and uses the linear photocoupler to form an optocoupler feedback circuit. By adjusting the duty of the control terminal to change the duty cycle, the purpose of precision regulation is achieved, and signal reception and transfer are also achieved [11].

4. Analysis of experimental results

4.1. Verifying output voltage

As shown in Figure 3, the waveform of the output voltage during operation, the maximum output voltage +12V requires 0.006ms to stabilize the output. The time of stable operation for the transformer is 13ms, and the time from the start to the stable operation of the power supply is not exceeded 14ms [9], indicating that the power supply has good dynamic response characteristics [12-13].

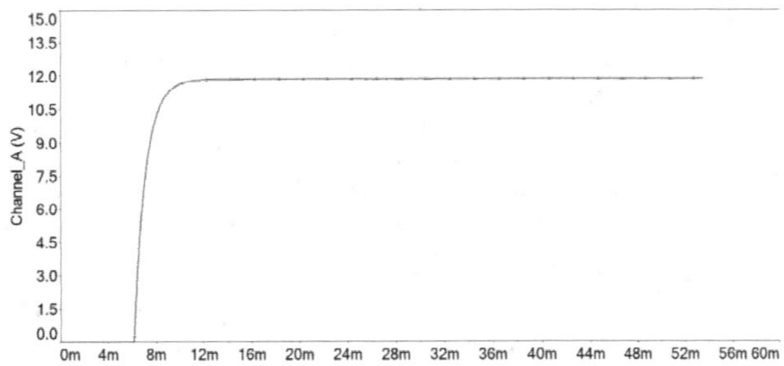

Figure 3. Output voltage.

4.2. Verifying output voltage ripple

The data for the output voltage is led to excel, and the waveform of the output voltage ripple is sampled as shown in Figure 4. The peak-to-peak ripple value of the output terminal is 11.853-11.860V, the ripple voltage is 17mV, and the ripple coefficient is γ=0.017/11.85×100=0.14%. It is known that the power supply has good ripple characteristics and good stability [12-13].

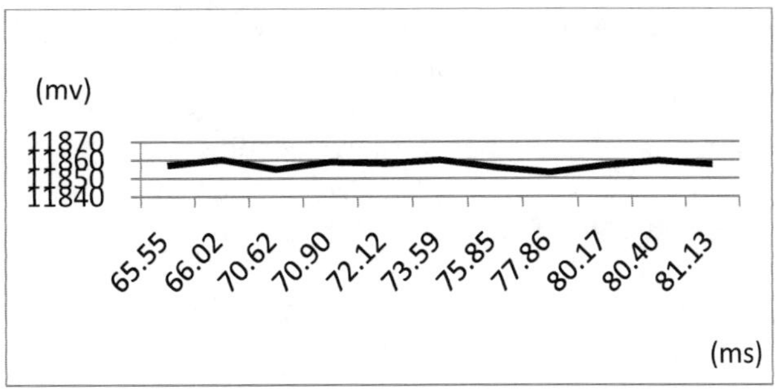

Figure 4. Output ripple voltage.

4.3. Load capacity

The output voltage waveform of the simulation plus equivalent load 50kΩ-50Ω changes is shown in Figure 5. After the equivalent load, the output voltage changes to 11.854-11.862V, and the fluctuation is 8mV. The voltage regulation rate is α=0.008/11.85=0.068%, and load regulation rate is β = (0.001 + 0.002) / 11.85 = 0.025%. It is fully indicating that the power supply has a strong load capacity [12-13].

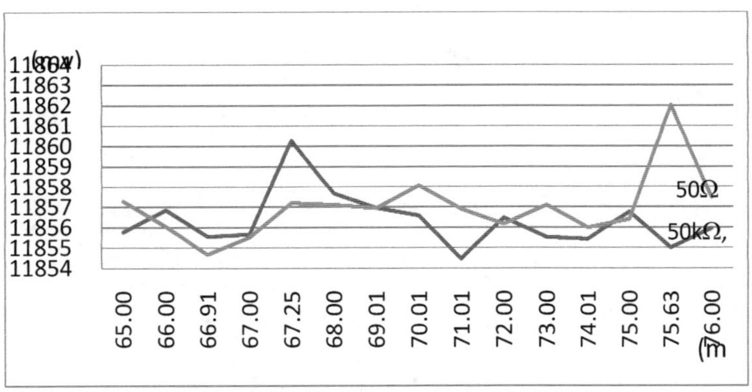

Figure 5. Voltage waveform with load.

IOP Conf. Series: Materials Science and Engineering **644** (2019) 012011 doi:10.1088/1757-899X/644/1/012011

5. Conclusion

The design of the single-ended forward switching power supply is mainly divided into two parts, namely circuit design, and magnetic circuit design. However, the design of the magnetic circuit is relatively complicated and difficult to grasp. The magnetic components of the high frequency switching power supply are mainly transformers and filter inductors. This paper fully introduces the calculation of the number of turns of the forward transformer, the method of winding, the low leakage inductance, and the core reset circuit, etc. The method of designing the magnetic circuit is clear, and the design of the high frequency switching power transformer and the design of the filter inductor achieve high efficiency, small size, low cost, and improved reliability.

Based on the single-ended forward switching power supply developed by TOP246Y, this paper introduces the characteristics of TOP246Y and the design of each part of the peripheral circuit. According to the simulation results by using National Instruments Multisim, the switching power supply can stably output the DC voltage of 11.85V (about 12V) when the load is changed from 50Ω to 50kΩ. The turn-on delay time is about 13ms, and the ripple coefficient is maintained at 1.4%. The voltage rate adjustment is about 0.068% as well as the load regulation β is about 0.025%, which verifies the excellent dynamic response characteristics of the power supply, unique ripple characteristics, and stable load carrying capacity. Through the analysis of the results, the feasibility of the theory was verified, and excellent results were obtained.

6. References

[1] D.H. Zhang, L. Le, B.G. Orr, Low Volt. Appar. 8, 32 (2006)
[2] W. Zhang, R.M. Li, Shanxi Elec. Tech. 3, 8 (2017)
[3] I. P. Abraham, Switching power supply design (Mc Graw-Hill, New York, 1998)
[4] N. Abeywickrama, Y.V. Serdyuk, S.M. Gubanski, IEEE T. Power Delivery 23, 1432 (2005)
[5] Z.Y. Sha, Y.M. Wang, J.Y. Ge, L.Z. Zhao, Power Elec. 37, 69 (2003)
[6] T.H. Zhao, J. Liu, Switching power supply design technology and application (Posts & Telecom, Beijing, 2007)
[7] D.G. Cheng, Elec. Comp. & Device App. 8, 13 (2001)
[8] J.G. Chen, Power Elec. 41, 60 (2007)
[9] D.L. Chen, W.Z. Chen, Y.G. Yan, J. Nanjing Uni. Aero. & Astro. 29, 231 (1997)
[10] S. Hariharan, Power Elec. Design 10, 38 (2005)
[11] M. Han, Auto. App. 3, (2018)
[12] Y. Song, (2010). Design of single-ended forward high-frequency switching power supply. (master's thesis). Xidian University, Xi'an, China
[13] L. Dixon, Unitrode Corp. 3, 356 (2001)

IOP Conf. Series: Materials Science and Engineering **644** (2019) 012012 doi:10.1088/1757-899X/644/1/012012

Design of Adjustable Switching Power Supply with Current Type Single-Ended Flyback Output

Chia-Wei Huang, Jun-Yu Du, and Wei-Tai Hsu

Department of Electrical Engineering, Zhaoqing University, 526061, China
E-mail:wthsu1@gmail.com

Abstract. This study is based on the UC3845 controller, designed a flyback switching power supply, which has a wide input voltage range, high output voltage accuracy, and high load regulation. According to the characteristics of flyback switching power supply and UC3845, adding filter circuit, a feedback circuit, protection circuit, etc., develop a single-ended flyback switching power supply with adjustable output, stable and reliable, and high efficiency. According to the simulation results, it can be seen that the designed switching power supply has high output voltage accuracy, and the input voltage in the range of 200V~240V, the accuracy of the output voltage is kept within 1.3%; the voltage adjustment is less than 1%, The output ripple factor is kept below 0.09%, and the output voltage is adjustable.

1. Introduction

The single-ended flyback switching power supply has been widely used in the power supply of automatic control and intelligent instrument because of its advantages of simple structure and few components [1]. The duty cycle of the power MOSFET is regulated by pulse width modulation (PWM) technology to stabilize the output voltage. This paper used a high-performance integrated control chip UC3845 as a pulse width modulation device. Its main features are an internal oscillator, high-precision error comparator, cycle-by-cycle current sampling-comparison, low start-up current, high-current totem pole output, etc., which are ideal devices for driving MOSFETs [2].

2. Switching power supply circuit design

The whole circuit mainly is included the rectifier and filter circuit, converter (including power switch tube, high-frequency transformer, etc.), control circuit, and a feedback circuit. The circuit diagram is shown in Figure 1.

The circuit set takes the input of 220V AC as an example. After passing through the bridge rectifier circuit and filter circuit, the output is turned to a DC close to 311V. By starting resistor R2 and capacitor C14, the UC3845 is brought to the start condition and can maintain stable operating conditions [3]. When the UC3845 is operating normally, a drive pulse can be provided to drive the MOSFET. The DC voltage can be output through a flyback transformer and rectification filtering [4]. When interference affects the output circuit, feedback adjustment is required to ensure voltage stability. In this paper, the internal error amplifier is forced to output a high level by directly pulling down the voltage of pin 2. The voltage at the output of the error amplifier is regulated by an optocoupler to adjust the duty cycle [5]. That is, the output signal is used as a feedback signal. The feedback circuit formed by the TL431 and the optocoupler is used to regulate the duty cycle of the output pulse. It can also protect the control chip UC3845 from achieving the purpose of stabilizing the output voltage.

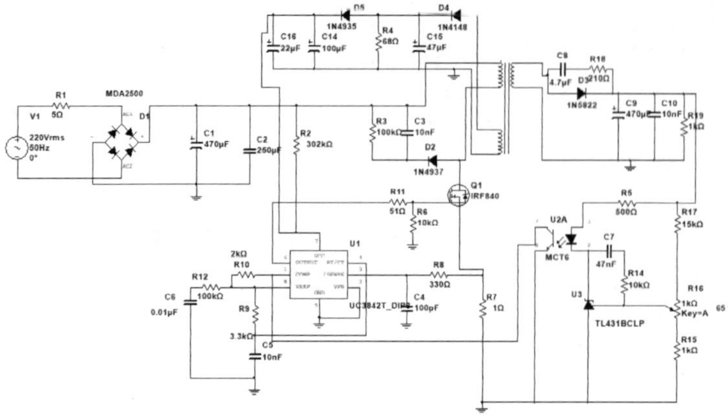

Figure 1. Circuit schematic.

3. Circuit design around the chip UC3845

3.1. Power supply circuit

The UC3845 power supply is divided into two phases: the start-up phase and the normal phase. In Figure 1, the starting resistor R2 and capacitor C14 constitute a starting circuit. At start-up, the filtered DC voltage is supplied to the capacitor C14 via the start-up resistor R1. The voltage of C14 must reach the threshold of 8.4V of UC3845's starting voltage. Otherwise, an undervoltage lockout (UVLO) function is enabled. During regular operation, the UC3845 provides the drive pulse that is driven by pin 6 to operate the switch. In this process, since the current supplied by R2 is not enough to maintain the operating current of UC3842, capacitor C1 is discharged, and the capacitor's voltage is gradually reduced. The auxiliary power supply must provide the operating voltage before the capacitor's voltage drops to the cut-off operating voltage of UC3842. The UC3842 was not able to work when the voltage drops to the cut-off operating voltage on C1 [6], while the auxiliary power supply consists of the auxiliary winding, the rectifier circuit, and the filter. Therefore, for regular operation, the starting capacitor C14 must be able to provide enough energy, that is, a sufficiently large capacitance of the capacitor [3]. Capacitor C16 is used to filter out high frequency superimposed signals [7]. According to the chip's datasheet, the formula for starting resistor is

$$R_2 = \frac{V_{in} - V_c}{I_{OUTPUT}}. \tag{1}$$

3.2. Current sampling circuit

This circuit is used to collect the current of the power switch tube to the pin 3, and the pin 3 and the output of the error amplifier inside the UC3845 are determined the duty ratio of the output. Therefore, this circuit plays a feedback role and acts as a current limiter [8]. For sampling resistor R8 its peak circuit is

$$I_{R8} = \frac{V_{COMP} - 1.4V}{3R_8}, \tag{2}$$

where V_{COMP} is the voltage at the output of the error amplifier. In order to eliminate current spikes and prevent false triggering of the chip, an RC filter circuit is required. Therefore, a filter capacitor C4 needs to be added to this circuit.

3.3. Feedback circuit design

The circuit is used voltage feedback composed of the photocoupler and TL431. The voltage feedback circuit has high precision and is favorable for the stability of the output voltage [9]. For the characteristics of the UC3845, there are two types of voltage feedback circuits. One is to provide a stable power supply to the optocoupler through the pin 8 and connect the emitter of the optocoupler to the pin 2 of the UC3845 to provide a comparison value for the error amplifier in the chip. It results in its output value, which controls the duty cycle of the output [3]. This article is applied another way to control the duty cycle. As shown in Figure 1, pin 2 of the UC3845 is grounded, and the error amplifier is forced to output the high voltage. The output of the error amplifier is divided by the optocoupler to adjust the duty cycle of the output as shown in Figure 2.

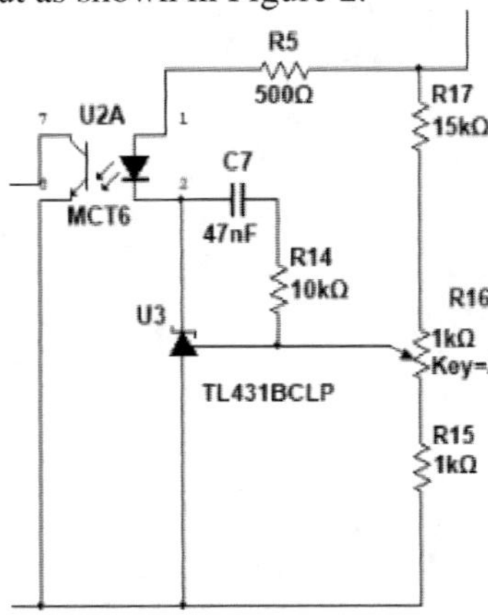

Figure 2. Feedback circuit.

The role of the TL431 is to provide a reference voltage to the primary side of the optocoupler. According to the characteristics of the TL431, the reference voltage can be changed by varying the resistance to which it is connected [8]. The voltage at the Comp is changed by adjusting the voltage on the secondary side of the optocoupler. Thus, the duty cycle of the output voltage, by which the output waveform can be adjusted. The equation of adjusted output voltage is

$$V_O = V_{REF} \times \frac{(R_5 // R_{15}) + R_{17}}{R_{15} + R_{16}} \tag{3}$$

where V_O is the secondary side output voltage of the photocoupler, and V_{REF} of TL431 is 2.5V.

3.4. Primary buffer circuit of the transformer

Due to the fast turn-off and turn-on of the power switch MOSFET, the back electromotive force generated by the inductor is large, causing serious damage to nearby circuits [9]. In order to prevent the destruction of its peak voltage, a buffer circuit composed of a diode and a resistor-capacitor circuit on the primary side of the transformer is shown in Figure 3 [10]. Since the current generated by the coupled inductor of the transformer is large, the values of R3 and C3 should be as large as possible within a reasonable range. For diodes, models with better EMI characteristics and higher reverse voltage should be selected.

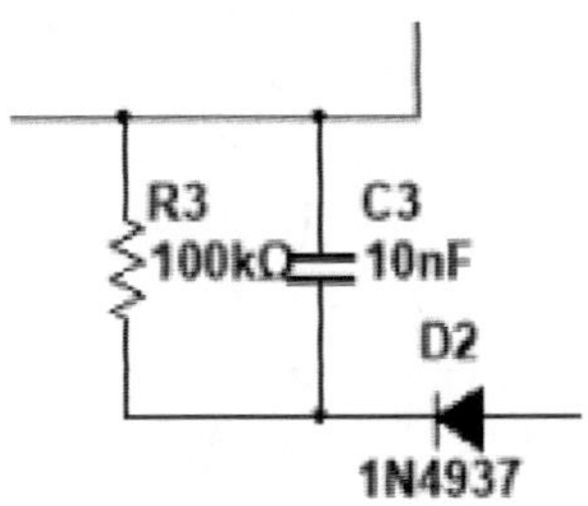

Figure 3. Primary buffer circuit diagram.

3.5. Secondary buffer circuit of the transformer

For the secondary side of the transformer, a sharp waveform is generated at the moment when the diode D5 is turned off. If the spike exceeds the withstand voltage of the diode D5, the diode may be broken down. In order to protect the diode and improve EMI, it is necessary to add the buffer circuit composed of R18 and C8, as shown in Figure 4 [10]. During the adjustment of the buffer circuit, it is necessary to pay attention to the overheating of R18 and C8.

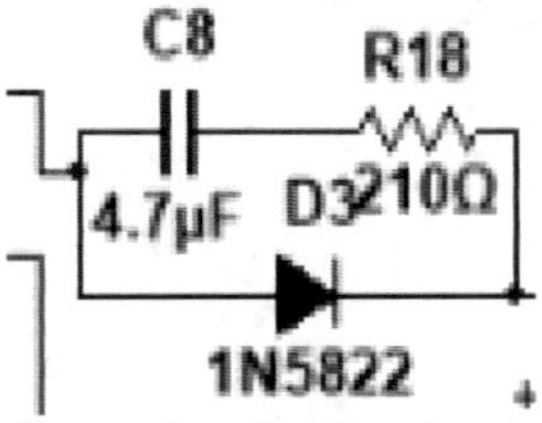

Figure 4. Secondary buffer circuit diagram.

4. Simulation Results and Discussion

This paper used Saber to model, simulate and analyze the circuit schematic. The overall circuit of the modelling and simulation is shown in Figure 5.

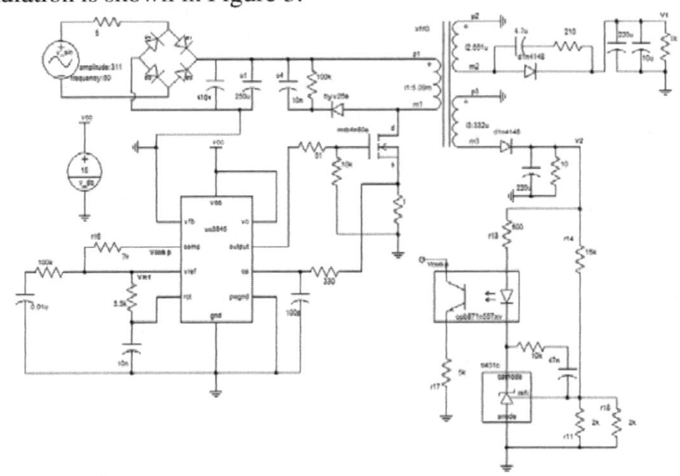

Figure 5. Simulation model by using Saber.

Figure 6 is shown the output voltage waveform when the input voltage is 220V. As shown in the figure, when the power is turned on, the waveform starts to rise sharply, and the final output voltage is stable at 24.103V. According to the accuracy formula of the output voltage is

$$\delta = \frac{|U_1 - U_O|}{U_O} \times 100\% \tag{4}$$

where U_1 is the actual output voltage and U_O is the ideal output voltage.

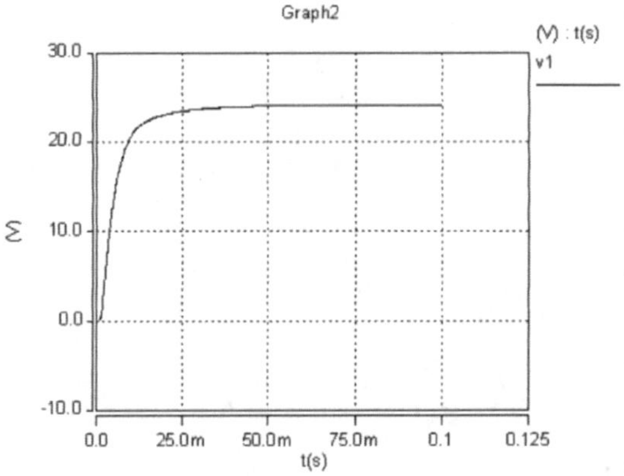

Figure 6. Output voltage waveform.

IOP Conf. Series: Materials Science and Engineering **644** (2019) 012012 doi:10.1088/1757-899X/644/1/012012

Therefore, the accuracy is 0.425%, and it is reached the expected design goal. By adjusting the range of the input voltage, the output-voltage accuracy corresponding to different input voltages is obtained, as shown in Figure 7. The input voltage is within 200~240V, and the accuracy of the output voltage can be kept within 1.3%. It shows that the voltage stability of the switching power supply is good, the feedback voltage works normally, the PWM regulation is accurate, and the input voltage is wide.

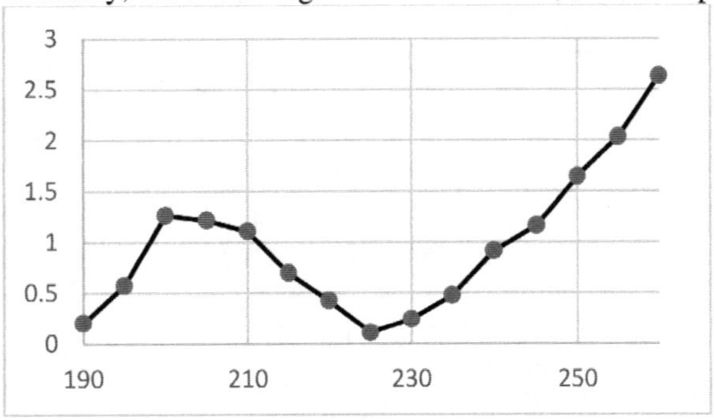

Figure 7. Output-voltage accuracy.

The ripple of switching power supply can be calculated by calculating the ripple factor, as shown in Figure 8. From the simulation diagram, the peak-to-peak value, that is, the output ripple voltage is 0.022V, and the calculated ripple coefficient is

$$\gamma = \frac{0.022}{24} \times 100\% = 0.09\%.$$

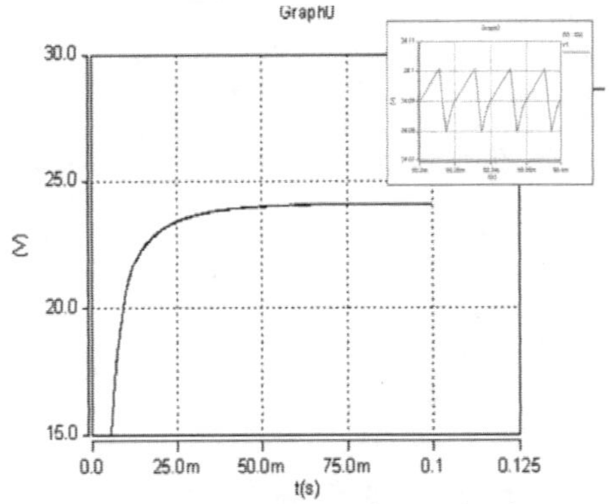

Figure 8. Output-voltage ripple.

In the voltage feedback circuit, adjusting the resistance of the TL431 can change the value of the output voltage. The output voltage can be adjusted by changing R11 and R18 in the simulation model. Through simulation analysis, it is concluded that when the parallel resistance of R11 and R18 is reduced, the output voltage is increased. For the convex wave appearing in front of the waveform, it is mainly used for the secondary side buffer circuit of the transformer to attenuate the convex wave, thereby making the waveform smoother, as shown in Figure 9.

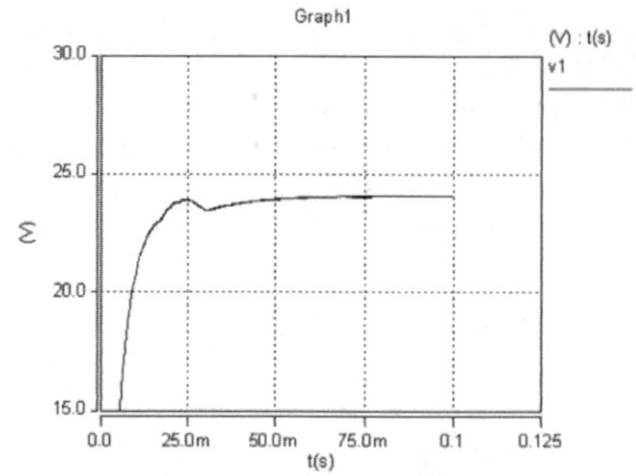

Figure 9. Adjustment of the voltage after feedback.

Figure 10 and Figure 11 are shown for different values of the output voltage. In Figure 10, the actual voltage of the output 9V is 9.07V, and the actual voltage of the output 12V is 11.93V. In Figure 11, the actual voltage of the output 5V is 5.5372V, and the actual voltage of the output 3.3V is 3.2764V.

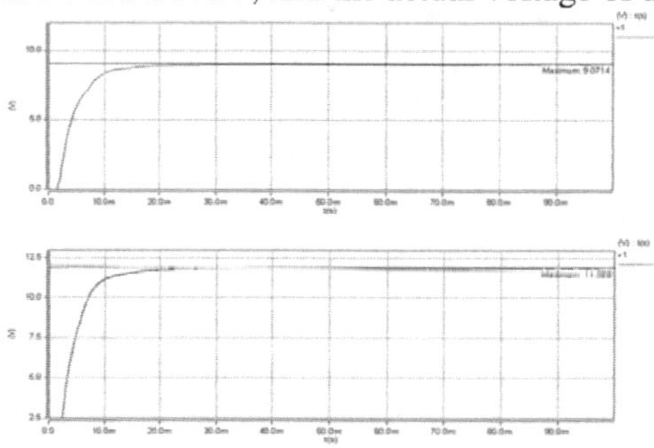

Figure 10. 9V, 12V output voltage waveform.

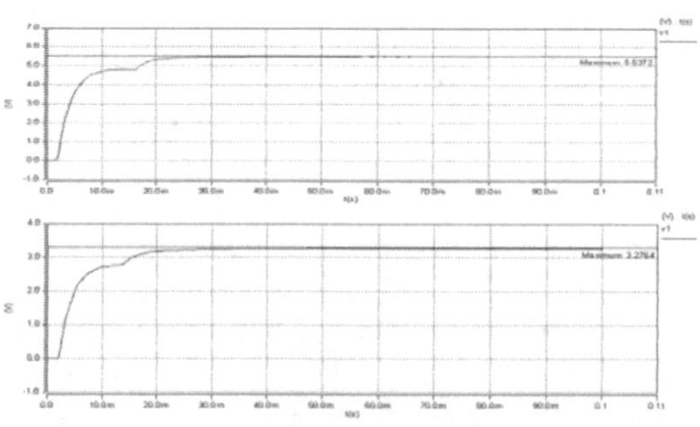

Figure 11. 5V, 3.3V output voltage waveform.

According to the data in the figure, the output voltage is relatively flat, and the response is faster at 9V or higher. For voltages below 9V, the output voltage waveform is affected by the feedback circuit. However, due to the limitation of the feedback circuit parameters, the waveform error below 9V is large. Thus, the overall waveform is relatively flat.

5. Conclusion

This paper is based on the switching circuit designed by UC3845 control chip. For the basic topology of switching power supply, the flyback converter circuit and its control method, the working mode of UC3845 chip and its surrounding sub-circuit and protection circuit are compared systematically. According to the simulation results, the design circuit of this paper has high output-voltage accuracy. For the input voltage in the range of 200V~240V, the accuracy of its output voltage is kept within 1.3%. For high-frequency switching, the strong ripples can cause surge voltage or current, resulting in

IOP Conf. Series: Materials Science and Engineering **644** (2019) 012012 doi:10.1088/1757-899X/644/1/012012

burning of electrical appliances. This design is excellent in suppressing ripple, and the ripple is controlled within 0.1%. This result is lower than 0.5% of the current industry. It is seen that power quality is excellent.

6. References

[1] X.L. Fang, Mod. Elec. Tech. 35, 174 (2012)
[2] Y.B. Xin, Switching power supply basis and application (2009)
[3] X.Q. Zhu, (2013). Design and Implementation of the UC3842 multiple-output flyback switching power. (master's thesis). Human University, Human, China
[4] Z.Z. Xia, B. Wang, J. Li, Power Supply Technol. Appl. 11, 6 (2008)
[5] X.L. Guan, F.L. Liu, S. Chi, J. Wei, Telecom Power Tech. 27, 53 (2010)
[6] Y.W. Zhang, (2011). Design and Implementation of Multi-output Adjustable Switching Power Supply. (master's thesis). University of Electronic Science and Technology of China, Chengdu, China
[7] J.F. Gong, Power Supply Technol. Appl. **13**, 25 (2010)
[8] Z.H. Zheng, Y.K. Ma, J. Hanshan Norm. Univ. (2017)
[9] S.Y. Wu, (2006). Research on Single-Ended Flyback Switching Power Supply Based on TOPSwitch-II. (master's thesis). Nanchang University, Nanchang, China
[10] D.L. Jiang, X.F. Zhang, Z.Y. Lu, Power Supply Technol. Appl. 8, 1 (2005)

IOP Conf. Series: Materials Science and Engineering **644** (2019) 012013 doi:10.1088/1757-899X/644/1/012013

Design of Photovoltaic Inverter Based on STM32 Microcontrollers

Wei-Tai Hsu, Jing-Feng Fang, and Chia-Wei Huang

Department of Electrical Engineering, Zhaoqing University, 526061, China
E-mail:wei353328@gmail.com

Abstract. In this paper, the STM32 microprocessor is used as the central control core, and a 500W photovoltaic inverter is designed. The inverter adopts a two-stage conversion structure. The high-speed timer of the STM32 microprocessor generates high-resolution PWM and SPWM pulses and drives the first-stage DC/DC convertor after driving the chip through UCC27324 and IR2111 respectively. The rectification and the second-stage DC/AC inverter convert the DC signal into a utility frequency. In order to reduce the higher harmonics derived from the DC/AC inverter, the circuit also adopts the principle of LC filter design. Finally, in the Multisim simulation, the operation is stable. The error of the output AC voltage is controlled within 3%. It is said that the inverter has the advantages of high reliability, low harmonic content, and high efficiency.

1. Introduction

Photovoltaic (PV) inverter is the core device for energy conversion of the photovoltaic power generation system, which plays a decisive role in the safety, energy conversion efficiency and stability of the system. Therefore, in order to improve the power generation efficiency and improve the control ability of the photovoltaic power generation system, it is of considerable significance to study a high performance photovoltaic off-grid inverter.

System and Structure

The structural design of the photovoltaic off-grid inverter is shown in Figure 1. It consists of several parts: STM32 microprocessor, conversion circuit, and sampling circuit. The STM32 microprocessor (MCU) generates pulse width modulation (PWM), and sinusoidal pulse width modulation (SPWM) signals to be amplified by the driving circuit to control the MOSFET switch in the conversion circuit, thereby achieving the purpose of boosting and rectifying the input voltage and converting to the DC voltage. The output voltage wave of the conversion circuit is a high-voltage SPWM pulse wave, and a smooth sinusoidal AC voltage wave is obtained after passing through the LC low-pass filter circuit. In order to stably output sinusoidal alternating current, the voltage and current sampling circuit samples the voltage and current signals outputted by the DC/DC converter into a low-voltage DC signal, and then connects to the I/O port of STM32 and converts the data signal and the PWM signal by the built-in ADC. It results in obtaining a modulated PWM signal [1].

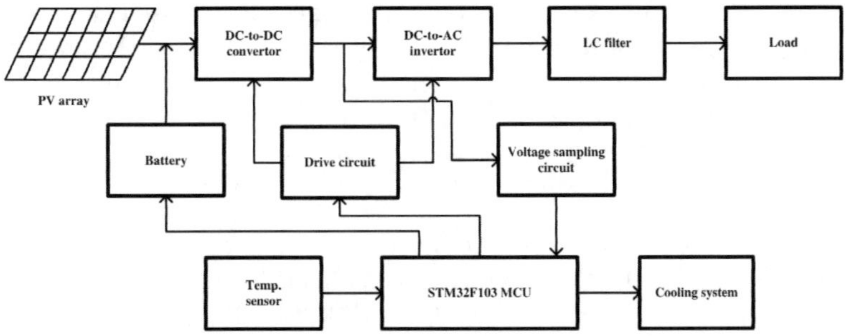

Figure 1. Schematic diagram.

Due to the output voltage of the PV array and its instability, in order to provide continuous and stable power supply to the load in this design, the STM32 controls the battery for additional power supply when the PV output voltage is insufficient. The MOSFET of the inverter is in the working stage of high-frequency [2], the heat emitted by the system may cause the inverter system to work at high temperature, and the damage to the electronic device is significant. In order to improve the lifetime of the inverter, the cooling system consisting of a temperature sensor is also designed.

2. Hardware design

2.1. Boost converter circuit

PV modules are easily interfered by various external factors. For this reason, the photovoltaic output voltage fluctuates greatly and needs to be converted to a stable bus voltage by boosting [3]. Therefore, the boost converter circuit shown in Figure 2 is designed. The output current and voltage are 2A and 310V respectively. For this diode D1, it has a strong forward surge withstand capability (SWC) of 30A and a high reverse breakdown voltage of 1000V. The switch adopts an N-channel MOSFET 2SK3070S with a working frequency of 50 kHz, a maximum forward withstand voltage of 600V, and an on-resistance of 0.3 Ω. In order to ensure that the boost converter circuit operates in the continuous mode of the inductor current, the value of the inductor L1 is selected to be ten mH, and the ripple voltage is required to be no more than 1%.

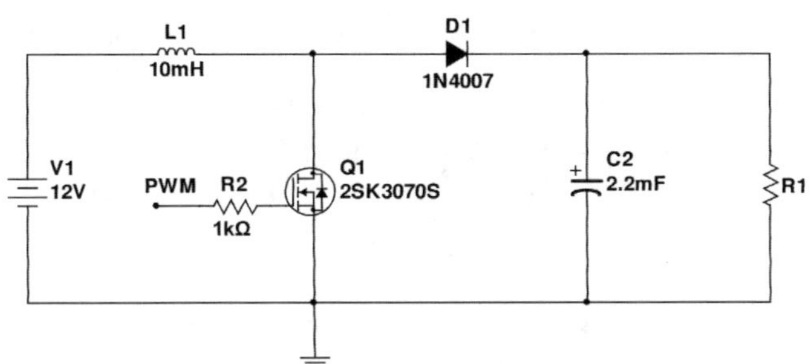

Figure 2. Boost converter circuit.

2.2. Single-phase full-bridge inverter circuit design

In order to meet the design requirements for the 500W inverter, the power switch tube IRF840 is selected. As shown in Figure 3, the inverter circuit is composed of four IRF840s to form four bridge arms. Q3 and Q6 form a pair of bridge arms driven by the same group of pulses, and Q4 and Q5 form a pair of bridge arms operated by the same group of pulses [4]. The two sets of complementary drive pulses are PWM pulses that are sinusoidally varied using a duty ratio to drive the bridge arms to convert the DC to the sinusoidal AC power source.

Due to the inherent characteristics of PWM technology, the output voltage of the inverter circuit contains a lot of higher harmonics. Therefore, a low-pass filter is needed at the output to reduce the harmonic content and smooth the sinusoidal output voltage. The passive low-pass filter network designed in this paper consists of a first-order LC low-pass filter and an EMI filter [5], which filters the SPWM output waveform after power amplification. According to the principle of the LC low-pass filter, the cutoff frequency of the designed LC filter circuit is

$$f_0 = \frac{1}{2\pi\sqrt{LC}}. \tag{1}$$

Since the AC output frequency is 50 Hz for the inverter, the upper limit of the cutoff frequency f_z of the LC filter is taken as 100 Hz. Considering the speed characteristics of the power components, the frequency of SPWM should not be too high. The f_s is 17.5KHz in the design, and the cutoff frequency of the filter can be calculated according to the equation (1):

$$f_0 = \sqrt{f_z f_s}. \tag{2}$$

The cutoff frequency is 1323 Hz from the equation (2), and the capacitance C2=4.7 μF and the inductance L1=2.5 mH are taken. The parameters for the post-stage EMI filter can take the conjugate inductor of 20μH and the large capacitor (104).

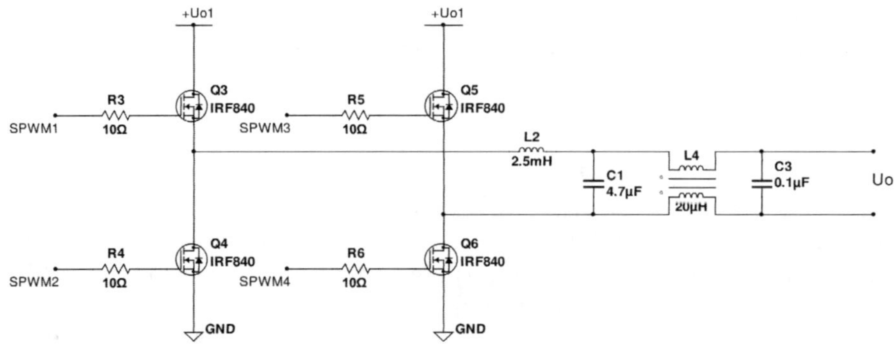

Figure 3. Single-phase full-bridge inverter circuit.

2.3. Drive circuit design

Since the I/O output of STM32 is 3.3V logic level and cannot directly drive the FET, the PWM drive circuit shown in Figure 4 is designed to drive the boost converter circuit. This design uses a dual MOSFET driver UCC27324 to build the drive circuit. The UCC27324 has a simple peripheral circuit and strong driving capability. It is a low power dissipation MOSFET with a supply voltage of VDD of 12V and 3.3V logic level. The output of PWM pulse for STM32 can be connected to the input with the protection resistor. Then the gate of MOSFET powered by STM32 turns on and off.

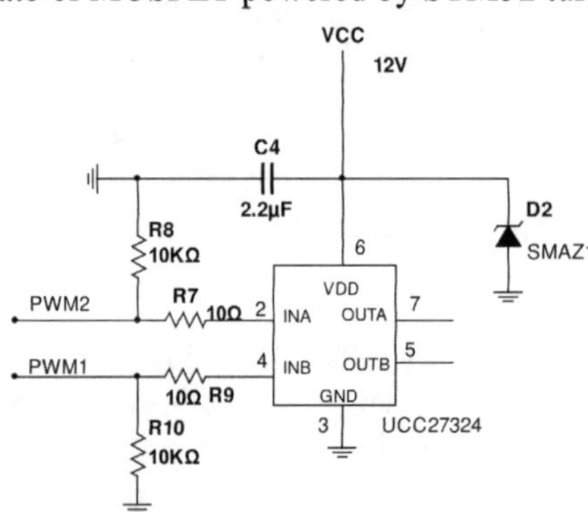

Figure 4. PWM drive circuit.

At the same time, the SPWM driving circuit shown in Figure 5 is also designed. The pins of STM32 can only work normally at a low level. In order to electrically isolate it from the high voltage side, the

SPWM control of the input driving circuit is performed. The signal is first subjected to electrical high voltage isolation by a high-speed optocoupler 6N136. The 6N136 uses a system's supply voltage of +3.3V. The VCC provides LED on-current, and the output side is powered by an auxiliary supply of 12V. The two half-bridge power switch tubes are driven by the driver chip IR2111S designed to drive the half-bridge circuit. The IR2111S has generated two 750ns complementary SPWM signals. The current capability is related to the characteristic of the switch.

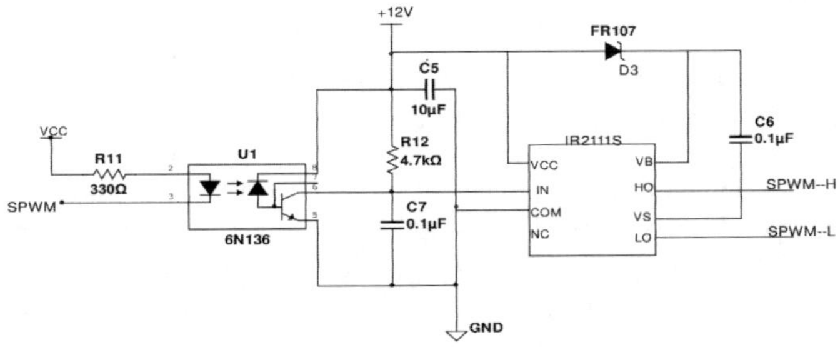

Figure 5. SPWM drive circuit.

2.4. Sampling circuit design

This design is controlled the stable output of AC 310V by regulating the high-voltage DC bus, so the sampling circuit is needed to sample the current and voltage of the high-voltage DC bus. The output signal of the sampling circuit is sent to the ADC of STM32. It is also necessary to consider the electrical isolation between the high side and the control side. Therefore, in the sampling circuit design, the optocoupler isolation amplifier HCPL-7840 is used for current and voltage sampling. The HCPL-7840 is a dedicated current/voltage detection optocoupler that is immune to external magnetic fields, has good linearity, small size, and low cost. Since the specification of the inverter is 500W, considering the back-end inverter efficiency, the boost DC is not exceeded 2A at most. The high-power manganese copper wire resistance of 0.1Ω is selected as the current sampling resistor, and the output current detection circuit is shown in Figure 6. If the output current is I_OUT, according to equation (3), the output current range is reached 3A under the premise that the output detection circuit AD_I is not exceeded 2.5V, which meets the design requirements.

$$AD\_I = 8 \times R37 \times I\_OUT \tag{3}$$

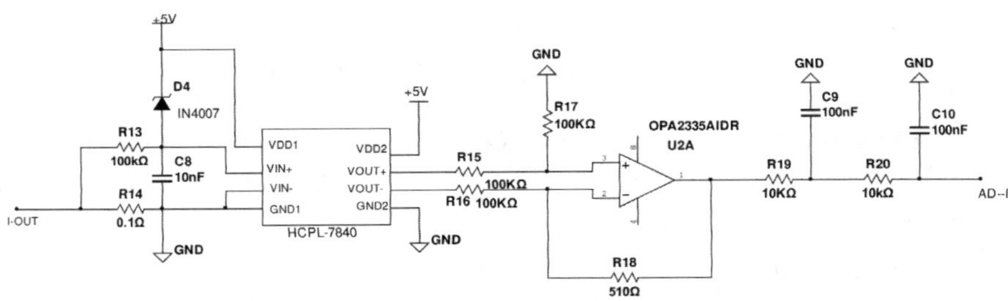

Figure 6. DC detection circuit for the output.

When designing the output current and voltage detection circuit for the high-voltage DC bus of 310V, it is necessary to select the appropriate resistors for a voltage divider and a current sampling and consider the internal 8x amplification relationship of the linear optocoupler HCPL-7840 and the STM32 internal A/D converter with voltage reference of 2.5V. According to equation (4), when the output voltage is 310V, the voltage detection circuit outputs 1.264V, which is within the range of voltage reference and has a positive voltage margin to meet the design requirements. The output voltage detection circuit is shown in Figure 7.

$$AD\_U = \frac{R_{22}}{R_{21} + R_{22}} \times 8 \times 310 \tag{4}$$

IOP Conf. Series: Materials Science and Engineering **644** (2019) 012013 doi:10.1088/1757-899X/644/1/012013

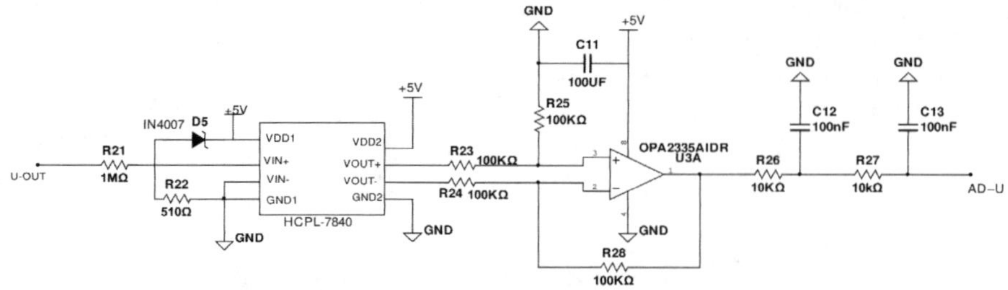

Figure 7. DC voltage detection circuit for the output.

In the design of the inverter, most of them are constructed with high-frequency switching tubes, which lead to the temperature rise of the inverter as the operating temperature of the inverter increases. When the temperature is too high, the cooling fan is started. The temperature sensor NTC-10K is selected, which is small and placed directly in the heat sink to make the temperature measurement closer to the inverter temperature. The circuit diagram is shown in Figure 8. The first half is the bridge measurement circuit, the middle is the amplifier circuit dominated by the instrumentation amplifier AD620, and the second half is the voltage clamp and overcurrent protection circuit.

The specific model of the negative temperature coefficient sensor is MF52-103/3435, and the temperature range is from -55°C to 125°C. When the temperature is equal to 25°C, the resistance is 10kΩ. As the temperature increases, the resistance gradually decreases. In this design, it is only necessary to start the cooling fan when the temperature of the inverter heat sink is higher than 45°C, and it is not necessary to know the accurate temperature value in other cases of the inverter. In the circuit design, the instrumentation amplifier is powered by a single power supply. The AD620 outputs the voltage signal only when the temperature of the heat sink is greater than 25°C. Below this value, the output is 0V. This design eliminates the need for the voltage boost circuit on the output side of AD620, simplifying circuit design. Also, the capacitor C14 is filtered out the differential mode noise in the input signal of AD620. The capacitors C15 and C16 have filtered the common mode noise in the input signal, making the temperature data acquisition more accurate. The AD620 has a high common-mode rejection ratio, low noise, and low input offset voltage. It is mainly used in high-precision data acquisition systems. The calculation formula of gain G is

$$G = \frac{49.4K\Omega}{R_{37}} + 1. \tag{5}$$

When setting the gain, it is considered that the output of AD620 should not be higher than 3.3V within the normal operating temperature range of the inverter. Otherwise, the AD620 and the microcontroller are damaged. The current limiting resistors R38 and R39 are designed to limit the overcurrent of AD620 and STM32, respectively, and function as current limiting protection. The output voltage of AD620 is calculated as:

$$V_{th} = 2.5 \times \left(\frac{R_{30}}{R_{29} + R_{30}} - \frac{R_{32}}{R_{31} + R_{32}}\right). \tag{6}$$

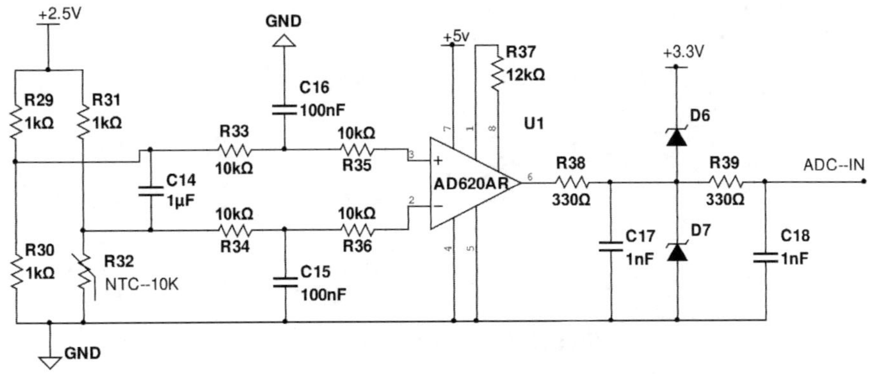

Figure 8. Temperature detection circuit for the inverter.

3. Simulation Results and Discussion

3.1. Boost converter circuit simulation

In NI Multisim, the boost converter circuit is built for the simulation. When the input voltage is +12V, the simulation output is shown in Figure 9. It can be seen from Figure 9 that the output voltage of the boost converter circuit is about 304V. The ripple is detected after the output voltage is stable. The period of ripple voltage is about 2ms, and the range of ripple voltage is from 100mV to 200mV, which is much smaller than the design requirement (less than 1%).

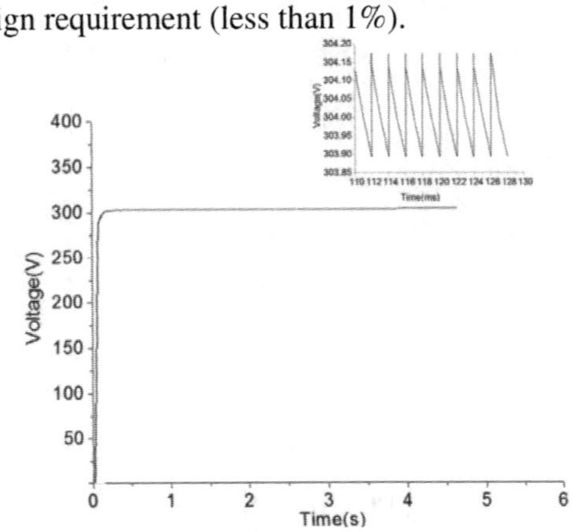

Figure 9. Output ripple voltage.

3.2. Single-phase full-bridge inverter circuit simulation

The single-phase full-bridge inverter circuit is built in NI Multisim for the simulation. The simulation results are shown in Figure 10, Figure 11, Figure 12 and Figure 13. Figure 10 is the SPWM waveform amplified by the optocoupler driving IR2111 chip, channel A is a driving signal of the upper arm, and channel B is a driving signal of the lower arm. It can be seen from Figure 10 that for a short period of time, the two driving signals are all low level, that is, the dead time is in this period, and the function of this time period is to ensure that the switching tubes of the upper and lower arms are not turned on at the same time.

IOP Conf. Series: Materials Science and Engineering **644** (2019) 012013 doi:10.1088/1757-899X/644/1/012013

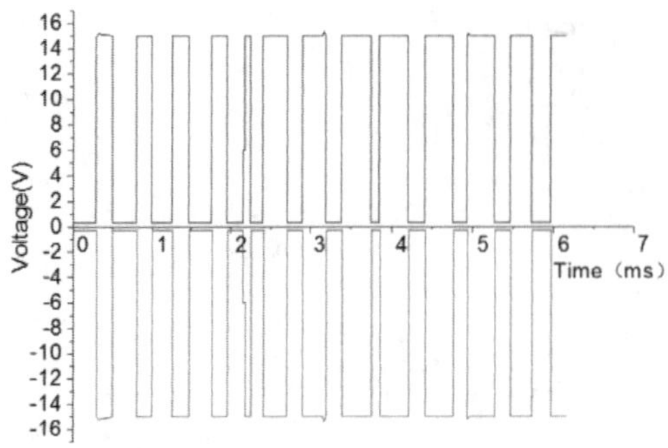

Figure 10. SPWM waveform.

This study is performed Fourier transform analysis on the output voltage waveform of the inverter circuit to obtain the phase, amplitude distribution map and Fourier analysis table of voltage harmonics at different frequencies, as shown in Figure 11. In Figure 11(a), it can be seen that the DC output voltage of the inverter is 0.00778453V and the total harmonic coefficient (THD) is 5.59376%. These two experimental data show that the filter design has a strong inhibitory effect on harmonics. It is greatly reduced the power pollution and improved the power quality. At the same time, it is also improved the stability, lifetime and efficiency of the use of electrical equipment. The distribution of harmonics is shown in Figure 11(b) and (c). The larger amplitude of each harmonic appears at the integral multiple of the fundamental frequency, and its amplitude increasing with frequency is gradually decreased. The amplitude after the ninth harmonic is almost close to zero, and the SPWM inverter is effectively suppressed the low-order harmonics. Therefore, in general, the voltage of low-order harmonic is minimal and is ignored [6].

Dc component:	0.00778453				
No. Harmonics:	9				
THD:	5.5938%				
Grid size:	256				
Interpolation Degree:	0				
Harmonic	Frequency	Magnitude	Phase	Norm. Mag	Norm. Phase
0	0	0.00778453	0	0.315362	0
1	50	0.0246845	-111.71	1	0
2	100	0.00116436	-51.276	0.047169	60.4328
3	150	0.000623475	-48.904	0.0252578	62.8046
4	200	0.000274026	-36.072	0.011011	75.6369
5	250	0.000197636	-45.1	0.00800651	66.6084
6	300	0.000149518	-30.011	0.00605718	81.6978
7	350	0.000103466	-32.197	0.00419154	79.5115
8	400	9.74716E-05	-25.253	0.0039487	86.4555
9	450	7.36076E-05	-18.323	0.00298194	93.3859

(a) Fourier analysis

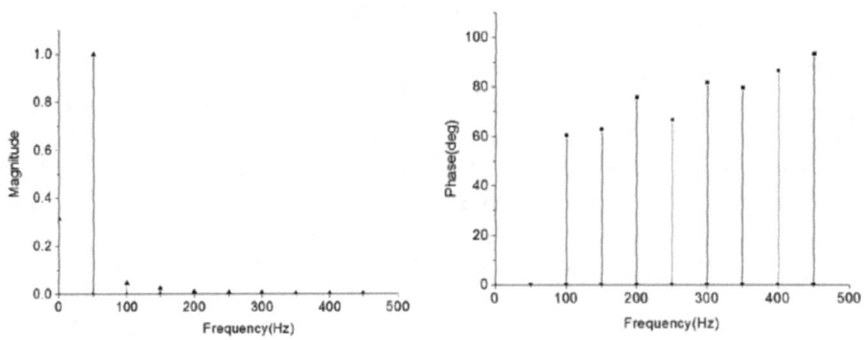

(b) Magnitude (c) Phase

Figure 11. Fourier analysis of output voltage for the inverter circuit.

Figure 12 is a diagram of the input and output power of the inverter. From the figure, the input and output power of the inverter designed in this paper is a zero-crossing linear relationship. In Figure 12, the inverter efficiency is about 96.5% on average. The data of these simulation results are shown that the designed inverter has the characteristics of high photovoltaic conversion efficiency and strong output stability.

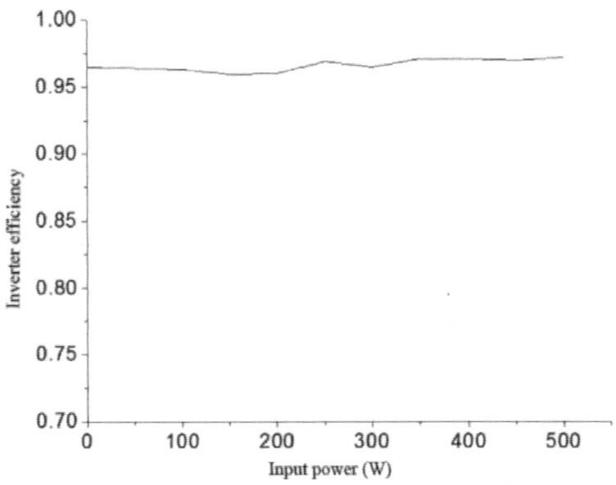

Figure 12. Input power versus inverter efficiency.

Figure 13 is shown the AC output voltage waveform for the inverter. The peak voltage of the sine wave for the inverter is 304V. The average voltage of the sine wave is about 215V, and the frequency is $f = 1 / (2 \times 10.000\text{ms}) \approx 50$ Hz. The value of the simulated voltage is smaller than the value of the ideal voltage. There is a certain impedance in the circuit that consumes energy. Since the electrical equipment is worked normally at 220±10V, the output error of the inverter designed in this paper is acceptable, and the inverter is suitable for popularization.

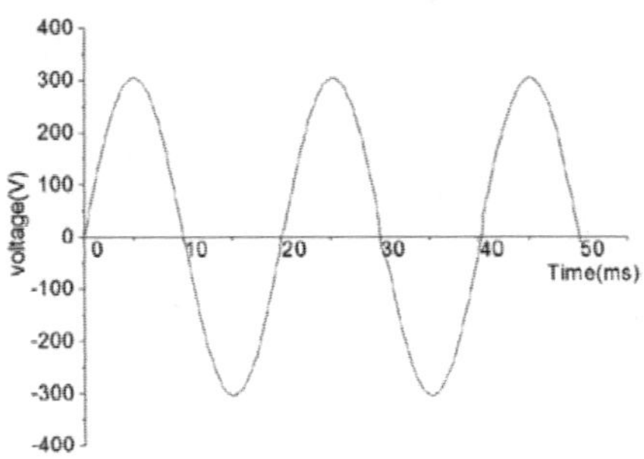

Figure 13. Output voltage waveform for the inverter.

4. Conclusion

The inverter is used the STM32 as the main controller to realize sinusoidal alternating current that converts PV DC into AC source. In Multisim simulation, the ripple coefficient is less than 0.1%, which largely avoids the generation of surge voltage and current, and greatly improves the efficiency of the inverter. Due to the harmonic, aging phenomena in cables insulation is accelerated, and the service life of the cable is shortened. The output is analyzed by Fourier to obtain a THD of 5.59%, which proves that the circuit has a strong suppression effect on these harmonics. The inverter conversion efficiency of this design is 96.5%, which is higher than 95% of the current market. Also, the two-stage conversion structure is adopted in this design to make the whole inverter have the advantages of simple structure, small volume, convenience-driven, and low cost. From the above, by introducing a protection circuit, the inverter is protected in time and effectively, improving safety and reliability.

References

[1] C.Y. Yan, C.J. Zhang, W.Y. Yan, M.L. Jia, K. Zhang, Trans. China Elec. Soc. 27, 59 (2010)
[2] K. Sun, Y.J. Lu, H.F. Wu, Y. Xing, L.P. Huang, Proc. CSEE 33, 37 (2013)
[3] K.L. Yu, A.H. Lin, Y.C. Zhou, Tech. Inno. App. 12, 23 (2012)
[4] X.H. Fan, M.C. Zhu, S.L. Nie, W.Q. An, Power Elec. 47, 52 (2013)

[5] M.G. Yang, (2018). *Application research of independent photovoltaic power generation technology*. (master's thesis). Jilin Institute of Chemical Technology, Jilin, China
[6] C.L. Liu, J. Xi'an Mining Inst. 2, 179 (1993)

IOP Conf. Series: Materials Science and Engineering **644** (2019) 012014 doi:10.1088/1757-899X/644/1/012014

Design of Switching Power Supply for Electric Vehicles with UC3842 Chip

Wei-Tai Hsu, Yong-Ke Chi, and Chia-Wei Huang

Department of Electrical Engineering, Zhaoqing University, 526061, China
E-mail:wei353328@gmail.com

Abstract. This article is applied to UC3842 as a current mode PWM generator component, which is used to construct a single-ended flyback switching power supply with DC48V input and DC12V output. For the analysis of the instability caused by the peak current control, and the slope compensation is $m=0.75m_2$, the duty cycle is no longer interfere with the stability of the system, which can improve the anti-interference ability of the power supply.

1. Introduction

Electric vehicles are currently zero-emission vehicles. As green vehicles, they will bring huge changes to society. DC brushless motors rely on excellent control modules and power supply performance. High-quality and high-performance systems of power supply occupy an important position in the entire electric vehicle system. In the switching power supply, the pulse width modulation (PWM) control circuit has two types of voltage mode and current mode. The current control mode has the disadvantage of the voltage control mode. Based on the voltage control mode a current feedback loop is added. The inductor current is no longer an independent variable so that the power supply circuit responds to the changes of input voltage and the load. The response is fast, the anti-interference performance is strong, and the loop stability is better [1]. When the peak current control works at the duty ratio D>0.5, it results in problems such as open-loop unstable system, subharmonic oscillation, and poor anti-interference ability. Therefore, the slope compensation must be introduced [2]. The flyback converter has a simple circuit structure, a high operating frequency, a stable output voltage, and its transformer can be used as an output filter inductor in addition to isolation. There is no need to add additional output inductors, so adjustments can be made to follow changes in input voltage and load.

2. Performance indicators of flyback switching power supply design

Electric vehicles are generally powered by lead-acid batteries. They are available in 24V, 36V, and 48V versions, but most of the electric vehicles use DC voltages of around 12V.

Based on UC3842 PWM control technology, this paper designs the switching power supply with basic parameters such as input voltage DC 36~48V, output voltage 12V/1A, switching frequency 100KHz, duty ratio less than 60%, power efficiency 80%, and ripple voltage less than 500mV.

IOP Conf. Series: Materials Science and Engineering **644** (2019) 012014 doi:10.1088/1757-899X/644/1/012014

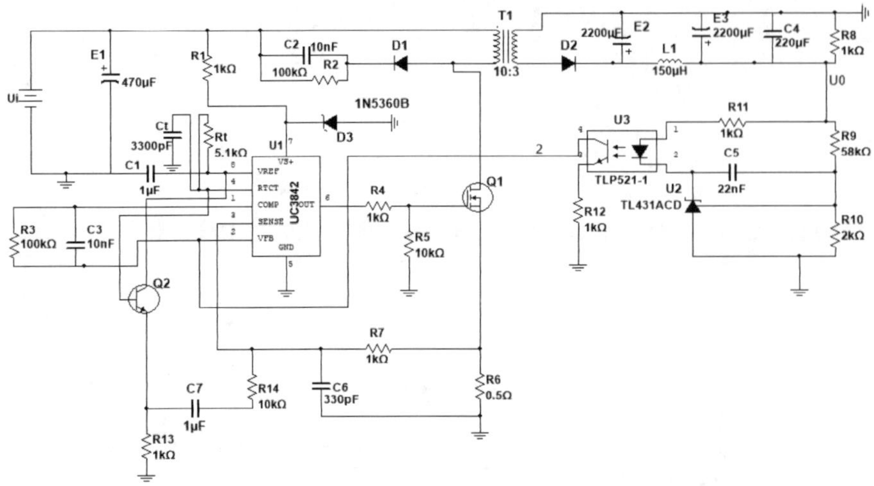

Figure 1. Schematic of flyback switching power supply.

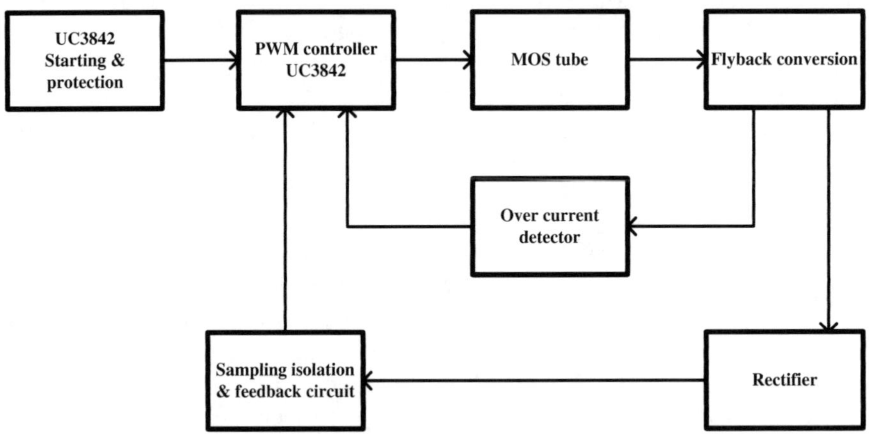

Figure 2. Block diagram of flyback switching power supply.

Figure 1 is a circuit diagram of a flyback switching power supply designed in this study [3], which can be divided into UC3842 driving circuit, flyback DC conversion circuit, PWM control circuit, voltage feedback and output filter circuit, ramp compensation circuit, etc. The block diagram is shown in Figure 2.

3. The Circuit design of flyback switching power supply

3.1. Drive circuit of UC3842

The driving circuit is composed of a voltage dividing resistor R1 and a Zener diode D3. The input voltage is supplied with a voltage drop across the resistor R1 to provide a chip driving voltage. In order to prevent the input voltage from being too high and damage the chip, the Zener diode D3 is connected to the power supply terminal of the chip to stabilize the voltage.

3.2. Protection circuit of switching tube

Due to the turn-to-turn capacitance of high-frequency transformer and the recovery time of output rectifier diode, the turn-off voltage of the MOS tube is spiked, which may damage the switch. Therefore, the clamp circuit is designed with R2, C2, and D1. When the MOS tube is turned off, the energy in the leakage inductance of the transformer is transferred to C2 and consumed on the resistor R2, thereby effectively suppressing the turn-off voltage spike of the MOS tube.

IOP Conf. Series: Materials Science and Engineering **644** (2019) 012014 doi:10.1088/1757-899X/644/1/012014

3.3. Feedback circuit of DC conversion

As shown in Figure 3, the flyback topology is: When the switch Q1 is turned on, the input voltage is applied to the primary side N_1 of the transformer, and the transformer stores energy. According to the polarity of the transformer, the induced electromotive force (emf) on the secondary side N_2 is positive (up) and negative (down), the diode D is cut off, and no current flows through the secondary side. When the switch Q1 is turned off, the polarity of the induced electromotive force on N_2 is negative (up) and positive (down), and the diode D is turned on. During the turn-on of the switch, the energy stored in the transformer is released to the load through the diode D [4].

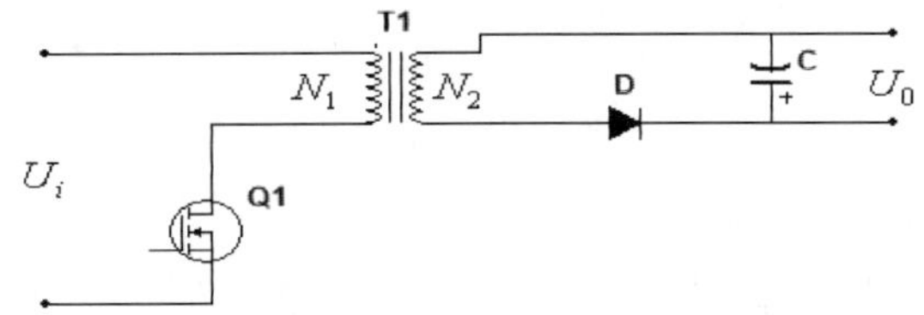

Figure 3. Schematic of flyback topology.

In the constant current state, the output voltage of the flyback converter is determined only by the turn ratio N of the primary and secondary windings, the duty ratio D, and the input voltage, regardless of the load resistance. The expression is

$$U_O = \frac{U_i N_2 D}{N_1(1-D)}. \quad (1)$$

That is, when the input voltage changes within a certain range or the switching power supply is disturbed, the output voltage can be stabilized within the certain range by changing the duty ratio of the switching tube.

3.4. Output circuit after rectification and filtering

The switching power supply performs high-frequency chopping on the DC input and then transmits it through the high-frequency transformer. Therefore, there must be high-frequency noise interference at the output, and the MOS tube also causes high-frequency noise during the high-frequency switching process. Thus, L1, E2, and E3 are used to form an LC filter circuit to filter out the interference of high-frequency noise.

3.5. Voltage feedback circuit

The voltage feedback circuit is mainly composed of TL431 and TLP521. The output voltage is divided by R9 and R10 to the sampling voltage, which is compared with the 2.5V reference voltage provided inside the TL431. When the sampling voltage is equal to the reference voltage, the current flowing through the optocoupler diode does not change, and the voltage feedback to the pin2 of UC3842 is also unchanged, and the duty ratio of UC3842 for the square wave output is also unchanged. The output voltage of the switching power supply is stable (12V). When the output voltage is high, the sampling voltage increases, the cathode potential of TL431 decreases, and the current flowing through the optocoupler diode become larger. Then the voltage of the triode collector that is feedback to the pin2 of UC3842 also becomes larger, and the output of UC3842 is also increased. The duty ratio for the square wave output becomes smaller, thereby stabilizing the output voltage (12V). When the output voltage is low, the sampling voltage is reduced, the cathode potential of TL431 is increased, and the current flowing through the optocoupler diode becomes smaller. Then the voltage of the triode collector that is feedback to the pin2 of UC3842 becomes smaller, then the duty ratio of UC3824 for the square wave output becomes large, which stabilizes the output voltage (12V) [5].

IOP Conf. Series: Materials Science and Engineering **644** (2019) 012014 doi:10.1088/1757-899X/644/1/012014

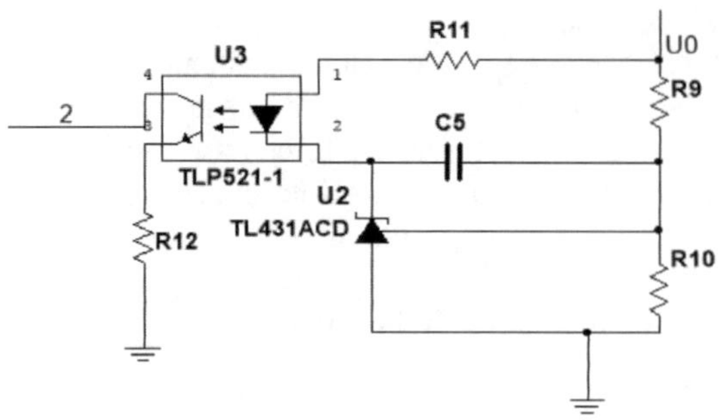

Figure 4. Voltage feedback circuit.

3.6. Slope compensation circuit

Peak current control has excellent load-regulation characteristics and anti-interference ability for the input and is easy to implement current limiting and overcurrent protection, so it is fully applied in switching power supply. In the state of peak current control, when there is a disturbance ΔI_1, the current waveform of inductor changes as shown in Figure 5 and Figure 6.

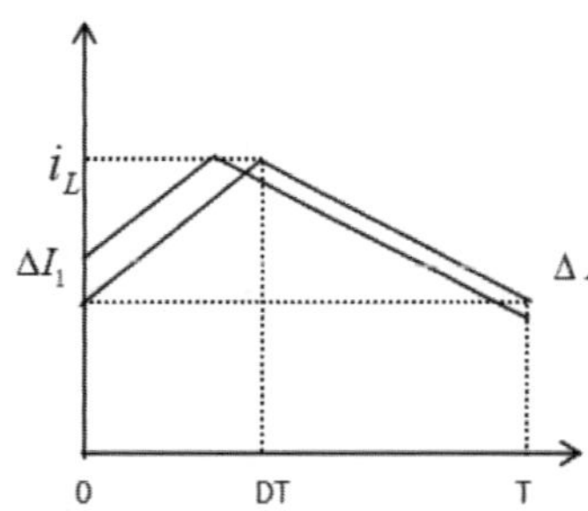

Figure 5. D<0.5, $\Delta I_2 < \Delta I_1$.

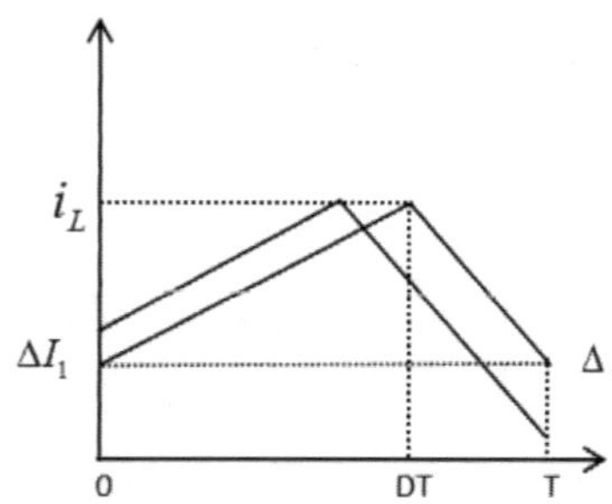

Figure 6. D>0.5, $\Delta I_2 > \Delta I_1$.

Let the rising slope of inductor current be m_1, and the falling slope is m_2, then $\Delta I_2/\Delta I_1 = D/(1-D) = m_2/m_1$. After n cycles, the disturbance error is $(m_2/m_1)^n$. When D<0.5, $\Delta I_2 < \Delta I_1$, the disturbance error decreases, and after n cycles, the disturbance error decays to 0, and the system is stable. However, when D>0.5, the disturbance error increases. After n cycles, the disturbance error gradually enlarges, which leads to the system out of control, and the anti-interference ability of power supply is poor, which can not work stably [6]. Therefore, the slope compensation must be performed; that is, a signal of a fixed slope is superimposed on the current feedback signal, as shown in Figure 7. Let the slope of the added ramp waveform be $-m$, then $\Delta I_2/\Delta I_1 = (m_2-m)/(m_1+m)$. In order to maintain the stability of the system, there is $(m_2-m)/(m_1+m)<1$. And because $m_1=(1-D)m_2/D$, the ramp slope of the system can be stabilized by adding slope compensation: $m>(2D-1)m_2/2D$, i.e., $m>0.5m_2$.

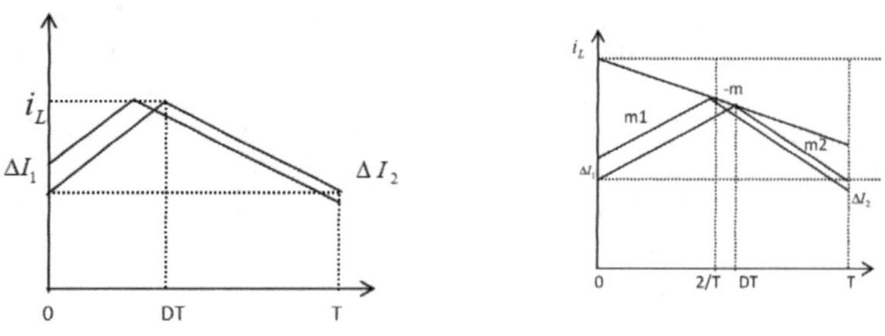

Figure 7. Slope compensation.

IOP Conf. Series: Materials Science and Engineering **644** (2019) 012014 doi:10.1088/1757-899X/644/1/012014

Normally, the method of slope compensation is to use the sawtooth wave generated by the pin4 of UC3842 to be superimposed on the sampling of current signal pin3 directly by the divider of the resistor, as shown in Figure 8 [7]. The slope compensation circuit composed of RC is simple, but the shunting action of R1 and R2 inevitably affects the charging time of the capacitor, thus affecting the operating frequency of the circuit, and the smaller the R1 and R2, the more significant the effect. Because of this drawback, a triode is added by introducing an emitter follower. The equivalent resistance of the compensation network is increased to reduce the impact of the compensation network on the operating frequency. It is more adaptable than the resistance compensation. The function of the capacitor C8 is to filter out the DC component in the sawtooth wave, as shown in Figure 9 [8].

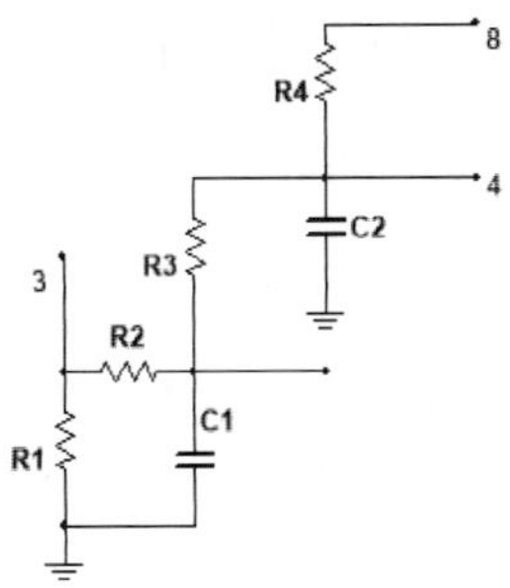

Figure 8. RC compensation.

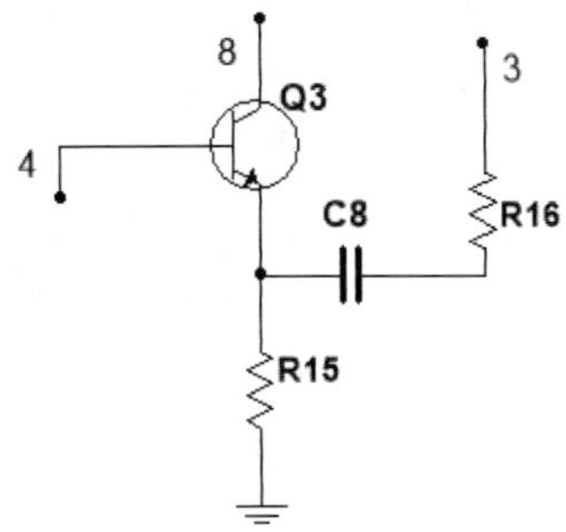

Figure 9. Emitter follower compensation.

4. Simulation results and discussion

The simulation of input and output is shown in Table 1.

Table 1. Input vs. output.

Input voltage/V	48	45	42	39	36
Output voltage/V	11.985	11.874	11.777	11.727	11.722

When the load current is constant, the voltage-regulation coefficient is $S_r=(\Delta U_o/\Delta U_i)\times(U_i/U_o)$, as shown in Figure 10.

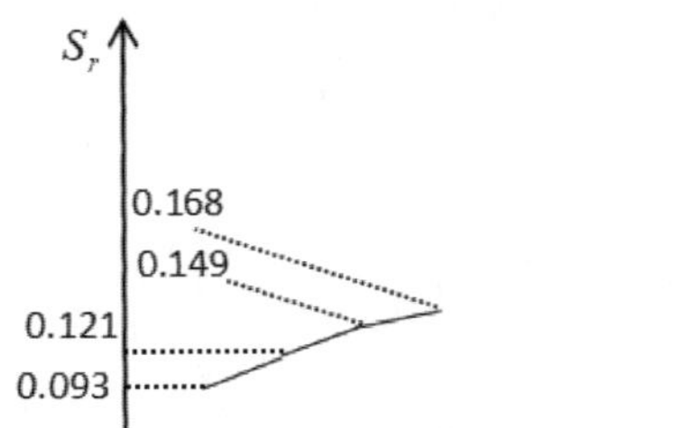

Figure 10. Voltage-regulation coefficient for different input voltage.

It can be seen from Figure 10 that the voltage-regulation coefficient can be controlled within 0.17, and the output voltage stability is useful in the face of input voltage variation. In the peak current mode control, the switching power supply is stable when the duty ratio D<50%, but when the duty ratio D>50%, the power supply is prone to subharmonic oscillation and runaway.

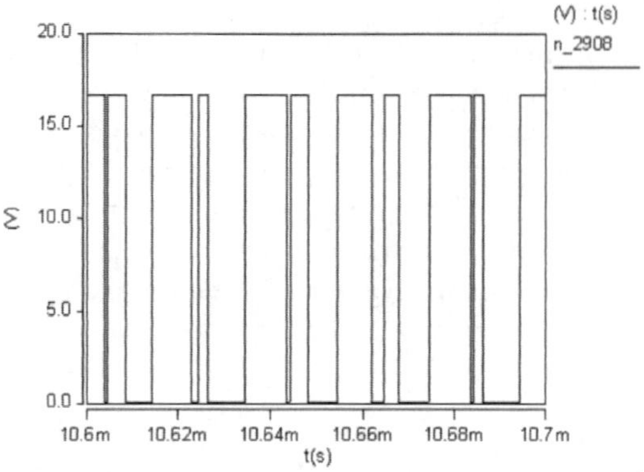

Figure 11. m=0, the waveform of the switch tube.

According to the test data, when the duty ratio is D=58%, no slope compensation is performed, as shown in Figure 11. The subharmonic oscillation occurs in the waveform of the switching tube, which makes the system's anti-interference ability weak and the output stability is poor.

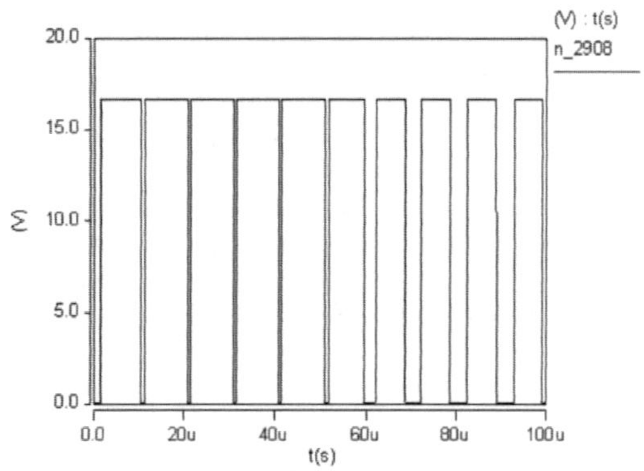

Figure 12. $m<0.5m_2$, the waveform of the switch tube.

By designing the emitter follower, the slope compensation is $m<0.5m_2$. As shown in Figure 12, the anti-interference ability of the system becomes stronger, but subharmonic oscillation occurs at around T=60μs, which is caused by insufficient slope compensation.

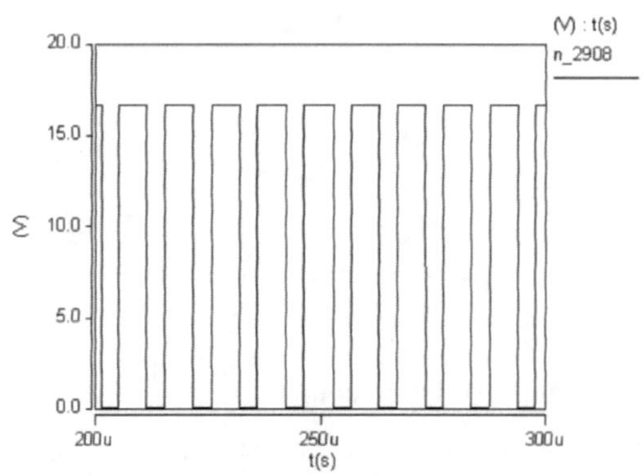

Figure 13. $m=0.75m_2$, the waveform of the switch tube.

When the slope compensating is $m=0.75m_2$, as shown in Figure 13, the switching tube drive waveform is stable, eliminating subharmonics, achieving significant compensation effects, stronger anti-interference ability, and high output voltage stability. Therefore, when the duty ratio is D>0.5, the slope compensation must be performed. When the slope compensation more significant than 50% of the falling slope of the inductor current is added, the power supply can be operated within the variation range of the duty ratio of 0<D<1. It improves the anti-interference ability, eliminates subharmonic oscillation, and ensures stable operation of the power supply.

5. Conclusion

In this study, the UC3842 was used as the current mode PWM controller to construct the single-ended flyback switching power supply with DC 48V input and DC 12V output. By using Saber simulation, when the input voltage is changed, the voltage-regulation coefficient is always controlled within 0.17. When the duty ratio is larger, the voltage-regulation coefficient is smaller, indicating that the switching power supply has a larger duty cycle and it is more stable. By introducing a relatively simple structure of the emitter follower compensation method, the impact of direct slope compensation on the operating frequency of the system is significantly reduced. Therefore, the compensation method is more adaptable than the RC compensation circuit. When the circuit operates at an excessive duty ratio, the subharmonic hazard is eliminated, the transformer is protected, and additional energy loss is prevented. By introducing the slope compensation circuit, the range of the duty ratio is expanded resulting in the utilization ratio of the transformer is improved. The miniaturization, weight reduction and popularization of the switching power supply are promoted.

6. References

[1] Z.P. Zhang, Research and design of electric vehicle DC/DC converter. (master's thesis), Shandong University, Shandong, China, (2014)
[2] Z.Z. Xia, Y.Z. Li, X.P. Tao, Power Elec., 42**,** 71 (2008)
[3] X.P. Fang, L.B. Guo, C.J. Li, X.J. Sun, J. Shandong Univ. Sci. Tech., 30, 99 (2011)
[4] L. Lu, Research on flyback switching power supply control loop. (master's thesis), FiberHome Technologies Group, Wuhan, China, (2012)
[5] C.L. Zhu, L.P. Xiong, H.H. Wei, Elec. Design Engr., 25, 148 (2017)
[6] Y.S. Shi, J.F. Zhang, Q.J. Ning, L. Niu, Chinese J. Liquid Cry. Displays, 27, 217 (2012)
[7] C.L. Zhu, L. Gao, X.D. Zheng, Elec. Design Eng., 26, 177 (2018)
[8] S.L. Liu, J. Liu, J.M. Zhong, Power Elec., 39, 78 (2005)

IOP Conf. Series: Materials Science and Engineering **644** (2019) 012015 doi:10.1088/1757-899X/644/1/012015

Design and Testing of a New Vascular Stent with Enhanced Fatigue Life

Kuang-Lei Ho, Ming-Ya Hung, Jiong-Hong Chen, Yu-Min Jian, and Hao-Ming Hsiao

Department of Mechanical Engineering, National Taiwan University, Taipei, 10617, Taiwan
E-mail: hmhsiao@ntu.edu.tw

Abstract. Vascular stents could suffer from repetitive motions due to pulsatile blood pressure and daily activities. Stent fatigue resistance has thus become a critical issue for stent design. In this paper, an intriguing stent design concept aimed at enhancing the fatigue life was investigated. The concept was to re-distribute stresses more uniformly by tapering the stent strut width. Finite element models were developed to evaluate the mechanical integrity and fatigue safety factor of the stent under various loading conditions. Simulation results show that the fatigue safety factor of this novel stent design increased by 4 times that of a conventional stent. Conceptual stent prototypes were cut by a pulsed-fiber laser, followed by a series of expansions and heat treatments to gradually shape the stent to its target size. A rotating bending fatigue tester was built for this study and stent fatigue tests were conducted for proof of concept. Experimental results show that this stent design concept successfully enhanced the fatigue life as designed. Its fatigue cycle number jumped to 6~7 times that of a conventional stent, which agreed well with the trend predicted by FEA simulations. The findings of this paper provide an excellent guide to greatly improve stent fatigue life.

1. Introduction

Stenting procedure has received great attention from the medical society since its introduction. However, these vascular stents could suffer from various repetitive motions due to pulsatile blood pressure and daily body activities after implantation, especially in the indications of peripheral arteries. Such motions oscillate the stents repeatedly, leading to the potential risk of stent fatigue failure. Such failure has drawn much attention in recent years and, therefore, stent fatigue resistance has become a critical issue for stent design.

Nitinol is an excellent candidate for medical devices due to its unique properties. The peripheral stents are perhaps the most celebrated applications of the nitinol material [1-4]. They are super-elastic, i.e., crush recoverable, exert a gentle chronic outward force (COF) to the artery wall, and are more physiologically compatible than balloon-expandable stents in many indications such as carotid, superficial femoral, popliteal, iliac, and renal arteries. Implantation of self-expanding nitinol stents has demonstrated good clinical results for the treatment of peripheral arterial diseases; however, a fracture rate of up to 50% has been reported in some cases [5]. These fractures are likely due to in vivo cyclic deformations of vascular motion. Stent failure may cause loss of radial support to the arteries, focal restenosis, thrombus formation, vessel perforation or dissection by the protruded stent struts [6-9].

In this paper, a simple yet intriguing stent design concept that significantly enhances the fatigue life, first proposed by Duerig et al. in their U.S. patent and further explored by Hsiao et al., was

IOP Conf. Series: Materials Science and Engineering **644** (2019) 012015 doi:10.1088/1757-899X/644/1/012015

investigated [10-12]. Finite element analysis (FEA) models, widely used in the medical society today [13-16], were developed to evaluate the mechanical integrity and fatigue safety factor of this novel stent under various loading conditions. A rotating bending fatigue tester was built and stent fatigue tests were conducted to validate this design concept. The findings of this paper provide an excellent guide to the optimization of future stent design for the enhancement of stent fatigue performance.

2. Methods

2.1. Tapered-strut stent design concept

A stent is a combination of a series of nested rings interconnected with bridging connectors. Figure 1(a) defines a vascular stent and its three important elements (crown, strut, and connector). Among them, the most critical element is the crown which controls key clinical attributes of a stent such as acute fracture, crimp profile, and long-term fatigue life. The strut thickness is believed to play a critical role in restenosis, whereas the strut width and length dictate the stent radial strength and scaffolding, respectively [17, 18]. During the stent crimping and deployment procedures, the maximum stresses usually occur at the most highly curved crown. In contrast, there are little to no stresses on the straight portion of the strut, which connects two opposing crowns. Therefore, modifications on the stress-free strut has great potential to achieve our goal, with minimal negative impact on other stent clinical attributes.

This design concept is to shift the highly-concentrated stresses away from the crown and re-distribute them along the stress-free strut by tapering the strut width. This design tweak of narrowing the strut width at the midpoint of the struts (Figure 1(b)) could reduce the burden on the crown and enhance the stent fatigue life.

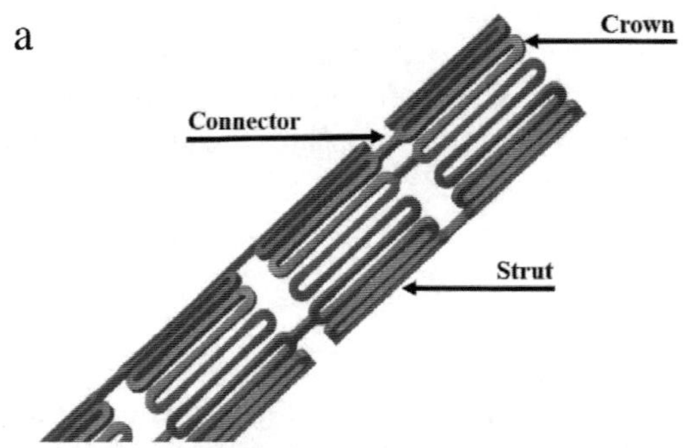

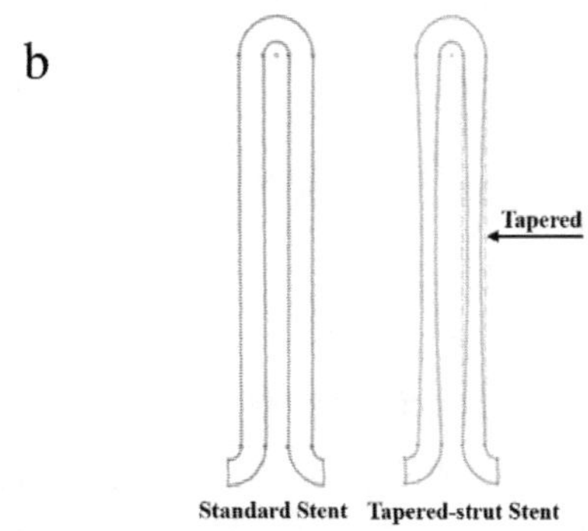

Figure 1. (a) Definitions of three important stent elements, (b) Tapered-strut design concept of narrowing the strut width at its midpoint (2-D sketch).

2.2. Finite element model

Finite element analysis has emerged as an important tool for the optimization of stent designs. It is able to give insights into various aspects of the stent behavior, which may consequently improve the clinical outcome [19-22]. The making of a self-expanding nitinol stent requires an alternating series of expansion and heat treatment procedures as it is formed into its final shape and dimension. The repeated heat treatment in each incremental expansion relieves stresses within a stent, thus allowing it to be expanded to the next size without fractures. After implantation, the stent is then subjected to various modes of loading and may eventually fracture during its functional life.

In this paper, finite element models were developed to evaluate the mechanical integrity and fatigue resistance of a stent subjected to various loading conditions consistent with the current clinical practice. These loading conditions include multiple stent expansions and their corresponding heat treatments during manufacturing, crimping of the stent inside a catheter and its final release into an artery, and simulation of fatigue resistance under systolic/diastolic blood pressure cycles. Finite element analysis was performed using the ABAQUS/Standard finite element solver (Dassault Systems Simulia Corp., Providence, RI, USA) with the user-defined super-elastic material subroutine UMAT [23]. The nitinol material behavior for the ABAQUS UMAT subroutine inputs were obtained from Pham et al. [24]. The material ultimate strain and fatigue endurance limits used in the fatigue life analysis were 10% and 1%, respectively [25]. These procedures were simulated by finite element analysis with the following steps:

Step 1: Expand the stent to 4.0 mm ID and heat-treat.
Step 2: Expand the stent to 6.0 mm ID and heat-treat.
Step 3: Crimp the stent inside the 2.0 mm ID catheter delivery system.
Step 4: Release the stent into the 5.0 mm ID artery.
Step 5: Simulate stent pulsatile fatigue resistance under systolic/diastolic pressure cycles by applying ±3 % stent diameter oscillation (4.85 ~ 5.15 mm).

2.3. Goodman fatigue life analysis

Goodman life analysis has been widely used for assessing the device fatigue resistance and provides an indication of device long-term fatigue resistance. It is also recommended by the U.S. Food and Drug Administration (FDA) that Goodman life analysis to be conducted to determine the stent fatigue safety factor subjected to physiologic loading up to 4× 〖10〗 ^8 fatigue cycles [5].

In Goodman fatigue life analysis, the mean stress/strain and the stress/strain amplitude are typically used to assess the fatigue safety factor of a material or device. Modified strain-based Goodman life analysis has been widely used for the nitinol stent due to the unique material characteristics [5, 25, 26]. It states that fatigue failure will occur if the strain state in the device satisfies the following relation.

$$\left(\frac{\varepsilon_a}{\varepsilon_e}\right) + \left(\frac{\varepsilon_m}{\varepsilon_u}\right) \geq 1 \qquad (1)$$

where ε_a is the strain amplitude applied to the device, ε_e is the modified material endurance limit for non-zero mean strain, ε_m is the mean strain applied to the device, and ε_u is the material ultimate strain. The Goodman diagram is a plot of the normalized strain amplitude $\varepsilon_a/\varepsilon_e$ (on the y-axis) vs. the normalized mean strain $\varepsilon_m/\varepsilon_u$ (on the x-axis). The equation $\varepsilon_a/\varepsilon_e + \varepsilon_m/\varepsilon_u = 1$ represents the failure line on the Goodman diagram. The fatigue safety factor (FSF) is defined as the ratio of the strain amplitude against the modified endurance limit. An FSF of less than 1.0 indicates the stent fatigue failure:

$$\text{FSF} = \frac{\varepsilon_e}{\varepsilon_a} \qquad (2)$$

2.4. Stent laser cutting and heat treatment.

Manufacturing of a self-expanding nitinol stent requires laser-cutting of the designed pattern onto a nitinol hypotube, followed by an alternating series of expansions and heat treatments as it is gradually shaped into our desired dimension. A laser module consisting of a Rofin 100 W pulsed fiber laser, an Aerotech linear X–Y motor stage, and a Z - direction server motor were integrated, with Argon as the assisted gas. A 2-D stent drawing was first sketched on the X-Y plane and then coded into the 3-D cylindrical coordinate system. The design pattern was cut onto a seamless 2 mm ID nitinol hypotube (Minitubes, Grenoble, France) based on the coded geometry by the laser module.

After laser cutting, multiple heat treatments were applied to shape the stent to the final dimension and relieve its residual stresses. Steel rods with diameters of 4.0 mm and 6.0 mm were used to gradually expand the stent at each heat-treatment stage. The high temperatures associated with laser cutting and heat treatments may create spatter, oxide layers, and other debris that must be removed with further surface finishing. Surface finishing was accomplished in two major steps, sand-blasting and electro-polishing, to form a titanium-rich oxide layer (mainly TiO2) on the surface of the stent, which not only increases its corrosion resistance but also improves its surface conditions.

IOP Conf. Series: Materials Science and Engineering **644** (2019) 012015 doi:10.1088/1757-899X/644/1/012015

2.5. Rotating bending fatigue testing

The combination of numerical simulations and bench testing usually leads to a better and faster evaluation process for new medical devices. In order to prove our design concept, a rotating bending fatigue tester, similar to the setup for small diameter solid round wires, was built for stent-deployed hollow round tubes instead. This machine design is based on the rotating beam principle by looping a specimen of pre-determined length through an arc. The bending stresses are determined from the geometry of the loop thusly formed. The specimen functions as a simple beam symmetrically loaded at two opposing points. When rotated a half revolution, the stresses in the specimen originally below the neutral axis are reversed from tension to compression and vice versa. Upon completing one revolution, the stresses are again reversed so that the specimen passes through a complete cycle of flexural stresses (tension and compression). This relatively harsh environment makes it possible to test a large number of specimens quickly to provide a statistically significant distribution of fatigue life at a given stress level. A constant temperature can be maintained by immersing the specimen in a constant temperature fluid bath.

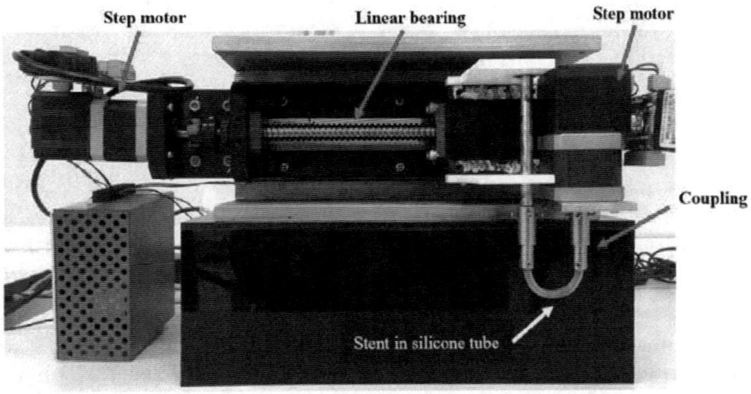

Figure 2. Rotating bending fatigue tester for small diameter hollow round tubes with stent deployed inside (front view).

The rotating bending fatigue tester consists of three parts: (1) a linear bearing module was used to control the displacement and thus the arc formation with high movement precision and load tolerance, (2) three-phase stepper motors were used to provide the tube rotation which generates cyclic stresses at the working speed of 1,000 rpm, with a motor accuracy of 10-4 degrees, (3) RS232 (UTN405) was used for the development of the control module, with Microsoft Visual Studio 2015 as the platform (Figure 2). Since the stent is a hollow cylindrical structure that cannot be mounted directly on the tester as the wire does, we chose a 3 mm ID transparent silicone tube as the medium with a stent deployed inside for rotations. The span of the two rotating couplings was 3 cm which formed the arc radius of 1.5 cm. Upon completing one revolution, tested stents experienced through a complete cycle of combined tension and compression.

3. Results and discussion

A tapered-strut design concept to enhance the stent fatigue life was investigated. In order to quantize such enhancement, the strut width at the midpoint was varied from 100% to 60% of its original dimension in 10% increments. FEA simulation was first performed on a standard nitinol stent with a constant strut width to establish the baseline information for key attributes: maximum strain at expansion and crimping, maximum strain when released inside an artery, and fatigue safety factor subjected to systolic/diastolic blood pressures. During expansion, the maximum strain occurred at the inner surface of the crowns, while there were little to no strains on the struts. Among all five

IOP Conf. Series: Materials Science and Engineering **644** (2019) 012015 doi:10.1088/1757-899X/644/1/012015

simulation steps, the crimping step appeared to be the most critical stage of all, as the highest strains were recorded.

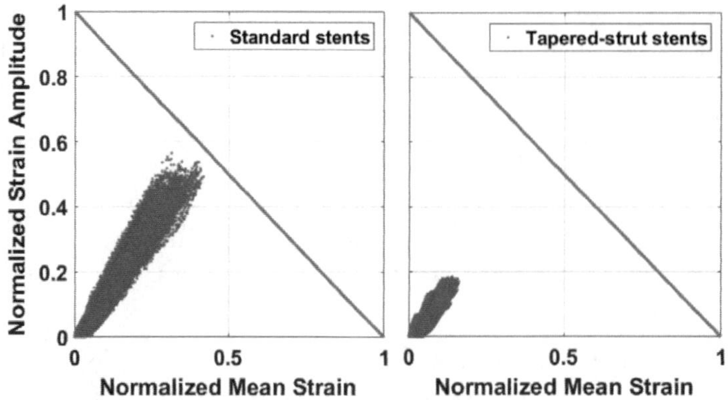

Figure 3. Goodman fatigue life analysis of standard nitinol stent and 40% tapered-strut counterpart stent.

For the tapered-strut counterpart stent, its fatigue safety factor jumped to 4.2 times that of the standard stent (from 1.11 to 4.66), as shown in Figure 3. This is amazing considering that such a small design tweak could result in multiple times of increase in stent fatigue resistance. This interesting phenomenon could be attributed to the stress/strain re-distribution after the modification of strut width at the midpoint of the strut, which effectively shifts the stresses/strains away from the crown and towards the strut. In addition, since the stent strut was loaded predominantly in bending, the narrowed strut became easier to bend, thereby taking more energy than usual during deformation. In Figure 4, the none-blue color of the contour plot extended deeper into the mid strut region along the edges for the tapered-strut stent. Its stress distribution was spread out more uniformly than that of the standard case, with lower intensity level near the crown region. From the energy standpoint, this is a more efficient way to manage energy by subjecting a higher volume of the stent structure to the same loading.

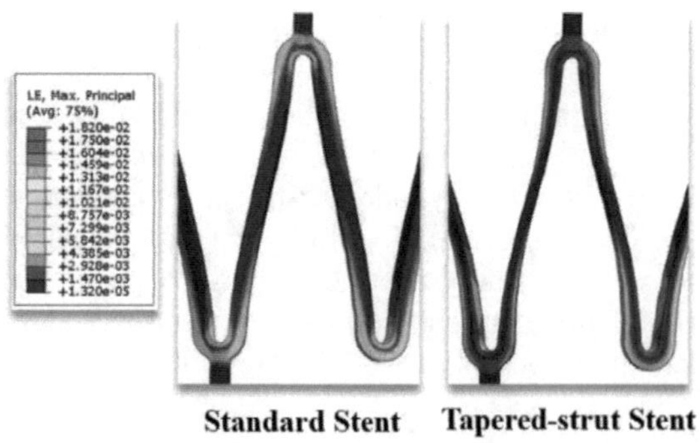

Figure 4. Contour plot comparison of strain distribution at 5.0 mm expansion between standard stent (left) and tapered-strut stent with 40% strut width reduction at the midpoint (right).

Figure 5 shows the simulation results of the fatigue safety factor as a function of the narrowed strut width at the midpoint in 10% increments. The fatigue safety factor increased sharply as the strut tapered degree increased, especially between 30% and 40%. A 40% taper in the strut width significantly boosted the fatigue safety factor by 4.2 times.

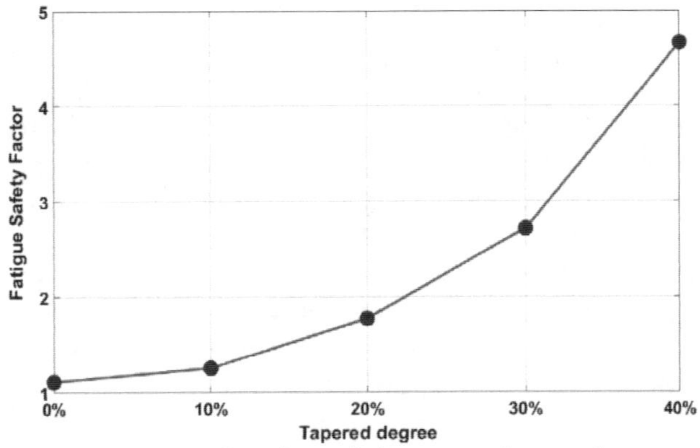

Figure 5. Fatigue safety factor as a function of the tapered-strut degree at the midpoint in 10% increments.

Table 1. Experimental data of rotating bending fatigue tests.

	Cycles	
Tapered Degree	0%	40%
Sample #1	6464	21022
Sample #2	8243	66938
Sample #3	5374	40657
Mean	6693	42872
Standard Deviation	1182	18810

A rotating bending fatigue tester was built and stent fatigue tests were conducted to prove our design concept. Fatigue tests were manually stopped at approximately every 1,000 cycles for visual inspection. The final cycle number was recorded when the first stent fracture was observed. Three stents were tested in each group (standard stent vs. 40% tapered-strut stent), as shown in Table 1. The average cycle number at break of the standard stent was 6693 cycles, while that of the 40% tapered-strut stent jumped to 42872 circles. This is an impressive 6~7 times increase when compared to the standard stent (Figure 6). This significant increase in the fatigue cycle number at break confirmed the rising trend predicted by earlier FEA simulations.

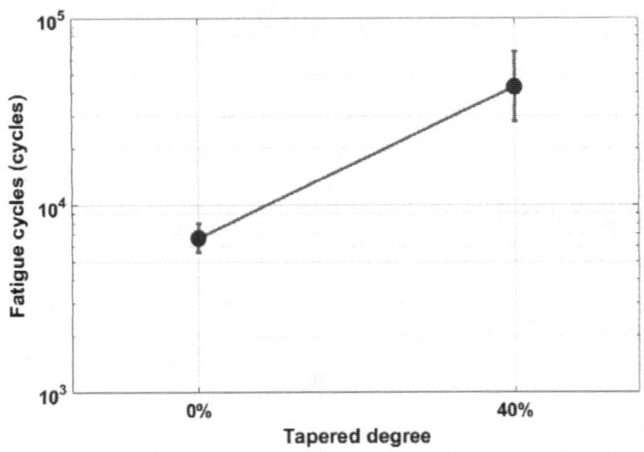

Figure 6. Fatigue cycle comparison between standard nitinol stent and 40% tapered-strut counterpart stent.

IOP Conf. Series: Materials Science and Engineering **644** (2019) 012015 doi:10.1088/1757-899X/644/1/012015

4. Conclusion

A simple yet intriguing design concept for the enhancement of stent fatigue life was investigated. The concept of this tapered-strut stent design was to shift the highly concentrated stresses/strains away from the crown, thus allowing the stress-free strut to carry more loads by narrowing its strut width at the midpoint. Simulation results show that when the strut width was reduced to 60% of its original dimension (or 40% taper) at the midpoint of the strut, the fatigue safety factor could be increased by up to 4.2 times that of its standard counterpart. A rotating bending fatigue tester was built and stent fatigue tests were conducted for proof of concept. Experimental results show that this stent design concept successfully enhanced the fatigue life as designed by 6~7 times that of a standard stent, which agreed well with the rising trend predicted by FEA simulations. This design has great potential for further fatigue enhancement once fully optimized and could help future stent design to achieve vast fatigue enhancement.

Acknowledgments

This research was supported by the Ministry of Science and Technology in Taiwan through Grants MOST-105-2221-E-002-092-MY3 and MOST-106-2221-E-002-130-MY2. The authors are grateful for the support and help.

5. References

[1] C.T. Dotter, R. Buschmann, M. K. McKinney, J. Rösch, Radiology 147, 259-260 (1983)
[2] T. Duerig, A. Pelton, D. Stöckel, Mater. Sci. Eng. A 273-275, 149-160 (1999)
[3] S.W.K. Cheng, A.C.W. Ting, J. Wong, Cardiovasc. Surg. 9, 133-140 (2001)
[4] L. Cho, M. Roffi, D. Mukherjee, D.L. Bhatt, C. Bajzer, J. S. Yadav, J. Invasive Cardiol. 15, 198-200 (2003)
[5] A.R. Pelton, V. Schroeder, M.R. Mitchell, X.Y. Gong, M. Barney, S. W. Robertson, J Mech Behav Biomed Mater 1, 153-164 (2008)
[6] J. Li, Q. Luo, Z. Xie, Y. Li, Y. Zeng, Heart Vessels 25, 333-337 (2010)
[7] D. Scheinert, S. Scheinert, J. Sax, C. Piorkowski, S. Bräunlich, M. Ulrich, G. Biamino, A. Schmidt, J. Am. Coll. Cardiol. 45, 312-315 (2005)
[8] X. Gong, D. Chwirut, M. Mitchell, B. Choules, ASTM Int 6, 1-10 (2010)
[9] J.M. Gibbs, C.S. Peña, J.F. Benenati, Tech. Vasc. Interv. Radiol 13, 37-42 (2010)
[10] T.W. Duerig, J. Burpee, M. Mathis, US Patent, US6190406 (2001)
[11] H.M. Hsiao, M.T. Yin, Biomed Microdevices 16, 133-141 (2014)
[12] H.M. Hsiao, L.W. Wu, M.-T. Yin, C.H. Lin, H. Chen, Comput Mater Sci 86, 57-63 (2014)
[13] F. Migliavacca, L. Petrini, V. Montanari, I. Quagliana, F. Auricchio, G. Dubini, Med. Eng. Phys. 27, 13-18 (2005)
[14] M. De Beule, R. Van Impe, B. Verhegghe, P. Segers, P. Verdonck, Technol. Health Care 14, 233-241 (2006)
[15] K. Takashima, T. Kitou, K. Mori, K. Ikeuchi, Med. Eng. Phys. 29, 326-335 (2007)
[16] V. Dehlaghi, M.T. Shadpoor, S. Najarian, J Mater Process Technol 197, 174-181 (2008)
[17] H.M. Hsiao, Y.H. Chiu, K.H. Lee, C.H. Lin, Comput Aided Des 44, 757-765 (2012)
[18] H.M. Hsiao, K.H. Lee, Y.-C. Liao, Y.C. Cheng, Micro Nano Lett 7, 430-433 (2012)
[19] J. McGarry, B. O'donnell, P. McHugh, J. McGarry, Comput Mater Sci 31, 421-438 (2004)
[20] H.M. Hsiao, Y.H. Chiu, T.Y. Wu, J.-K. Shen, T.Y. Lee, Med. Eng. Phys. 35, 884-897 (2013)
[21] M. Azaouzi, A. Makradi, J. Petit, S. Belouettar, O. Polit, Comput Mater Sci 79, 326-335 (2013)
[22] M. Azaouzi, A. Makradi, S. Belouettar, Comput Mater Sci 72, 54-61 (2013)
[23] F. Auricchio, Int. J. Plast. 17, 971-990 (2001)
[24] T. M. Pham, M. DeHerrera, W. Sun, Mechanics of Biological Systems and Materials, 2, 1-10 (2011)
[25] Z. Lin, H.M. Hsiao, D. Mackiewicz, B. Anukhin, K. Pike, Adv. Eng. Mater 11B, 189-193 (2009)
[26] P. Adler, J. Allen, J. Lessar, R. Francis, J.ASTM Int. 4, 1-16 (2007)

IOP Conf. Series: Materials Science and Engineering **644** (2019) 012016 doi:10.1088/1757-899X/644/1/012016

Rapid Tooling by Using a Magnetic Powder System with Optimized Parameters

Chong-Ching Chang, Hsieh Chung Lo, and Yi-Ting Li
Graduate Institute of Mechatronic System Engineering, National University of Tainan, Taiwan
Email: changjeff0718@gmail.com

Abstract. For shortening customized products of development cycle and reduce costs, rapid prototyping technology is one of the ways for the industry. In the paper, it will be combined of 3D printing forming technology and the magnetic field lines control of molding technology, to develop a set of parameters with the best control of the magnetic powder mold system. First, after molding to 3D printing technology printing desired model, and then generate the cavity via magnetic molding technology, in order to shorten the mold manufacturing time. In which, in the powder molding equipment related parameter setting, use Taguchi method to find the relevant parameters to produce the desired combination of hardness and precision, such as voltage, density, magnetization time. The results are shown the best parameters combination of stiffness and precision, in precision design the best combination of parameters S / N ratio improved by 56%, and the surface hardness to design the best combination of parameters S / N ratio is improved by 21%. Reuse fuzzy inference system inference mechanism to simulate various voltage, density and magnetizing time formed by a surface hardness, inferences parameters of various combinations of different conditions, the surface hardness of the formed, inference values and experimental values results are compared error is about 2.45%.

1. Introduction

The molding tool industry is rather important in our country industry, according to current output value, the world placings of our country molding tool industry are to come out in front, and the molding tool industry takes up above 25% of the manufacturing industry. In the manufacture process of molding tool, because the size of variety of models are different and shape also has the dissimilarity, they must manufacture a molding tool of a variety of different sizes, respond various different models. For example shoe insoles, the different shoe insoles sizes need the molding tool of matching with the dissimilarity, if the same style shoe insoles have many different sizes, needing to manufacture various molding tools, the cost is opposite to raise, and manufacture time more long, the procedure is more complicated, so need the manufacture methods of changing the tradition. Hsu [1] put forth to shorten product development time and decrease cost, promote the diversification product to the market quickly, combine a 3D printer to print out to model technique and magnetic force to control the magnetic powder system, development has the environmental protection to also repeat the rapid molding tool system of using, solve a traditional molding tool to the environment pollution and the cost problem. However, because of market at size accuracy need more and more fine, the methods of making use of the Taguchi methods can find out the quality characteristic of influence of each factor upon, reducing the time and manpower cost, finding out the best parameter to design to product and to promote the quality characteristic [2-3].

Chang [6] mentioned rapid prototyping (RP) is a technology directly from the part via the CAD design, the use of stacked laminated manufacturing technology. Lin [7] refers to rapid prototyping (RP) systems in the major advanced countries have gradually developed a wide variety of rapid prototyping systems. Generally it can be classified with a liquid processes, powder process, plastic extrusion process, paper lamination process, surface exposure process, 3D printing process. Guo [8] use of rapid prototyping (RP) combining computer-aided manufacturing of the 3D stereoscopic graphic design generated by the production of molds and prototype the size of the error can be easily mastered. Yan [9] using the way of materials Precision accumulation, namely the point piled into the surface, 3D entity formed by the surface accumulation. KJ Suganth Kumar et al. [10] referred to the magnetic molding is an environmentally friendly and economical way to take advantage of keeping the ball in the field to make a mold for casting metal forming mold. JL Lin et al.[11] use Taguchi method combines fuzzy inference theory, and put forward a set of algorithm of multiple quality problems.

Vijayalaksmi [12] proposed water supply system of the region desired by the user, a fuzzy inference system prediction. In the past this team [13] using the magnetic lines of force to control the magnetic stereotypes method combined with rapid tooling technology to produce a new type of shaping powder mold system, this team [14] also use the Taguchi method with the L18 orthogonal table magnetic control fast sole control of the magnetic powder of depression mechanical process parameter optimization.

This research existing 3D printer to model the rapid molding tool technique is foundation, developing a set of control the rapid molding tool system of magnetic powder by the 3D printing forming and the magnetic force. First, the 3D printer prints a prototype, making use of the electromagnet produce electricity magnetic field, the line of magnetic force controls the magnetic powder cladding molding tool of hard mode (Taiwan patent number:I319705), removing prototype molding tool. Finally, turning off electricity of electromagnet, stopping a function of electromagnetism field, will make the magnetic powder return to not-hard mode of plasticity molding tool [4-5]. This method can reduce development cost and time of molding tool of different sizes, promote the molding tool manufacture efficiency. In the system design, find out each parameter that the surface degree of hardness to the system mold and size accuracy influences and design the best surface degree of hardness and parameter combination of accuracy. The related data the parameter under a variety of conditions use fuzzy inference establish a database, reducing caused by human factors measurement error experiment in the future. Improve the magnetic powder to sticky to the problems of finished products, and through dissimilarity of flow a design to inquire into the results of finished product then raising the finished product rate of finished product.

2. Design and experimental methods

This paper will develop a rapid tooling system device of magnetic controlling equipment development process shown in Fig.1. There are included two parts: 1. To use Taguchi method to design the best combination of accuracy and precision parameter and 2. Fuzzy inference theory is to establish different parameter combinations of hardness repository. In this paper, developed the magnetic powder rapid tooling system to control its operation principle for the use of 3D molding equipment for printing a list of the required prototype device, the purpose of the magnetic charge and magnetic devices combine to achieve magnetic control electromagnet magnetic molding of Fig.2. Users can via adjust parameters of the charge magnetic device and the mold of quickly and accurately obtain production.

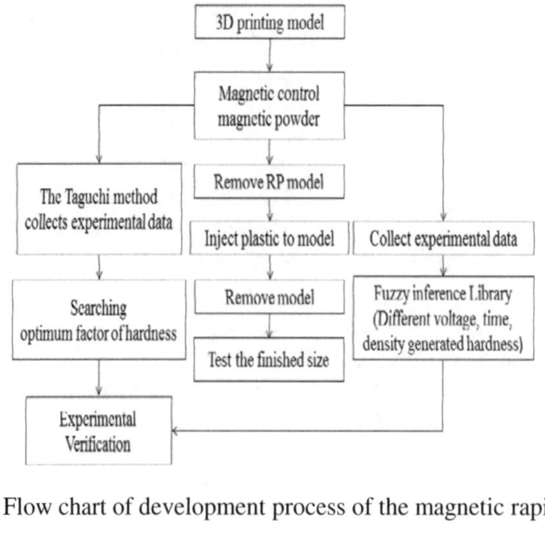

Figure 1. Flow chart of development process of the magnetic rapid tooling.

IOP Conf. Series: Materials Science and Engineering **644** (2019) 012016 doi:10.1088/1757-899X/644/1/012016

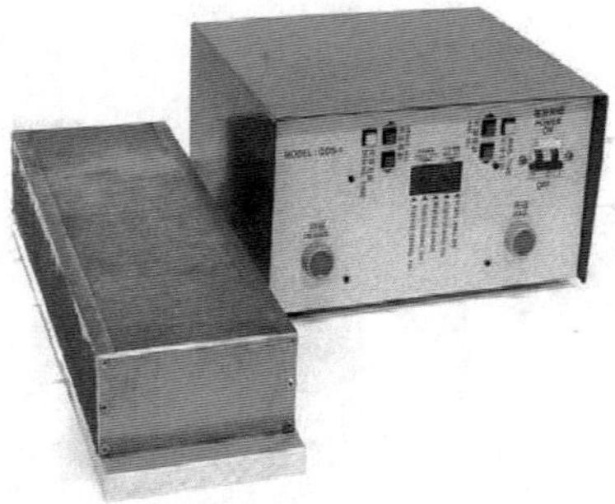

Figure 2. Electromagnet and magnetization control system equipment.

Taguchi quality method to design the best combination of parameters desired hardness and precision, such as voltage, density, magnetization time, and then use the fuzzy inference system inference mechanism to simulate the surfaces of various voltage, density, magnetization time formed the die face hardness, achieving simple, fast and low-cost production of magnetic rapid tooling system. Taguchi quality method used in this paper is made in order to find the best combination of parameters, Taguchi quality method is a systematic quality and highly efficient way to import numerical experiments. Surface hardness and precision molding is appropriate, it will directly affect the magnetic changes produced by the different parameters of the finished product will not be good deformed or damaged, the flow chart shown in Fig.3.

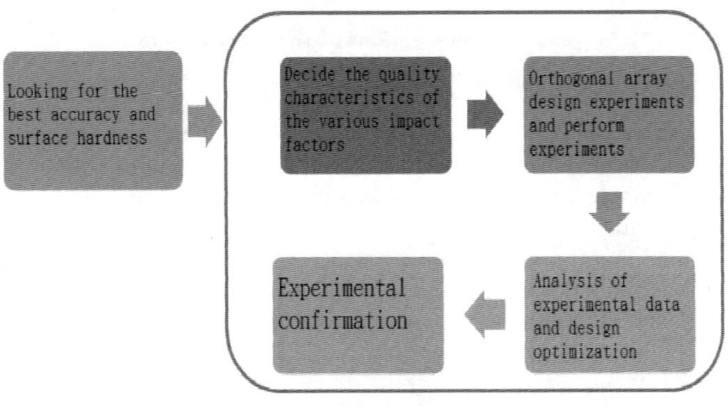

Figure 3. Taguchi experimental method flow chart.

Taguchi experimental method selected control factors used in the paper. They are: A. Voltage strength (80V, 120V, 160V and 200V), B. The particle size (40 ~ 70 um, 0.5mm, 1mm and 3mm), C. Density (1.8 kg / m^3, 1.9 kg / m^3, 2.0 kg / m^3 and 2.1 kg / m^3), D. Magnetizing time (0.3s, 0.5s, 0.7 s and 1.0s), E. Box Type (quasi-metal box and plastic box (ABS material)), where the parameter A ~ D are set to four levels, the parameter E there are two standards. In the Taguchi method orthogonal array experiments we choose the orthogonal table shown in Table 1. The surface hardness test be made by 16 sets in Taguchi method, measuring the surface hardness of the five points shown in Fig.4.

Table 1. Orthogonal table 16 sets in Taguchi method.

Exp.	A	B	C	D	E
1	1	1	1	1	1
2	1	2	2	2	1
3	1	3	3	3	2
4	1	4	4	4	2
5	2	1	2	3	2
6	2	2	1	4	2
7	2	3	4	1	1
8	2	4	3	2	1
9	3	1	3	4	1
10	3	2	4	3	1
11	3	3	1	2	2
12	3	4	2	1	2
13	4	1	4	2	2
14	4	2	3	1	2
15	4	3	2	4	1
16	4	4	1	3	1

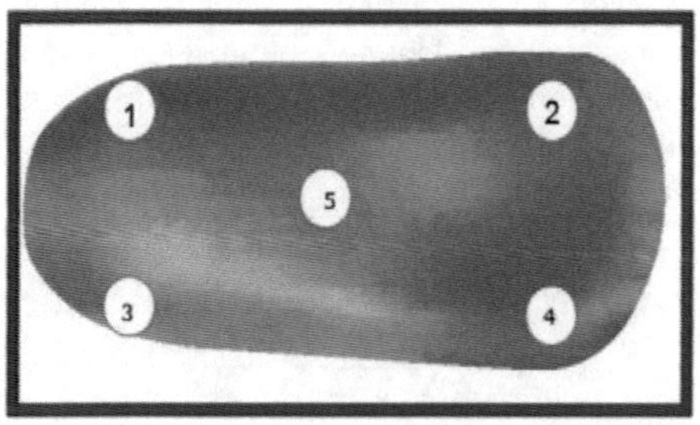

Figure 4. Magnetic mold surface measurement partition map.

Fuzzy systems under different conditions of voltage, density and magnetizing time, the use of guidelines inference mechanism of fuzzy inference system inference hardness output, and then establish a database, learning under different parameter combinations in the surface hardness. In this experiment, there are the voltage, density and magnetizing time as the three quality characteristics of the input variables, with surface hardness as output variables, shown in Fig.5. In the fuzzy interface, the present paper using the triangular function attributable to segmentation of fuzzy inputs and a total of four levels is defined, as shown in Fig.6. In the output section is nine output levels, as shown in Fig.6.

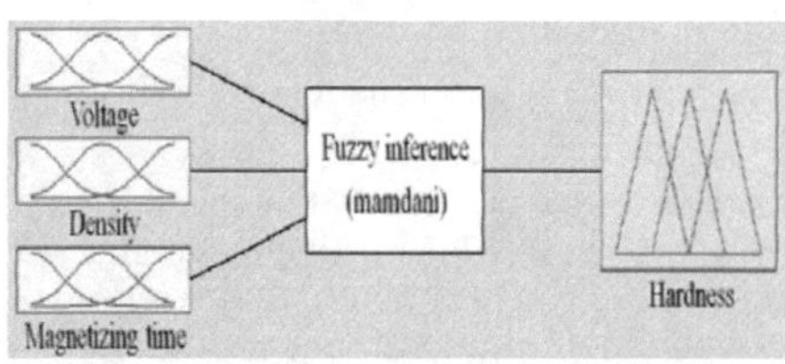

Figure 5. Fuzzy inference system interface.

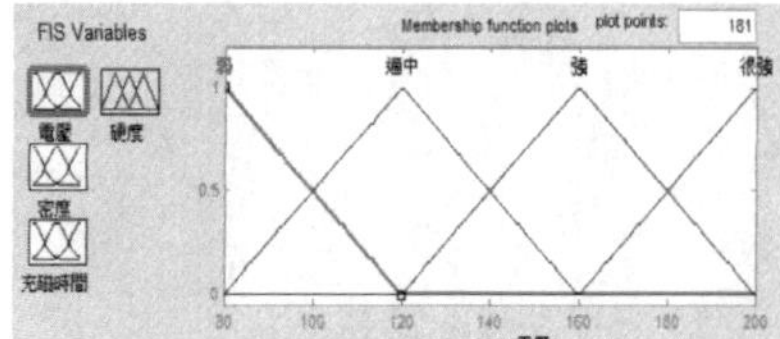

(a)

IOP Conf. Series: Materials Science and Engineering **644** (2019) 012016 doi:10.1088/1757-899X/644/1/012016

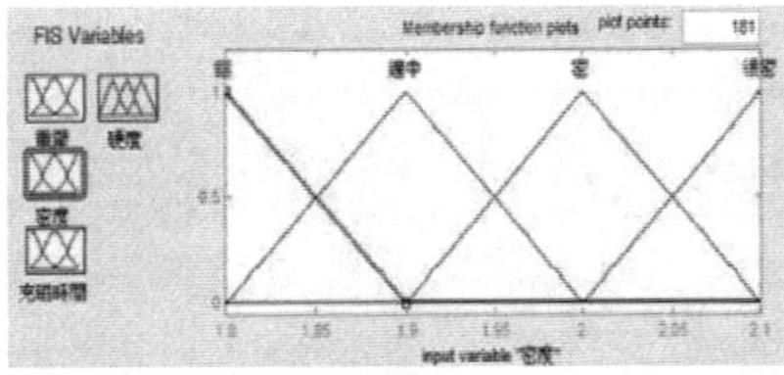

(b)

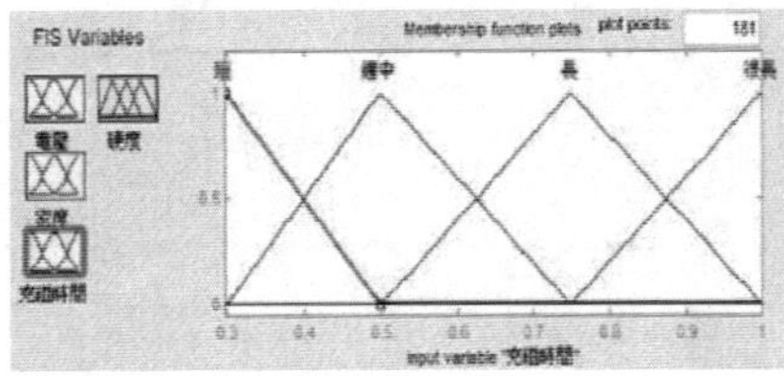

(c)

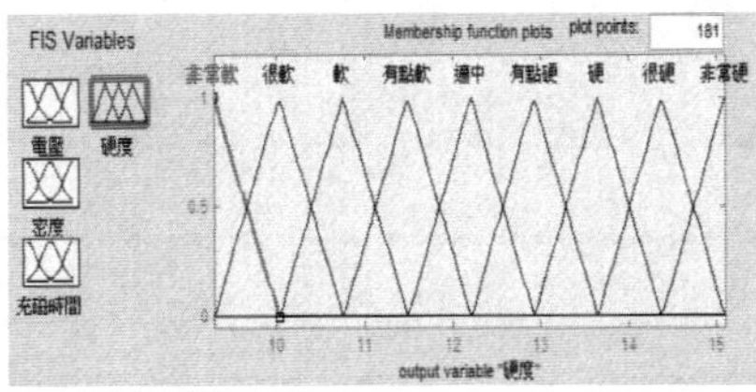

(d)

Figure 6. Fuzzy inference in the home function of the input(a) voltage, (b) density, (c) magnetizing time and (d) output hardness.

3. Results and Discussion

3.1. S / N ratio and quality characteristics

In the process of Taguchi Methods to determine the objective function and quality is the key characteristic of an analysis. The paper uses of large characteristics of the surface hardness and the experiment of Taguchi method S / N ratio analysis shown in Fig.7. Quality characteristics are continuous and the bigger is the better. The quality characteristics of the ideal value is treated as infinity, the experiments Taguchi method quality characteristics analysis shown in Fig.8.

In addition to, the paper uses Taguchi method in precision measurement (size error) using smaller characteristics, the Taguchi method in the experimental of dimensional accuracy S / N ratio analysis shown in Fig.9, when the quality characteristic is continuous and the smaller is the better. Quality characteristics of the ideal value are treated as zero, the experimental method of dimensional accuracy of Taguchi quality characteristics analysis diagram shown in Fig.10.

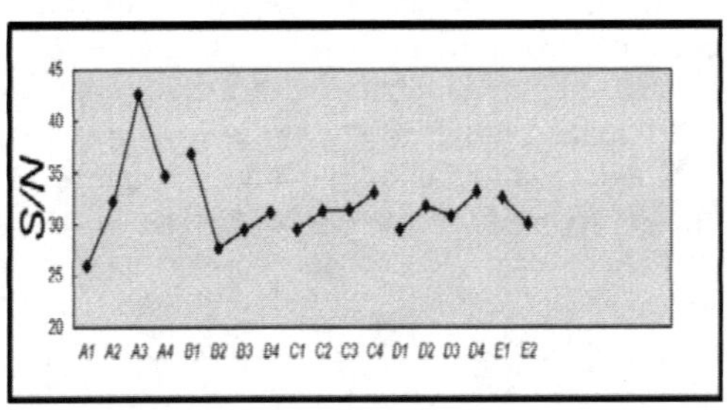

Figure 7. Surface hardness of S / N ratio analysis.

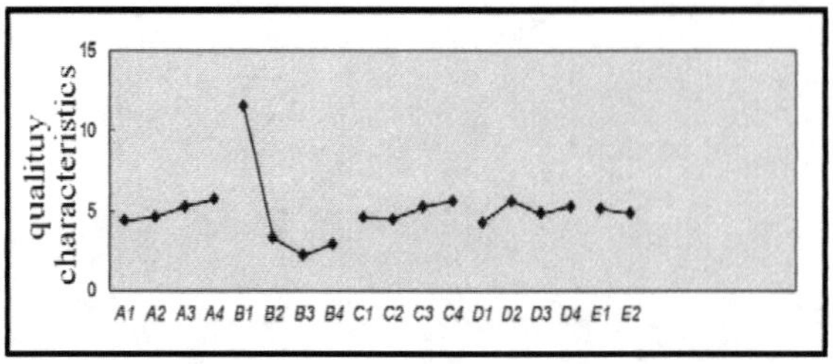

Figure 8. Surface hardness quality of characteristics analysis.

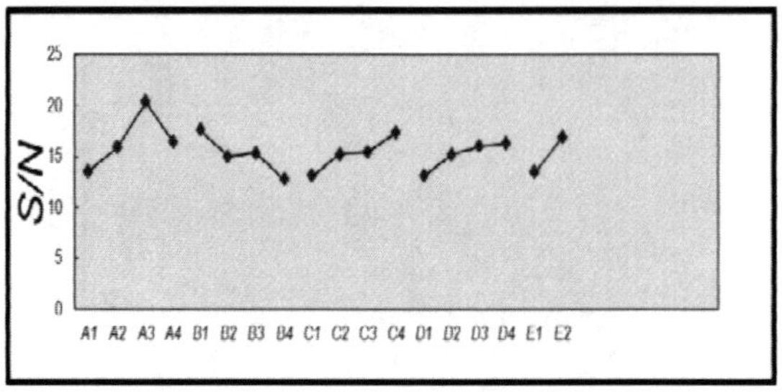

Figure 9. The dimensional accuracy of S / N ratio analysis.

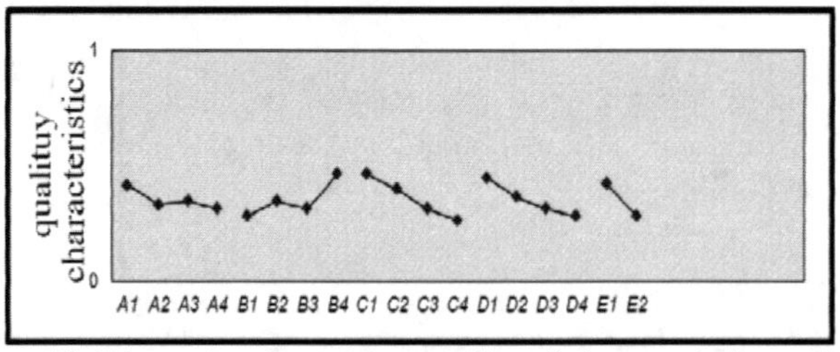

Figure 10. Dimensional precision quality of characteristics analysis.

The best combination of parameters design and verification experiment via Taguchi method for each factor analysis, surface hardness will be in the original design value A4, B2, C2, D2, E1 (voltage of 200V, particle size of 0.5mm, density of 1.9 kg / m3, magnetizing time is 0.5s and metal box) to modify the parameter combination of A3, B1, C4, D4, E1 (voltage of 160V, a particle size of 40 ~ 70 um, a density of 2.1 kg / m3, and magnetizing time is metal box) for the best combination of surface hardness level parameters.

In terms of accuracy, parameter combinations modify A3, B1, C4, D4, E2 (voltage 160V, particle size of 40 ~ 70 um, density of 2.1 kg / m3, the magnetizing time is 1s and plastic boxes) for the best accuracy parameter level portfolio, through the results of experimental verification of the optimum combination of parameters, as shown in Table 2. Table 2 shows the optimum parameters after improving design accuracy by S / N ratio of 6dB (about upping 56 per cent),and the hardness on the surface also improves 7.9dB(about upping 21%). In the results, can know the best parameters surface hardness and precision of different types of combination in that the box, it will do two boxes hardness test. First, the box is divided into A, B and C three parts and A, C in the case of the two sides, B is a point in the center of the box, At a voltage of 110V experimental conditions, the results shown in Fig.11 and Fig.12, after the figure can understand through magnetic, metal box will form with the direction of magnetic lines of magnetic force inconsistent, easy to make the model deformed shape, and in the use of plastic box can make uniform magnetic field lines down, improve the direction of the magnetic field lines do not mean to get a higher modulus surface accuracy.

Table 2. Experimental validation results.

	The original design portfolio	Best design portfolio (verification)		Improve
Reference standards	$A_4B_2C_2D_2E_1$	Surface hardness	$A_4B_2C_2D_2E_1$	-
		Surface accuracy	$A_4B_2C_2D_2E_2$	
Surface hardness (S/N Proportion)	10.7	16.7		+6dB (Rise 56%)
Surface accuracy (S/N Proportion)	37.4	45.5		+7.9dB (Rise 21%)

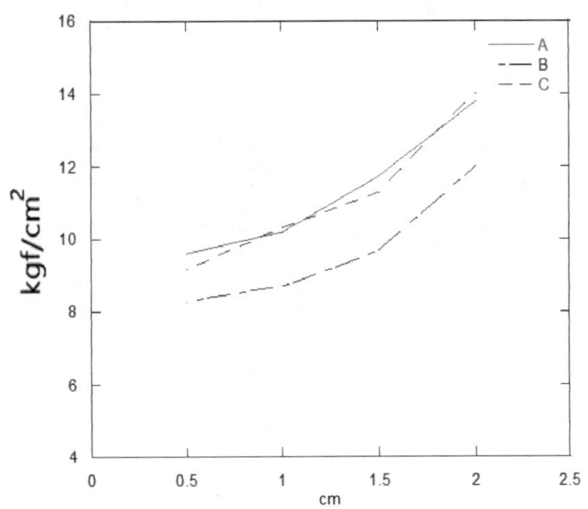

Figure 11. 110V Magnetic hardness experiment chart (metal box).

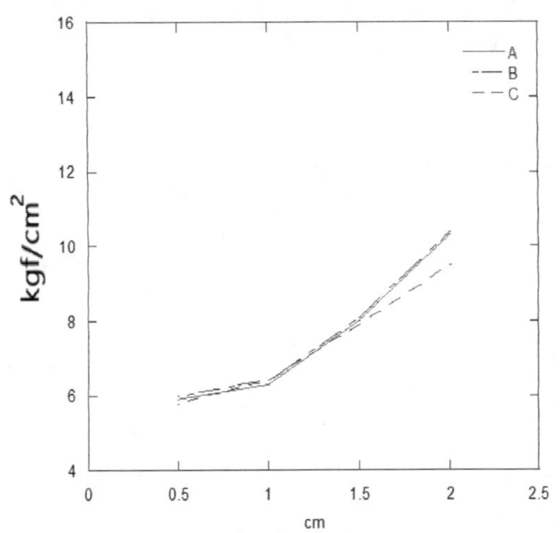

Figure 12. 110V Magnetic hardness experiment chart (Plastic box).

IOP Conf. Series: Materials Science and Engineering **644** (2019) 012016 doi:10.1088/1757-899X/644/1/012016

3.2. Fuzzy inference system

This study considers voltage, density and magnetizing time three parameters, the characteristics same degree of importance, and thus develop a fuzzy rule table, a total of 64 rules. Fuzzy inference result, the present experiment using Matlab software for fuzzy inference, the inference results shown in Fig.13 and Fig.14. The figure shows the relationship between voltages, density and magnetizing time three parameters, when the voltage and density parameters fixed, the surface hardness will increase with the magnetizing time enhanced surface hardness. While fixed parameters between the voltage and magnetizing the surface hardness of the magnetic density increases and enhanced. When the density and magnetizing time parameters fixed, as the voltage rises, the surface hardness increases, because when the electromagnet magnetic intensity reached saturation time, and then increase the voltage, the magnetic effect of increasing strength will become smaller and smaller, it was found from the experiment, electromagnet of this study, conditions will arrive at 160V magnetic saturation. 12 group via random sampling experiments verify the results in Table 3, the experiments found that the simulation results and fuzzy inference error is about 2.45 percent, therefore it is proven by fuzzy inference the surface hardness data generated by various parameter combinations of inferred and thereby establish a database.

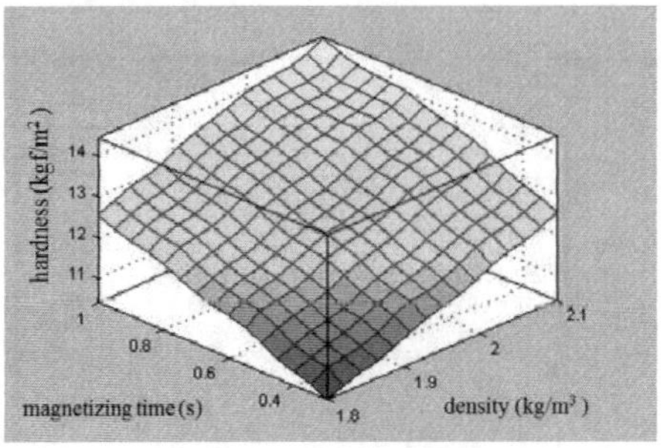

Figure 13. Fuzzy inference - magnetizing time and density graph.

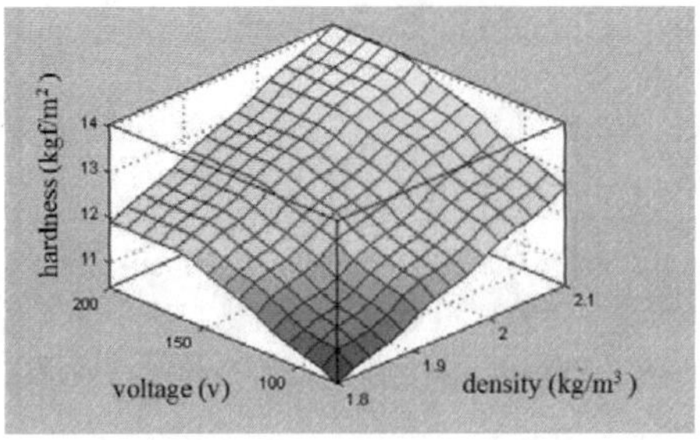

Figure 14. Fuzzy inference - voltage and density graph.

IOP Conf. Series: Materials Science and Engineering **644** (2019) 012016 doi:10.1088/1757-899X/644/1/012016

Table 3. Fuzzy inference experimental verification.

Voltage(v)	Density(kg/m^3)	Time(s)	Measured value	Inference value	Errors%
80	1.8	0.3	9.3	9.52	2.31
80	1.9	0.5	10.5	10.8	2.77
80	2	0.7	12.13	12	1.08
120	1.9	0.3	11	10.8	1.85
120	2.1	0.7	13.86	13.5	1.91
120	2	1	13.33	13.6	1.98
160	2	0.3	12.37	12.2	1.39
160	1.8	0.7	11.7	12	2.5
160	1.9	1	12.8	13.6	5.88
200	2	0.5	12.6	12.9	2.32
200	2.1	0.5	13.8	13.6	1.47
200	1.9	0.7	12.3	12.8	3.9
					Average =2.45%

Taguchi method and fuzzy inference results: In this study, the quality of the Taguchi method to find out the best design combination of surface hardness and dimensional accuracy, however, using a combination of parameters A3, B1, C4, D4, E1 (voltage 160V, particle size of 40 ~ 70 um, density of 2.1 kg / m3, magnetizing time is 1s and metal box) optimum surface hardness can be obtained, if the parameter combination of A3, B1, C4, D4, E2 (voltage 160V, particle size of 40 ~ 70 um, density of 2.1 kg / m3, the magnetizing time is 1s and plastic box) to obtain the best accuracy.

3.3. To improve the finished quality design

The use of the magnetic field lines control powder forming, because the molding surface is part of hard powder state, if perfusion silicone or resin material will generate magnetic sticking problems in the subsequent processing of finished products will increase the number of processing costs and high degree of difficulty, therefore, this study uses two methods to isolate powder and model surface, first using plastic wrap to cover up before molding surface materials needed reperfusion, isolating the magnetic and the material, after reperfusion make the finished molding material deformation, because the plastic wrap is part of an irregular shape, though the material can be effectively isolated from the powder, but could not immediately attached to the magnetic forming surface, it will have a greater size error. The second magnetic isolation mode, this study used a liquid rubber spray forming surface formed thereon is attached to isolated magnetic film as shown in Fig.15, the advantage is that at room temperature and normal humidity, normally hardened form elastomers, no special techniques and expensive equipment you can use. The temperature has little effect on the liquid rubber and can be used for vertical fill the gap seal, in the use of convenience, safety, material does not shrink, compression and tensile excellent elasticity, suitable for most of the material surface. From the experimental results, in addition to the use of spray liquid rubber powder can be effectively cut off from the outside with the material, size and surface finished to minimize the impact.

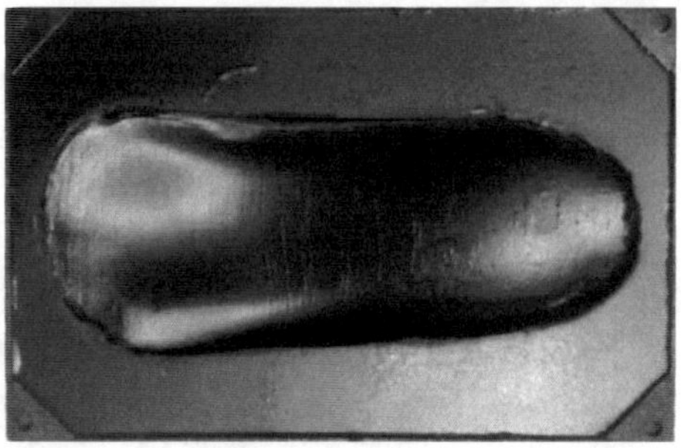

Figure 15. Magnetic isolate liquid rubber

4. Conclusion

Purpose of the paper was to develop a magnetic powder control of rapid tooling systems and this system features are by fast magnetic mold to quickly produce the products needed for rapid tooling. In addition, through rapid prototyping methods to control the magnetic basis of magnetic force, also to avoid the problems of the conventional mold and rapid tooling cost cannot be reused. Magnetic control in magnetic technology developers is to work with Taguchi method and fuzzy inference theory. The required magnetic mold system is quickly and accurately in the quality and design.

Development of this system is expected to accomplish the objectives set by:

1. Optimal parameter design: the use of Taguchi method can design the best parameters combination mold surface precision of A3, B1, C4, D4, E2 (voltage 160V, particle size of 40 ~ 70 um, density of 2.1 kg / m3, magnetizing time as 1s and plastic box) precision design parameter combination S / N ratio improvement 6dB (about 56 per cent rise), and the surface hardness of the optimum parameters combination A3, B1, C4, D4, E1 (voltage 160V, 40 ~ 70um particle size and a density of 2.1 kg / m3, magnetizing time is 1s and metal box) surface hardness design parameter combinations S / N ratio is improved 7.9dB (about up 21%).
2. Fuzzy inference repository established: fuzzy inference theory in combinations of different parameters of the conditions of various, formed by the surface hardness, pushing with the experimental error of about 2.45%.
3. Surface coating isolated magnetic surface coating: the use of liquid rubber surface coating powder sticking to overcome the problem and a smaller effect on the dimensional accuracy of the finished product.
4. Gate design to get a good product: The three kinds of gate design, it is found as a double gate and gate design placed on edge than the other two models available gate design products with a higher rate.

5. References

[1] X.R. Hsu, ME, Ming Chi Univ. of Tech., Master Dissertation(2013)

[2] P.J. Ross, Taguchi Techniques for Quality Engineering, McGraw-Hill, New York(1988)

[3] Taguchi, Introduction to Quality Engineering, Asian Productivity Organization, Tokyo(1990)

[4] C.C. Chang, Taiwan Patent No. M321703(2007)

[5] Wittmoser, International Giesserei-Fachmesse MIT Technical Forum, Dusseldorf(1968)

[6] K.H. Chang, Computer-Aided Engineering Design, 743(2015)

[7] Z. Lin, Mold technology information ， September, 1, (1997)

[8] C.C. Guo, ME, Nat. Taiwan Univ. of Sci. and Tech., Master Dissertation(1997)

[9] Y. Yan, et al., Tsinghua Science & Technology, 14, 1(2009)

[10] K.J. Suganth Kumar, S. Senthilkumaran, S. Rajendra Boopathy, and A. Ramesh, J. of Mat. Processing Tech., 235(2008)

[11] J.L. Lin, K. S. Wang, B. H. Yan, and Y. S. Tarng, J. of Sci. and Tech., 102, 48(2000)

[12] D.P. Vijayalaksmi and K. S. Jinesh Babu, International Conference, ICWRCOE(2015)

[13] Q.W. Cheng, GIMSE, Nat. Univ. of Tainan, Taiwan, Master Dissertation(2010)

[14] X.B. Yang, GIMSE, Nat. Univ. of Tainan, Taiwan, Master Dissertation(2010)

IOP Conf. Series: Materials Science and Engineering **644** (2019) 012017 doi:10.1088/1757-899X/644/1/012017

Development and Application of Image-base Plantar Measurement System

Hsien Chung Lo, Guo-Xiang Pan, Hui-Ling Chang, and Chong-Ching Chang

Graduate Institute of Mechatronic System Engineering, National University of Tainan, Taiwan
Email: changjeff0718@gmail.com

Abstract. The foot pain affects daily life function about 14-4b2% of adults over the age of 45 years old, since feet bear all the weight of the body. The most foot problems come from a great impact on the human body. It affects not only bone structures, but also the organs. Plantar information can be checked through the plantar pressure and arch index. The plantar pressure distribution is related to the walking balance, while the arch index can understand the foot bone structure. The paper aims to develop a system of plantar self-detection which is economical and easy to carry and operate. The system has built up two parts: one is the image capture equipment; another is the data analysis software. The portable scanning platforms can obtain high-resolution image, scanning work area is 216x356mm which the resolution is 1200dpi image. It only needs 10 seconds, using the wireless transmission to the plantar image through the pad for data analysis, calculated the plantar pressure distribution and arch index. The whole system price is less than one-tenth of the market selling pressure pad, providing a general system which is designed for plantar self-detection.

1. Introduction

Researches indicate that foot pain affects 14-42% of the adult population at ages above 45 [1]. Foot pain affects the functions of daily life. Additionally, as the population gradually becomes more aging, foot pain probably does not only produce poor balance, but also increase stumbling risk among the elderly people. It may also be combined with chronic diseases to cause worse harm. For example, 50% of patients with diabetes are hospitalized owing to diabetic feet [2]. Seemingly, foot pain is a problem unallowable to be neglected.

Both feet endure body weight and they act like the foundation of human bodies. It is the most direct connection between the body and the ground. It provides the functions of support and shock absorption. If there is any problem happening to feet, it will not only affect bone structure, but also impede organ functions. Foot problems can be known mainly through foot pressure and the arch index. The pressure distribution of feet let us know the balance of human bodies, while the arch index gives us the information to know the structure of human bodies.

If pressure is concentrated in a certain zone, it will not only affect the balance of human bodies, but also cause other diseases owing to poor blood circulation. Therefore, plantar pressure should be uniform to make good pressure distribution achievable. According to the judgment method published by the Cavanagh research team in 1987, the arch index is determined to be normal between 0.21 and 0.26 [3]. However, researches' pinpoint roughly 10-15% of the global population belongs to the group of a high arch [4] and 6.6% belongs to the group of a low arch [5].

Generally, people do not particularly care about foot problems. They often start paying attention after those problems happen. However, with the advancement of medical care and the rise of individualism, people begin paying attention to health problems [6]. Various medical test devices and customized products start appearing on markets. In 2015, T. Kaewwichit [7] developed a measurement system functioned with the automatic detection of plantar images. The plantar pressure measuring system is developed application software for image-based plantar pressure measurement and the database of foot correction [8].

Foot health care is bound to be a focus. It is extremely required to develop a set of testing software and equipment particularly designed for feet. This paper aims to develop an economical and portable foot self-detection system. This system is divided into two parts: the image acquisition instrument and the software of data analysis. High-resolution plantar images are obtainable through a portable scanning platform. Information such as foot pressure and arch index can be used to rapidly understand the information about feet through analysis results. It is available to understand the problems of body balance through the maps of foot pressure distribution and the bone structure problems of arch index. Moreover, it is an economical system helpful to implement foot self-detection.

2. Experimental method

2.1. Introduction to system functions

This system is divided into two parts: an instrument and the software of data analysis. The image acquisition instrument is a portable scanning platform (Fig. 1a) wherein the outer frame is mainly made of wood with a 10 mm tempered glass placed on the frame. The platform size is 36×48×16 cm functioned to obtain the images of users' feet through the scanner. The working area is 21.6x35.6 cm and the color image acquisition with 1200 resolution takes 10 seconds only. Scanning results are sent to a tablet for analysis through wireless transmission. Analysis software is functioned to analyze high-resolution image results through programs, calculate data such as arch index and foot pressure distribution, and provide a user-friendly operation interface for users. In the Figure 1b, We cooperated with MIRDC(Metal Industries Research & Development Centre in Taiwan) to commercialize the plantar pressure and arch index measurement system that wins the silver of Edison Awards 2018 honoring innovation, excellence in design, development, marketing & launch of new products.

Figure 1a. Portable scanning platform for plantar measuring system in the prototype.

IOP Conf. Series: Materials Science and Engineering **644** (2019) 012017 doi:10.1088/1757-899X/644/1/012017

Figure 1b. Commercialize optic 3D plantar pressure and arch index measurement system cooperated by MIRDC.

2.2. *Data collection method*

Foot cleanliness of experimental subjects should be firstly determined to avoid the influence on analysis results caused by dirt. Experimental subjects stand upright on the portable scanning platform with bare feet. Soles touch tempered glass with the principle similar to rubbing. When people naturally stand without being affected by external forces, the weight of whole body will be concentrated on soles. The lighter the color of the area with soles touching glass it is, the greater the pressure it is. On the contrary, the deeper the color it is, the smaller the pressure it will be. Therefore, experimental subjects should avoid unnecessary movement. Movement should ensure the body remains relaxed with eyes opened to observe the reference point located 3 meters ahead and the eye level as well as obtain the information about the plantar images of experimental subjects.

2.3. *Experimental process*

In this research, plantar images are firstly obtained by the portable scanning platform, and then these images are sent to a tablet for calculation through wireless transmission. The relevant background of experimental subjects is not required and the software can be simply operated through a touch panel. The analysis results of plantar data are displayed and stored in a tablet. Through this system, users can rapidly conduct data analysis. The flow chart is shown as follows:

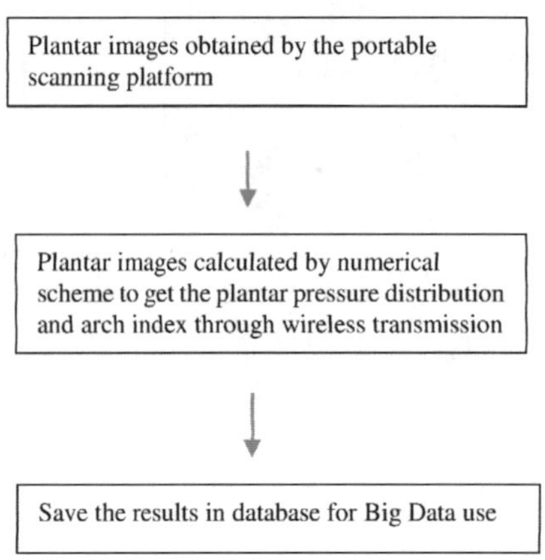

Figure 2. Flow chart for measuring plantar system.

3. Data analysis

3.1. Analysis of arch index

The information of plantar images is binarized and plantar positions are found through touching areas. Furthermore, the area touched by toes is removed and the remnant area is the plantar area required to determine arch index. The arch determination in this research is conducted in accordance with the judgment method proposed by the Cavanagh's team in a literature published in 1987 [3]. As shown in Figure 2, foot length is found through both the uppermost and the lowermost tangents of plantar edges. Foot length is divided into three equal parts, namely Zone A, B and C. The area of the Zone B means the area touched by foot arch divided by the total area touched by feet as shown in equation (1). It indicates the group of high foot arch is determined when the arch index is below 0.21; the group of normal foot arch is determined when the arch index is between 0.21 and 0.26 and the group of low arch is determined when the arch index is above 0.26. The arch index equation is shown as follows:

$$\text{Arch Index} = B/(A+B+C) \tag{1}$$

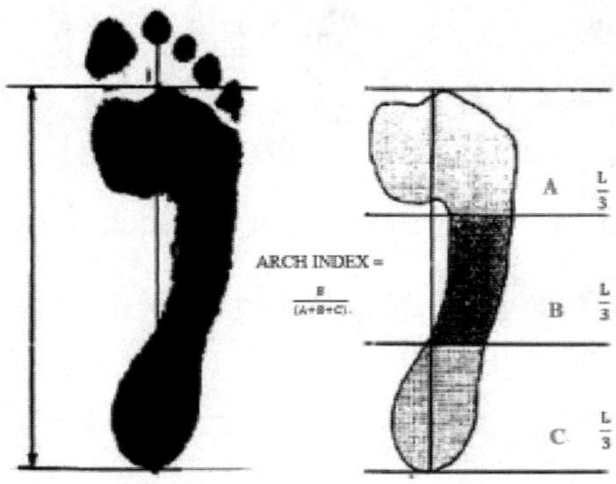

Figure 3. The segmentation map of arch index [3].

3.2. Analysis of plantar pressure

Plantar pressure is obtained with the weight of the area divided by the area of the zone. As shown in the equation (2), it is obtainable through image processing. The principle is meant to convert the color gradation of foot images into binary digit codes. By corresponding to the order from white to black, the values of color gradation are changing from large to small. The larger the color gradation value it is, the larger the area it is stressed [8]. Otherwise, the smaller the area it is stressed. By using the matrix, gray scale values are obtained through reading all the binary values in images. Zone weight is obtained with the weighting values of regional gray scales multiplied by actual weight. Zone areas are regional weight values multiplied by the actual area of images, so that the area of every zone can be calculated.

$$\mathbf{P}_K = \frac{\mathbf{W}_K}{\mathbf{A}_K} \tag{2}$$

P : Pressure distribution, N/cm^2
W: Weight loading, N
A : Area, cm^2
K: *k* zone

4. Results and discussion

The plantar images of the first experimental subject (Figure 4) are obtained by using a portable scanning platform. After data analysis is finished, the software show results by means of charts and tables. Through Table 1, experimental subjects can confirm their own analysis results by checking basic information to avoid confusion with other experimental subjects when storing data. When checking records, it is allowable to know the results of your own foot detection at different times and you can check them out later. Figure 5 provides a comparison with data and images allowing experimental subjects to understand the status of their own pressure distribution and body balance. The pressure profile of the first experimental subject shows that the pressure is concentrated on forefoot. Table 2 shows the arch index, arch status and foot length of both feet. Users can understand the problems of skeletal structure through arch index. The arch index of both left and right feet for the first experimental subject is above 0.26 according to the judgment method proposed by the Cavanagh team in 1987 [3] wherein the first experimental subject belongs to the group of low foot arch (flat foot).

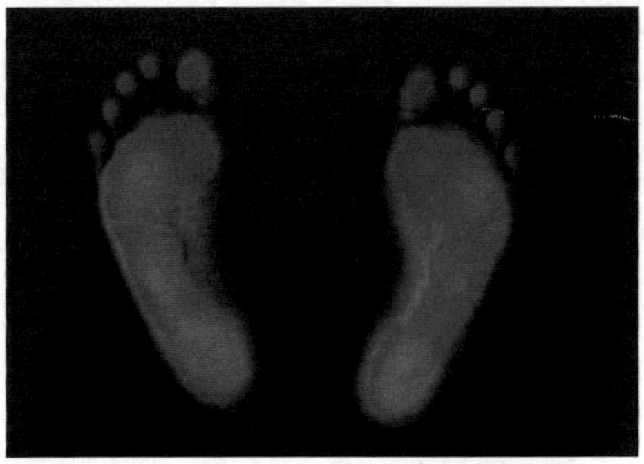

Figure 4. The plantar images of experimental subjects.

Table 1. Basic information and BMI value.

	Information
Name	No.1
Age	20
Height (cm)	160
Weight (kg)	60
BMI	23.4

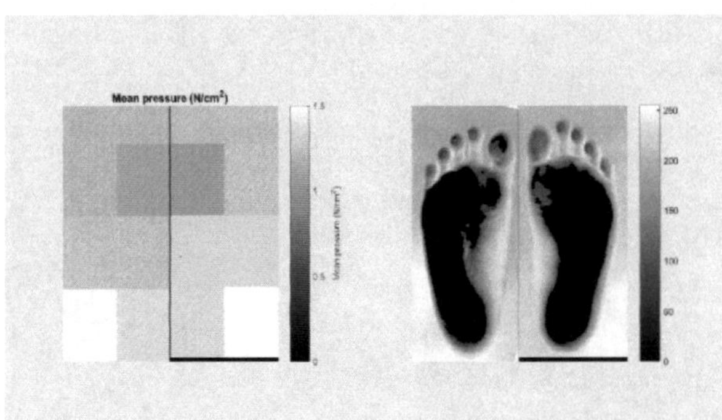

Figure 5. The map of plantar pressure distribution.

IOP Conf. Series: Materials Science and Engineering **644** (2019) 012017 doi:10.1088/1757-899X/644/1/012017

Table 2. Arch index and plantar length value.

	Left	Right
AI (Arch Index)	0.2729	0.2972
Arch type	Low Arch	Low Arch
Contact area (cm^2)	17.79	17.95
Plantar length (cm)	26.24	26.35
Plantar width (cm)	10.01	9.97

5. Conclusion

The main contributions in this research are to develop a foot self-detection system combined with plantar image capture hardware and a set of foot pressure and arch index analysis software. Users can obtain the results of high-resolution plantar images without using invasive instruments. Through the analysis software the both important information on foot pressure and arch index have been combined. The users can specifically understand the problems of his body balance and bone structure.

Because of scanning plantar image, it can avoid the fish-eye effect caused by photographing with accuracy improved. Analysis results are used to combine arch index and the information of plantar pressure and it is a system especially dedicated to foot self-detection. Users also can use measuring results to desire customized insoles and shoes. Besides, the price of this system is less than one tenth of those of the pressure pads currently on markets. It is economical and easily portable. It also provides the reference available for medical diagnosis or producing foot aids. The whole system is commercialized and won the silver of Edison Awards 2018 honoring innovation, excellence in design, development, marketing & launch of new products.

6. References

[1] H.B. Menz, Maturities, 91, 110(2016)

[2] J. Ahmad, Clinical Research & Reviews, 10, Issue 1, 48(2016)

[3] P.R. Cavanagh and M.M. Rodgers, Journal of Biotech, 20, 511(1987)

[4] W. Martha and H.J. Fan, Foot & ankle international, 2, 379 (1998)

[5] K. Gatzoulis, et al., Foot and ankle Surgery, 4, 75,(2009)

[6] R.R. Azevedo, E. S. da Rocha, P. S. Franco and F. P. Carpes, Physical Therapy in Sport, 24, 39(2017)

[7] T. Kaewwichit, GIMSE, National University of Tainan, Taiwan, PhD. Dissertation, (2015)

[8] C.N. Huang, M.Y. Lee and C.C. Chang, Computer Graphics and Applications, IEEE 31.2, 77(2011)

[9] W.C. Wang, K.H. Su and C.C. Chang, International Conference, MDEATED (2014)

[10] T. Kaewwichit, C.H Tseng, C.C. Chang, KMUTNB: International Journal of Applied Science and Technology. 7(2),77 (2014)

IOP Conf. Series: Materials Science and Engineering **644** (2019) 012018 doi:10.1088/1757-899X/644/1/012018

Stable and Robust Alternating Current Power Source Control Technique Applied to Ultraprecision Machining of Aluminum Alloy Materials

En-Chih Chang, Jerry Cheng, Yuan-Chieh Tsai, Meng-Han Tsai, Cheng-Syuan Li, and Rong-Ching Wu

Department of Electrical Engineering, I-Shou University, Kaohsiung City, Taiwan
Email: rcwu@isu.edu.tw

Abstract. This paper develops a stable, robust control technique for AC power sources in ultraprecision machining of aluminum alloy materials that maintains low total harmonic distortion output-voltage and fast transient under uncertain interferences. The proposed control technique associates the properties of the finite-time variable structure controller (FTVSC) with the gray Bernoulli model (GBM). The FTVSC provides finite system state convergence time with non-singular advantages, which is different from the asymptotic convergence of the traditional variable structure controller in infinite time. However, once there is a highly uncertain disturbance, there will be chatter around FTVSC. The chatter produces high voltage harmonics in the AC power source output, which reduces the stability of ultraprecision machining of aluminum alloy materials. The GBM is used to eliminate the chatter and produce a high performance AC power source. Experimental results on a high-performance AC power source show that the proposed stable and robust control technique can result in high quality AC output voltage against system uncertainties. Since the proposed control technique is easier to implement than the existing control technique, and has high-precision tracking, fast convergence speed and efficient calculation, this paper will cause an interest in researchers of related aluminum alloy materials applications.

1. Introduction

Alternating current (AC) power supplies have been used as key components in ultraprecision machining of aluminum alloy materials, such as forging, drawing and extrusion [1-4]. High-performance AC power supplies should include: (1) Output voltage waveform with low harmonic distortion for linear or non-linear cyclic loads. (2) Rapid transient when sudden load changes occur. (3) Steady state error should be as small as possible. In order to obtain requisitions, the proportional-integral controller performs well under constant load conditions; however, when the plant is disturbed by fluctuating loads, good performance cannot be met [5], [6]. Nonlinear control strategies such as repetitive control, deadbeat control and optimal state feedback, are often used [7-11]. However, they have complex algorithms and are difficult to implement. Variable structure controller (VSC) is a robust method for controlling nonlinear systems in which system uncertainties occur [12-17]; many VSCs have been developed for AC power supplies. A single sliding surface function is used to perform all control objects, but the transient and steady state are not good [18]. A multi-loop control is proposed through the improvement of the [18], but there is a steady-state error [19]. A discrete variable structure control scheme for the servo system is proposed [20], and it is modified for the use of the PWM inverter [21]. Although a fast transient response is obtained, the output voltage waveform is distorted because the step size of the discrete reference sinusoid is greater than the sampling period. In addition, a discrete feedforward sliding mode control scheme is developed that shows fast transient and good steady-state response [22]. Unfortunately, due to the rated load deviation, the responsibility of the VSC becomes heavier. A discretized version of the integral VSC scheme is introduced. Unless a smooth function is used, the chatter still exists [23]. A cheap 8-bit microcontroller for the outer loop replaces the classical analog dual-loop technique [24]. A first-order Takagi-Sugeno fuzzy controller is proposed to synthesize and analytically design a hybrid PWM and sliding-mode global nonlinear control for boost switching regulators [25]. A Type-2 fuzzy sliding mode controller is developed to handle the rule uncertainty when the operation is extremely uncertain and/or the membership level cannot be accurately determined [26]. A sliding mode fuzzy logic controller is proposed to obtain the better start-up and steady state performance [27]. As described above [18-23], the use of a linear sliding surface produces an infinite time convergence problem. Although satisfactory results for transients and steady-states are given in [24-27], they have high control complexity, require precise system parameters and need more calculations. In order to improve the convergence speed, a finite-time variable structure controller (FTVSC) using a nonlinear sliding surface is developed instead of a linear sliding surface [28-35]. Compared with the control based on the linear sliding surface, the FTVSC can drive the tracking error to converge to zero in a finite time [36-42]. However, in the event of highly uncertain disturbances, the chatter occurs and the AC power source output produces high harmonic distortion, resulting in thermal breakdown, which reduces the stability and reliability of ultra-precision machining of aluminum alloy materials. Based on past and present dynamic system data, a computationally efficient gray Bernoulli model (GBM) is used to describe and analyze future trends in serial numbers [43-48]. The GBM improves the accuracy of the classic grey prediction model and has been successfully applied in many areas of engineering [49-53]. Therefore, the mathematically simple and accurately prediction GBM is used to eliminate the chatter while the system uncertainty boundary is overestimated. Through the combination of FTVSC and GBM, the tracking error between the desired output and the actual output can be minimized, and the ultraprecision machining of the aluminum alloy material produces low total harmonic distortion, fast transient, reduced chattering and less steady state error. Experimental results are provided to verify the applicability of the proposed technique.

2. Modelling of AC power source

Figures 1 and 2 show the structure of two AC power sources for ultraprecision machining of aluminum alloy materials. In order to reduce the switching loss, Figure 2 is used, and its equivalent circuit is shown in Figure 3. The load can be linear/nonlinear to test the performance of the buck inverter. The output of the buck inverter is designed to provide 110 Vrms at 60 Hz frequency.

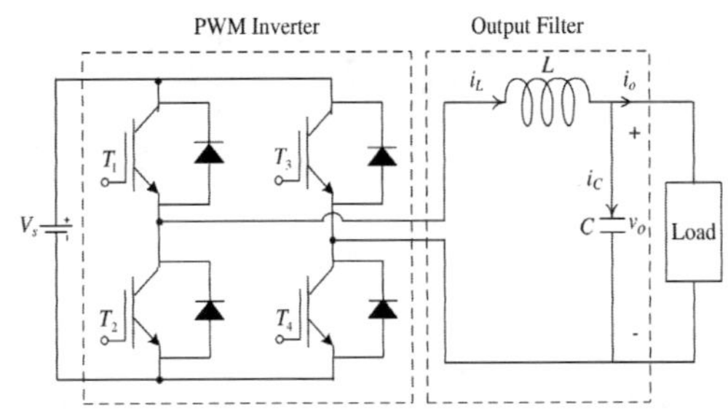

Figure 1. AC power source using PWM inverter.

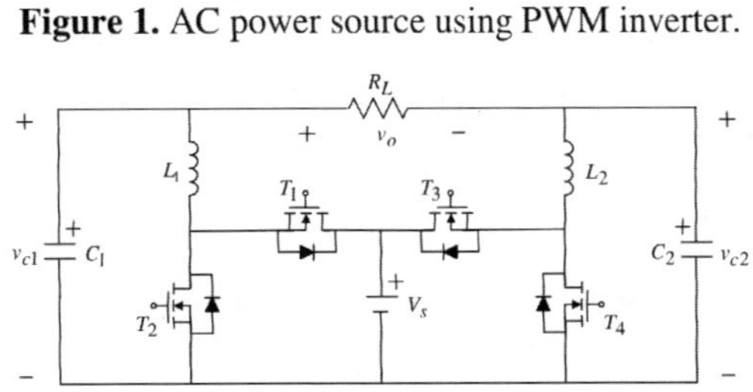

Figure 2. AC power source using buck inverter.

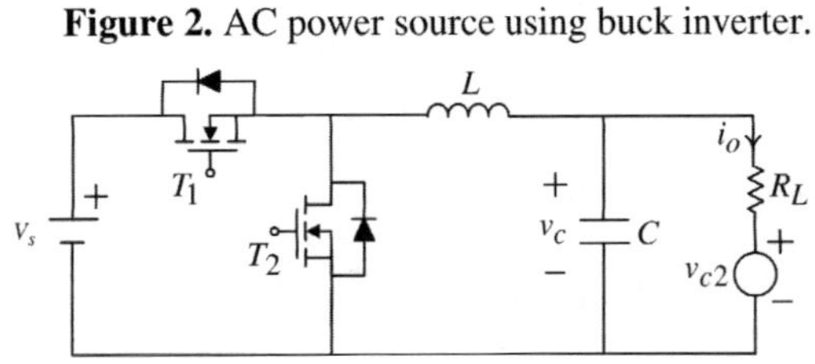

Figure 3. Equivalent circuit of buck inverter.

From the Figs. 2 and 3, the state space equation can be expressed as

$$\begin{bmatrix} \dot{v}_c \\ \ddot{v}_c \end{bmatrix} = \begin{bmatrix} 0 & 1 \\ -\dfrac{1}{LC} & -\dfrac{1}{R_L C} \end{bmatrix} \begin{bmatrix} v_c \\ \dot{v}_c \end{bmatrix} + \begin{bmatrix} 0 \\ \dfrac{V_s}{LC} \end{bmatrix} u + \begin{bmatrix} 0 \\ \dfrac{\dot{v}_{c2}}{R_L C} \end{bmatrix} \quad (1)$$

where $v_c = v_{c1}$, $L_1 = L$ and $C_1 = C$.

Set v_o be output voltage, v_r be the desired AC waveform, $e_1 = v_o - v_r$ stands for voltage error. Then, define $\dot{e}_1 = e_2$, $a_1 = b = 1/LC$, and $a_2 = 1/R_L C$. The parameter a_1 is known, but the parameter a_2 is uncertain because it depends on the load conditions. In fact, the load conditions are unpredictable, but the load variation range is limited by the inverter design. Therefore, the parameter a_2 are limited to $\bar{a}_2 - \Delta a_2 < a_2 < \bar{a}_2 + \Delta a_2$, where the bar on the character symbol stands for the nominal value and the symbol Δ represents the parameter change.

As can be seen from the (1), the control signal u must be designed to converge e_1 and e_2 to zero. Therefore, when GBM is used to reduce the chatter, the FTVSC can drive system tracking error to converge to zero for a finite time, thereby ensuring closed-loop stability and producing a higher performance AC output voltage. The design philosophy of this proposed control technique is to improve the classic VSC by introducing the FTVSC and the GBM to solve the problems of the infinite time convergence and the chatter. Finally, the AC power source will provide powerful performance for ultraprecision machining of aluminum alloy materials.

3. Design of control technique

A finite time variable structure controller with a non-linear sliding surface is defined as

$$s = e_1 + (1/\rho) \cdot e_2^{\xi} \quad (2)$$

where $\rho > 0$, $1 < \xi < 2$, and a sliding-mode reaching entry $\dot{s} = -\phi_1 |s|^{1+\kappa} sign(s) - \phi_2 |s|^{1-\kappa} sign(s)$ is used.

Therefore, the control law u can be written as

$$u(t) = u_{eq}(t) + u_{ft}(t) \quad (3)$$

with

$$u_{eq}(t) = b^{-1}[a_1 e_1 + a_2 e_2 - \frac{\rho}{\xi} e_2^{2-\xi}] \quad (4)$$

$$u_{ft}(t) = -b^{-1}\left[\phi_1 |s|^{1+\kappa} sign(s) + \phi_2 |s|^{1-\kappa} sign(s)\right] \quad (5)$$

where $\phi_1, \phi_2 > 0, 0 < \kappa < 1$, u_{eq} represents the equivalent control that determines the dynamics of the system, and u_{ft} indicates the sliding control that prevents system uncertainty. Therefore, the system will be driven to $s = 0$ and converged in a finite time; however, if highly uncertain disturbances are applied, the chatter will still occur and the system (1) will not be able to achieve accurate tracking performance. Therefore, the control signal $u(t)$ (3) is modified by adding a gray Bernoulli model (GBM) (u_{gbm}), which reduces the chatter in the AC power source system. The modeling steps for the GBM are as follows.

Step 1: Input the original sample data sequence

Represent the original data sequence as

$$X^{(0)} = \{x^{(0)}(k)\}, k = 1, 2, \ldots, n \quad (6)$$

where $x^{(0)}$ stands for the set of n original sample data.

Step 2: Use accumulated generating operation (AGO)

By taking the AGO on $x^{(0)}$, we can written the first-order AGO sequence as

$$X^{(1)} = \{x^{(1)}(k)\}, k = 1, 2, \ldots, n \quad (7)$$

where $x^{(1)}(k)=\sum_{i=1}^{k}x^{(0)}(i), k=1,2,\cdots,n$

Step 3: The grey differential equation for constructing GBM is

$$x^{(0)}(k)+\alpha z^{(1)}(k)=\beta[z^{(1)}(k)]^{\sigma} \tag{8}$$

In addition, the whitening differential equation is expressed as

$$\frac{dX^{(1)}}{dt}+\alpha X^{(1)}=\beta[X^{(1)}]^{\sigma} \tag{9}$$

where $z^{(1)}(k)=gx^{(1)}(k)+(1-g)x^{(1)}(k-1)$, $k=2,3,\cdots,n$, $g\in[0,1]$ denotes the generating coefficient of the background value, and σ is an any real number precluding $\sigma=1$.

Step 4: Solve the estimated parameters α and β by the least square method below.

$$[\alpha,\beta]^T=(D^TD)^{-1}D^TM \tag{10}$$

where $D=\begin{bmatrix}-z^{(1)}(2) & [z^{(1)}(2)]^{\sigma}\\ -z^{(1)}(3) & [z^{(1)}(3)]^{\sigma}\\ \vdots & \vdots\\ -z^{(1)}(n) & [z^{(1)}(n)]^{\sigma}\end{bmatrix}$ and $M=\begin{bmatrix}x^{(0)}(2)\\ x^{(0)}(3)\\ \vdots\\ x^{(0)}(n)\end{bmatrix}$.

Step 5: The solution of the (10) is generated as

$$\hat{x}^{(1)}(k)=[(x^{(0)}(1)^{1-\sigma}-\frac{\beta}{\alpha})e^{-\alpha(1-\sigma)(k-1)}+\frac{\beta}{\alpha}]^{1/1-\sigma} \tag{11}$$

where '^' denotes predicted value, $\sigma\neq 1$, and $k=1,2,...,n$.

Step 6: Use inverse accumulated generating operation (IAGO)

The data sequence $\hat{x}^{(0)}(k)$ using the IAGO can be estimated as

$$\hat{x}^{(0)}(1)=x^{(0)}(1) \tag{12}$$

$$\hat{x}^{(0)}(k)=\hat{x}^{(1)}(k)-\hat{x}^{(1)}(k-1),\ k=2,3,...,n \tag{13}$$

Therefore, the control law of the (3) is rewritten as

$$u(k)=u_{eq}(k)+u_{ft}(k)+u_{gbm}(k) \tag{14}$$

where the added compensation component is gray Bernoulli control u_{gbm}, which can eliminate the chattering.

$$u_{gbm}(k)=\begin{cases}0 & , \quad |\hat{s}(k)|<\tau\\ \psi\hat{s}(k)sign(s(k)\hat{s}(k)) & , \quad |\hat{s}(k)|\geq\tau\end{cases} \tag{15}$$

where $\hat{s}(k)$ represents for the predicted value of $s(k)$, ψ is a constant, and τ symbols the system boundary. The $sign$ function in the (15) is replaced by a continuous function, i.e., hyperbolic tangent function.

4. Experiments

The proposed system parameters are listed in Table 1. Figure 4 shows the simulated output voltage waveform of an AC power source controlled by the proposed technique under a rectifier load, including full-wave rectified with a parallel resistor 12 Ω and a capacitive filter 300μF. The output voltage has very low output voltage distortion (%THD is 0.91%) while the current suddenly rises. The simulated performance of a classic VSC (diode bridge with a capacitive filter 300μF and 12Ω resistive load) with a rectified load is shown in Figure 5. The simulated output voltage is hardly sinusoidal and produces high distortion (the %THD of the output voltage is 6.84%). To test the transient characteristics of AC power sources in ultraprecision machining of aluminum alloy materials, Figures 6 and 7 show the experimental output voltage obtained using the proposed technique and the classic VSC under step changes in load (from no load to full load) at a 90 degree

IOP Conf. Series: Materials Science and Engineering **644** (2019) 012018 doi:10.1088/1757-899X/644/1/012018

firing angle, respectively. A detailed examination of the waveforms shows that Figure 6 has a slight voltage dip and a fast recovery steady-state response. However, the classic VSC system as shown in Figure 7 produces unsatisfactory transient response and delayed recovery time. It has been well established that the proposed technique does produce higher tracking accuracy, lower harmonic distortion and faster convergence than classic VSC. Therefore, the finite time convergence of the system state is achieved by the FTVSC, and the chatter is reduced by the GBM. The performance of the proposed AC power source has been significantly improved, which is very suitable for the application of ultraprecision machining of aluminum alloy materials.

Table 1. System Parameters.

DC-link voltage	$V_s = 200$ V
Output voltage and frequency	$v_o = 110$ $V_{rms,}$ $f = 60$ Hz
Filter inductor	$L = 1.1$ mH
Filter capacitor	$C = 12$ μF
Resistive load	$R_L = 12$ Ω
Switching frequency	$f_s = 18$ kHz

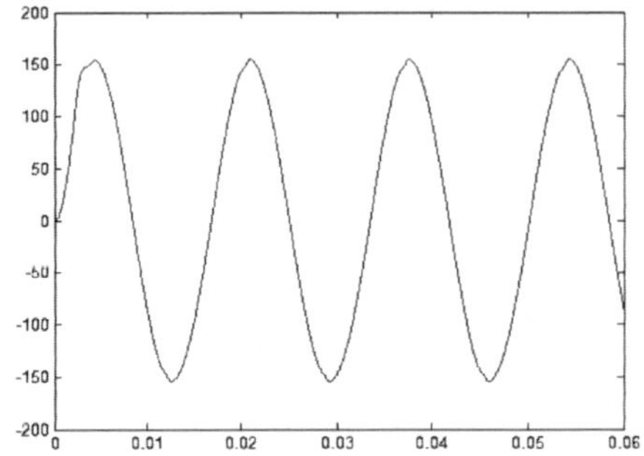

Figure 4. Output voltage for the proposed technique under rectifier load (100V/div; 5ms/div).

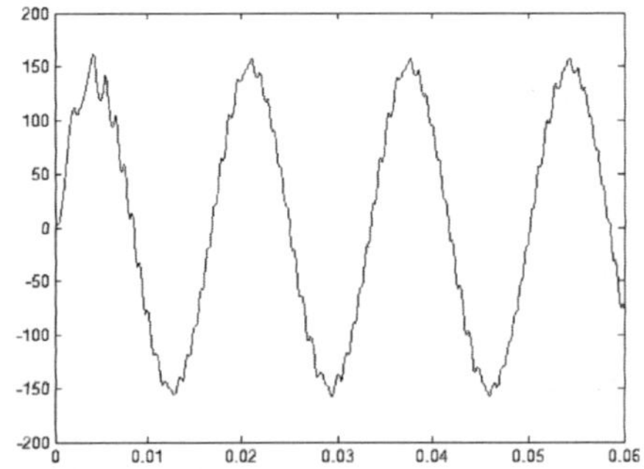

Figure 5. Output voltage for the classic VSC under rectifier load (100V/div; 5ms/div).

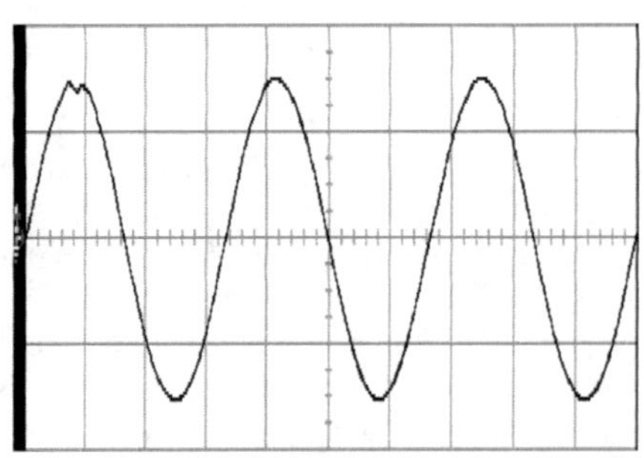

Figure 6. Output voltage for the proposed technique under step change in load (100V/div; 5ms/div).

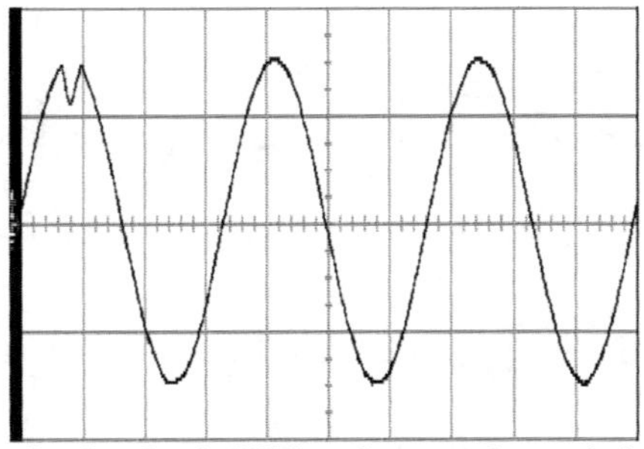

Figure 7. Output voltage for the classic VSC under step change in load (100V/div; 5ms/div).

5. Conclusions

This paper designs the FTVSC with the GBM to improve AC power sources in ultraprecision machining of aluminum alloy materials. The proposed technique not only has the robustness of the classic VSC, but also provides finite time convergence of the system state and reduces the chatter. According to theoretical analysis, simulation and experimental results, the effectiveness of the proposed technique is proved and suitable for ultra-precision machining of aluminum alloy materials. In addition, the proposed technique can be used for more complex three-phase AC power sources for machining applications in future research.

Acknowledgments

This work was supported by the Ministry of Science and Technology of Taiwan, R.O.C., under contract number MOST107-2221-E-214-006.

6. References

[1] H.A. Kishawy, and A. Hosseini, Machining Difficult-to-Cut Materials: Basic Principles and Challenges, Springer International Publishing, 2019

[2] Advanced Joining, Sustainable Manufacturing, Springer, New York, 2017

[3] K. Sundareswaran, Elementary Concepts of Power Electronic Drives, CRC Press, Boca Raton, FL, USA, 2019

[4] B.M. Wilamowski, and J.D. Irwin, Power Electronics and Motor Drives, CRC Press, Boca Raton, FL, USA, 2011

[5] J. Selvaraj, and N.A. Rahim, "Multilevel inverter for grid-connected PV system employing digital PI controller," IEEE Trans. on Industrial Electronics, vol. 56, no. 1, pp. 149-158, 2009

[6] J. Dannehl, F.W. Fuchs, and P.B. Thøgersen, PI state space current control of grid-connected PWM converters with LCL filters, IEEE Trans. on Power Electronics, vol. 25, no. 9, pp. 2320-2330, 2010

[7] T. Hornik, and Q.C. Zhong, A current-control strategy for voltage-source inverters in microgrids based on H^{∞} and repetitive control, IEEE Trans. on Power Electronics, vol. 26, no. 3, pp. 943-952, 2011

[8] D. Chen, J.M. Zhang, and Z.M. Qian, "An Improved Repetitive Control Scheme for Grid-Connected Inverter With Frequency-Adaptive Capability," IEEE Trans. on Industrial Electronics, vol. 60, no. 2, pp. 814-823, 2013

[9] J.B. Hu, and Z.Q. Zhu, Improved Voltage-Vector Sequences on Dead-Beat Predictive Direct Power Control of Reversible Three-Phase Grid-Connected Voltage-Source Converters," IEEE Trans. on Power Electronics, vol. 28, no. 1, pp. 254-267, 2013

[10] H. Yang, X. Lin, M.M. Khan, C. Chen, Y. Gang, and L.D. Zhou, "Robust deadbeat control scheme for a hybrid APF with resetting filter and ADALINE-based harmonic estimation algorithm, IEEE Trans. on Industrial Electronics, vol. 58, no. 9, pp. 3893-3904, 2011

[11] S.L. Jung, L.H. Ho, H.C. Yeh, Y.Y. Tzou, DSP-based digital control of a PWM inverter for sine wave tracking by optimal state feedback technique, Proc. IEEE PESC'94 (1994), pp. 546-551

[12] V.I. Utkin, Variable Structure Systems with Sliding Modes, IEEE Trans. Autom. Contr., vol. AC-22, no. 2, pp. 212-222, 1977

[13] U. Itkis, Control systems of variable structure. New York: Wiley, 1976

[14] K.D. Young, V.I. Utkin, and U. Ozguner, A control engineer's guide to sliding mode control, IEEE Trans. Contr. Syst. Technol., vol. 7, pp. 328-342, 1999

[15] G. Bartolini, L. Fridman, A. Pisano, and E. Usai, Modern Sliding Mode Control Theory, Springer-Verlag, Berlin, 2008

[16] L.G. Wu, P. Shi, and X.J. Su, Sliding Mode Control of Uncertain Parameter-Switching Hybrid Systems, Wiley, New York, 2014

[17] S. Vaidyanathan, and C.H. Lien, Applications of Sliding Mode Control in Science and Engineering, Springer, New York, 2017

[18] K. Jezernik, and D. Zadravec, Sliding mode controller for a single phase inverter, in Proc. IEEE APEC'90, 1990, pp. 185-190

[19] S.J. Chiang, T.S. Tai, and T.S. Lee, Variable structure control of UPS inverters," Proc. IEE-Elect. Power Applicat., vol. 145, no. 6, pp. 559-567, 1998

[20] C. Y. Chan, Servo-system with discrete-variable structure control, Syst. Contr. Lett., vol. 17, no. 4, pp. 321-325, 1991

[21] S. L. Jung, Y. Y. Tzou, Sliding mode control of a closed-loop regulated PWM inverter under large variations, in Proc. IEEE PESC'93, 1993, pp. 616-622

[22] S. L. Jung, Y. Y. Tzou, Discrete feedforward sliding mode control of a PWM inverter for sinusoidal output waveform synthesis, in Proc. IEEE PESC'94, 1994, pp. 552-559

[23] T. L. Chern, J. Chang, C. H. Chen, and H. T. Su, Microprocessor-based modified integral discrete variable-structure control for UPS, IEEE Trans. Ind. Electron., vol. 46, pp. 340-348, 1999

[24] E. Vidal-Ldiarte, L. Martinez-Salamero, F. Guinjoan, J. Calvente, and S. Gomariz, Sliding and fuzzy control of a boost converter using an 8-bit microcontroller, IEE-Elect. Power Applicat., vol. 151, no. 1, pp. 5-11, 2004

[25] S. Gomariz, E. Alarcon, F. Guinjoan, E. Vidal-ldiarte, and L. Martinez-Salamero, Piecewise PWM-sliding global control of a boost switching regulator by means of first-order Takagi-Sugeno fuzzy control, in Proc. Conf. IEEE Circuits Syst., 2001, vol. 3, pp. 715-718

[26] P.Z. Lin, C.M. Lin, C.F. Hsu, and T.T. Lee, Type-2 fuzzy controller design using a sliding-mode approach for application to DC-DC converters, IEE-Elect. Power Applicat., vol. 152, no. 6, pp. 1482-1488, 2005

[27] Y. Shi, and P.C. Sen, Application of variable structure fuzzy logic controller for DC-DC converters, in Proc. Conf. IEEE IECON'01, 2001, vol. 1, pp. 2026-2031

[28] M. Zak, Terminal Attractors for Addressable Memory in Neural Network, Physics Lett. A, vol. 133, no. 1,2, pp. 18-22, 1988

[29] X.H. Yu, and M. Zhihong, Model reference adaptive control systems with terminal sliding modes," Int. J. Contr., vol. 64, no. 6, pp. 1165-1176, 1996

[30] X.H. Yu, and M. Zhihong, On finite time mechanism: terminal sliding modes," in Proc. IEEE VSC'96, 1996, pp. 164-167

[31] X.H. Yu, and M. Zhihong, Terminal sliding modes with fast transient performance, in Proc. Conf. IEEE Decision and Contr., 1997, vol. 2, pp. 962-963

[32] D. Yue, S. Xu, and Y. Liu, Terminal sliding mode control of high-order systems, in Proc. Conf. IEEE Ind. Technol., 1996, pp. 811-815

[33] R. B. Fernandez, and J. K. Hedrick, Control of multivariable nonlinear systems by the sliding mode method, Int. J. Contr., vol. 46, no.3, pp. 1019-1040, 1987

[34] S. T. Venkataraman, and S. Gulati, Control of nonlinear systems using terminal sliding modes," in Proc. Conf. American Contr., 1992, pp. 891-893

[35] M. Zhihong, A. P. Paplinski, and H. R. Wu, A robust mimo terminal sliding mode control scheme for rigid robotic manipulators, IEEE Trans. Autom. Contr., vol. 39, pp. 2464-2469, 1994

[36] Z.P. Wang, Y.S. Mao, Z.H. Hu, and Y.X. Xie, A Sliding Mode Control Design based on the Reaching Law for Matrix Rectifiers, Journal of Power Electronics, vol. 16, no. 3, pp. 1122-1130, 2016

[37] H.B. Zhou, J. H. Song, and S.M. Song, Sliding mode guidance law considering missile dynamic characteristics and impact angle constraints, International Journal of Automation and Computing, pp. 1-10, 2016

[38] X. H. Yu, and M. Efe, Recent Advances in Sliding Modes: From Control to Intelligent Mechatronics, Springer, 2015

[39] X.H. Yu, J.X. Xu, Variable Structure Systems: Towards the 21st Century, Springer, 2002

[40] S.D. Xu, C.C. Chen, and Z.L. Wu, Study of Nonsingular Fast Terminal Sliding-Mode Fault-Tolerant Control, IEEE Trans. on Industrial Electronics, vol. 62, no. 6, pp. 3906-3913, 2015

[41] Z. Galias, and X.H. Yu, Dynamical Behaviors of Discretized Second-Order Terminal Sliding-Mode Control Systems, IEEE Trans. on Circuits and Systems II: Express Briefs, vol. 59, no. 9, pp. 597-601, 2012

[42] A. Mohammadpour, A. Ajami, S.G. Zadeh, M.P. Aghababa, and B.M. Mahboubi, Robust Terminal Sliding Mode Power Flow Controller Using Unified Power Flow Controller with Adaptive Observer and Local Measurement, IET Generation, Transmission & Distribution, vol. 8, no. 10, pp. 1712-1723, 2014

[43] J. L. Deng, Introduction to grey system theory, J. Grey Syst., vol. 1, pp. 1-24, 1989

[44] J. L. Deng, Control problems of grey system, Syst. and Contr. Lett., vol. 5, pp. 288-294, 1982

[45] H. C. Lu, Grey prediction approach for designing grey sliding mode controller, in Proc. Conf. IEEE Syst., Man, and Cybern., 2004, vol. 1, pp. 403-408

[46] K. L. Wen, F. H. Chen, and M. L. Huang, Existence of Grey Model GM(2,1), Journal of Grey System, pp. 209-214, 1998

[47] C. M. Chen, C. M. Hong, and Y. R. Hsieh, A Modeling Technique of Grey Prediction Model Using Discrete Difference Equation, Journal of Grey System, vol. 12, no. 1, pp. 65-72, 2000

[48] Y.T. Hsu, M.C. Liu, J. Yeh, and H.F. Hung, Forecasting the Turning Time of Stock Market Based on Markov-Fourier Grey Model, Expert Systems with Applications, vol. 36, pp. 8597-8603, 2009

[49] C.N. Wang, and V.T. Phan, An Improved Nonlinear Grey Bernoulli Model Combined with Fourier Series, Mathematical Problems in Engineering, pp. 1-7, 2015

[50] P.H. Hsin, and C.I. Chen, Application of trembling-hand perfect equilibrium to Nash nonlinear Grey Bernoulli model: an example of BRIC's GDP forecasting, Neural Computing and Applications, vol. 28, pp. 269-274, 2017

[51] S.F. Liu, and Y. Lin, Grey Information: Theory and Practical Applications, Springer, 2006

[52] S.F. Liu, Y.J. Yang, and J. Forrest, Grey Data Analysis: Methods, Models and Applications, Springer, 2017

[53] X. Ma, Z.B. Liu, and Y. Wang, Application of a novel nonlinear multivariate grey Bernoulli model to predict the tourist income of China, Journal of Computational and Applied Mathematics, vol. 347, pp. 84-94, 2019

IOP Conf. Series: Materials Science and Engineering **644** (2019) 012019 doi:10.1088/1757-899X/644/1/012019

The Optimization Wire Drawing Die Design for Aluminum A1050 and A1070

Jinn-Jong Sheu and Bao-Shan Wang

Department of Mold and Die Engineering, National Kaohsiung University of Science and Technology, No.415, Jiangong Rd., Sanmin Dist., Kaohsiung City 807, Taiwan
Email: jjsheu@nkust.edu.tw

Abstract. Wire drawing is a common used process to reduce the diameter of a small rod or a wire. The parameters of drawing die design such as die angle, bearing length, entrance fillet, and the drawing ratio of each pass are crucial to control the quality of a drawn wire. Taguchi method was adopted to optimize the drawing die geometry design. The lubrication conditions were taken into consideration to observe the effect of friction on the dimension accuracy of the drawn wires. The optimum combination of the drawing die design using 20% of reduction ratio is semi die angle 9 degrees, bearing length 1.2mm and entrance fillet 0.4mm. Two types of aluminum alloy 1070 and 1050 were tested using lubricants sodium sulfide (Na2S), graphite, and drawing oil, respectively. The lubrication effect of sodium sulfide is the best to achieve the lowest drawing load. The lubricant Na2S had the best performance in terms of lowest maximum drawing force for both of A1050 and A1070. The final diameters of the drawn wires were very close to the specification 4.0 mm. The CAE simulations were able to find the optimum design of die. The designed drawing machine was capable of carrying out the experiments and obtains consistent results.

1. Introduction

Drawing process is commonly used to reduce the wire diameter to a required specification for further forming processes or applications. Hsu et al. [1] used Taguchi method to get the optimum drawing wire quality and verified with experimental works. Tai [2] had adopted an $L_9(3^4)$ orthogonal array of Taguchi method to study the drawing process and found the optimum die designs. Ye [3] had proposed a tube drawing die design method to obtain the optimum surface quality of tubes. The effective strain distribution was used for the cost function of Taguchi method. Fan [4] had used $L_{27}(3^7)$ orthogonal array for tube drawing die design. The drawing rate is crucial for making a sound tube drawing process. Lin [5] had proposed curved profile drawing die design methods for better tube surface quality and reduced the pass number required.

Shao et al. [6] had carried out titanium alloy wire drawing tests with the lubricants of molybdenum disulphide grease and graphite. The moly grease had the better performance for the drawing process in terms of drawing force. Lee [7] had carried out the stainless wire drawing experiments successfully using the zirconia ceramics drawing die and the lubricants of molybdenum disulphide grease and graphite. Fan [8] had tested the moly grease, the graphite and the butter lubrications for the aluminium ironing process. The graphite lubricant seems had better results. Byonet al. [9] had tested two lubricants for wire drawing and found the powder type lubricant was able to fill the micro cavity of the wire surface and had better lubrication effect. In this paper, Taguchi method was proposed to study the aluminium wire drawing process with lubricants of sodium sulphide (Na_2S), drawing oil (90W-90) and graphite. The drawing loads, wire diameter, and the surface roughness were examined.

2. Drawing pass and drawing die design

2.1. Design of drawing pass

The aluminium wires of A1050, A1070, and A1100 with 5.0 mm diameter were drawn to 4.0 mm. The pass number of drawing was determined considering the reduction rate given in equation 1.

$$r = (A_0 - A_f)/A_0 \times 100\% \quad (1)$$

where A_0 is the original wire diameter, A_f is the final wire diameter of each pass. The toal reduction rate (r_t) of the test is 36%. The required drawing pass number (n) was calcuated with the assumption of equal reduction rate ,r, for each pass and given in equation 2.

$$r_t = [1 - (1 - r/100)^{1/n}] \times 100\% \quad (2)$$

Two passes of wire drawing was designed to obaint the better surface quality and prevent the wire broken during the experiment process. The reduction rate calculated was 20% for each pass and the wire diameter for each pass is 4.48, and 4.0mm, respectively.

2.2. *Design of drawing die*

The geometry and the dimensions of the drawing die designed were shown in Fig. 1 and Table 1, respetively. The approach fillet,R, the die angle,α, the entrance fillet, r,the bearing length, Lb , and the exit angle,β, were tahke into consideration.

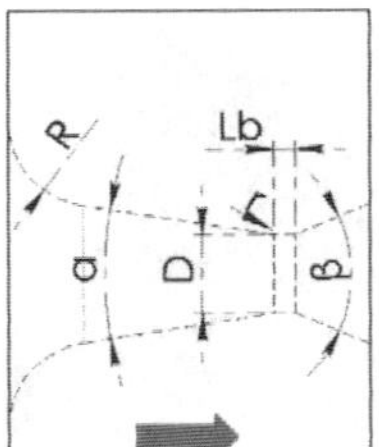

Figure 1. Dimensions of the drawing die orifice and the bearing length.

Table 1. Dimensions of the die orifice.

Approach fillet (R)	5mm
Semi die angle (α/2), degrees	5, 7.5, 9
Entrance fillet (r)	0, 0.2, 0.4 mm
Bearing length (Lb)	1.0, 1.2, 1.4 mm
Relief (β)	43°

2.3. *Taguchi method*

The wire drawing force was adopted for the cost function calcualtion and the goal of the optimization is the-smaller-the-beter. The orthogoanl array $L_9(3^4)$ was adopted for the layout of the expewriments. The considered thee level desgin factors A, B, and C are the bearing length, Lb , the die angle,α, and the entrance fillet, r,respectively. The experiment layout was given in Table 2. The calculation of the cost funtion for the-smaller-the better was givne in equation 3.

$$S/N = -10\log\left[\frac{1}{n}\sum_{i=1}^{n} y_i^2\right] \quad (3)$$

where y_i is maximum drawing force, n is the numbers of tests and equal to one for the CAE simulation, the S/N is the signal to noise ratio.

IOP Conf. Series: Materials Science and Engineering **644** (2019) 012019 doi:10.1088/1757-899X/644/1/012019

Table 2. Orthogonal Array $L_9(3\ )$for the drawing tests.

Try	A bearing length	B die angle	C entrance fillet
1	1	1	1
2	1	2	2
3	1	3	3
4	2	1	2
5	2	2	3
6	2	3	1
7	3	1	3
8	3	2	1
9	3	3	2

3. CAE Simulation and the experiment setup

3.1. CAE model

Die drawing die was designed using the CAD software and depicted in Fig. 2. The 3-D geometry of die was input to the CAE software, Deform 3D. The tetrahedral mesh of the A1050 wire was generated in the CAE pre-processor and shown in Fig. 3.The mesh numbers generated initially is 51,000. The Coulomb friction model was adopted and set to 0.08 and 0.12 for excellent and general lubrication of ceramic tools, respectively.

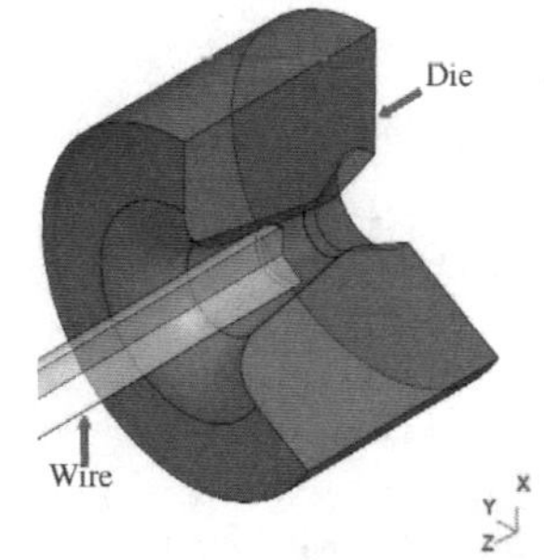

Figure 2. Three dimensional model of the drawing die designed in the CAD system.

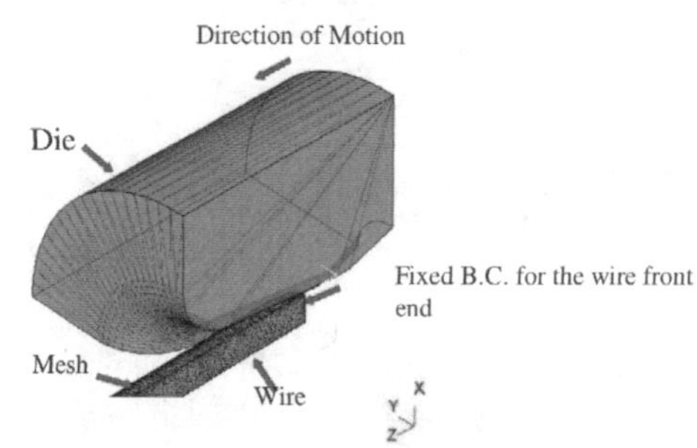

Figure 3. Mesh of the wire and the STL geometry of the drawing die.

3.2. Experiment setup of drawing

A wire drawing machine was designed and installed for the experiment tests. The servo motor (60w/110v/60Hz) installed on the wire drawing machine (Fig. 4) was able to stably draw the wire on the drum with a linear speed of 6 mm/s through the die orifice(shown in Fig. 5). The drawing force was measured using data acquisition system and the load cell (capacity 1,200N). The lubricants of sodium sulphide (Na_2S), drawing oil and graphite were dropped on the sponge in front of the die orifice.

IOP Conf. Series: Materials Science and Engineering **644** (2019) 012019 doi:10.1088/1757-899X/644/1/012019

Figure 4. Wire drawing machine designed for the experiment

First Pass Second Pass

Figure 5. Drawing orifice for the first and the second pass experiments

4. Results and discussion

4.1. Drawing force of the CAE simulation for A1050

The maximum drawing forces for general lubrication of the second pass as well as the S/N ratios for the 9 trials of the Taguchi design were shown in Table 3. The response charts for the different design factors were shown in Fig. 6, Fig. 7, and Fig. 8, respectively. The optimum combination for the drawing die design was bearing length 1.2mm, semi die angle 9 degrees, and the entrance fillet 0.4 mm. The prediction of the drawing force for the best design combination is 41.3Kgf. The prediction loads of the CAE confirmation simulation were 40.8 Kgf for friction coefficient 0.08. The 1% error of the confirmation simulation had verified the optimum design combination is feasible.

Table 3. Drawing force prediction and the S/N ratio for general lubrication condition.

Try no	1	2	3	4	5	6	7	8	9
Load,Kg f	48.4	43.2	42.0	44.0	44.4	40.8	46.4	43.2	42.4
S/N	-33.7	-32.7	-32.5	-32.9	-32.9	-32.2	-33.3	-32.7	-32.5

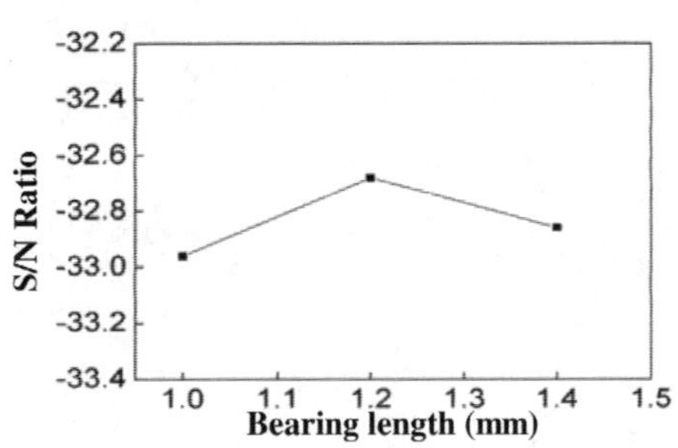

Figure 6. Response chart for the S/N ratio of bearing length.

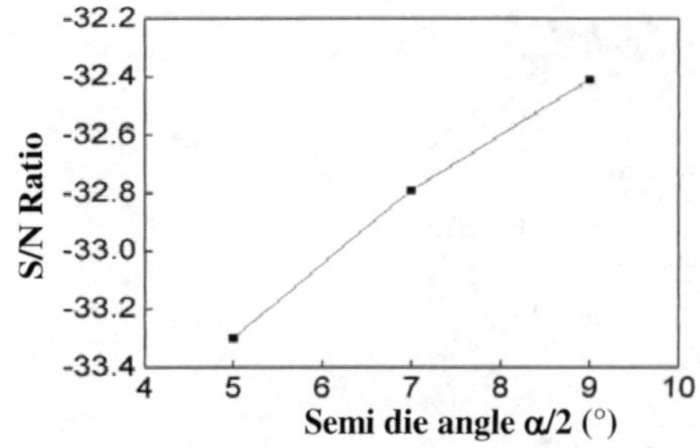

Figure 7. Response chart for the S/N ratio of die angle.

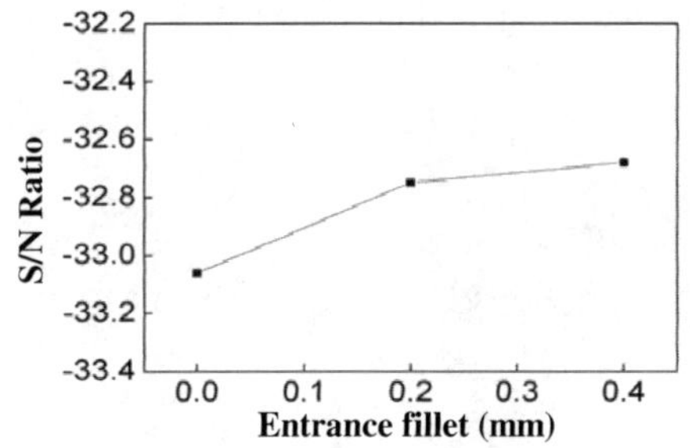

Figure 8. Response chart for the S/N ratio of entrance fillet.

5. Comparisons of the CAE simulations and the experimental results

5.1. Drawing force

The CAE simulations were carried out only for A1050 but the drawing experiments were tried for A105 and A1070. The drawing loads of the first and the second passes for the simulation and the experiments were compared in Fig. 9 and Fig. 10 for different materials. The curves with the labels of '0.08' and '0.12' in Fig. 9 were obtained via the CAE simulation with the friction coefficient 0.08 and 0.12, respectively. The curves with the label' no lubrication' in Fig. 10 were the experiment results of dry drawing without applying any lubricants. The drawing force increased dramatically in the beginning and reached a steady state value. The rank of the performance of the lubricants for both of the drawing passes was Na_2S, drawing Oil, and the graphite in terms of lower drawing force. The maximum drawing force of the experiments of the second pass for different lubrication were in the range of 31~36Kgf which are close to the CAE prediction with friction coefficient 0.08.

5.2 Surface roughness and the hardness

The measured maximum drawing forces, surface roughness, and the hardness for A1050, and A1070 were shown in Table 4, and Table 5, respectively. The maximum drawing forces for the A1050 were in the range of 34 to 40 Kgf for both of the pass one and pass two. The maximum drawing forces for the A1070 were in the range of 48 to 76 Kgf for both of the pass one and pass two. The maximum drawing loads for the second pass increased only a small amount that means the strain hardening phenomenon is not significant. The diameters of the final wire were very close to the specification that shows the drawing processes were successful.

For A1050, the surface roughness of the first and the second passes without lubrication were close to the surface roughness of die orifice (0.07mm). The performance of the graphite is the better among the three lubricants in terms of the surface roughness (the drawing loads showed no significant difference). The hardness measurements of the first and the second passes were changed insignificantly.

For A1070, the surface roughness of the first and the second passes without lubrication were close to the surface roughness of die orifice (0.07 mm). The performance of the drawing oil is the better among the three lubricants in terms of the surface roughness and the drawing loads. The measurements of the load and the

IOP Conf. Series: Materials Science and Engineering **644** (2019) 012019 doi:10.1088/1757-899X/644/1/012019

hardness for the first and the second passes were changed significantly which indicates the strong effect of strain hardening.

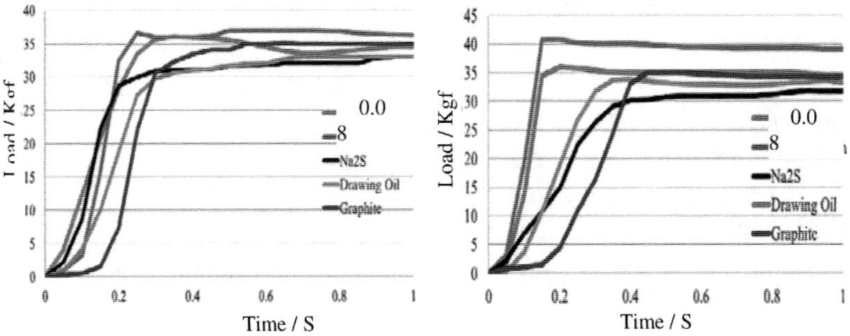

Figure 9. Drawing force of the first pass (left) and the second pass (right) for A1050.

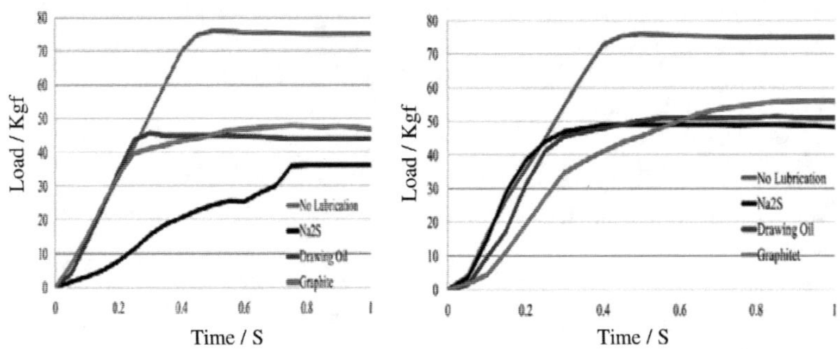

Figure 10. Drawing force of the first pass (left) and the second pass (right) for A1070.

Table 4. The drawing load, wire diameter, surface roughness, and the hardness of A1050 experiment.

Lubricants	Load(Kgf)	Wire diameter(mm)	Surface roughness (Ra)	hardness(Hv)
The first pass				
Deform(0.08)	35	4.48	-	-
No Lubrication	38	4.49	0.07	29.3
Na2S	35	4.48	0.07	30.9
Drawing Oil	34	4.48	0.17	31.2
Graphite	33	4.48	0.04	28.6
The second pass				
Deform(0.08)	35	4.0	-	-
No Lubrication	40	4.01	0.05	32.1
Na2S	31	4.0	0.13	31.9
Drawing Oil	33	4.0	0.08	31.8
Graphite	34	4.0	0.04	28.7

Table 5. The drawing load, wire diameter, surface roughness, and the hardness of A1070 experiment.

Lubricants	Load(Kgf)	Wire diameter(mm)	Surface roughness (Ra)	hardness(Hv)
The first pass				
No Lubrication	75	4.49	0.07	32.1
Na2S	35	4.48	0.07	33.4
Drawing Oil	43	4.48	0.13	30.3
Graphite	48	4.48	0.05	33.3
The second pass				
No Lubrication	76	4.01	0.1	44.3
Na2S	48	4.0	0.04	39.0
Drawing Oil	50	4.0	0.03	38.0
Graphite	58	4.0	0.17	45.3

6. Conclusions

In this research, the Taguchi method was adopted to obtain the optimum drawing die design combination via the CAE simulations for A1050 material. The optimum combination for the drawing die design was bearing length 1.2mm, die angle 18 degrees, and the entrance fillet 0.4 mm. The lubricant Na_2S had the best performance in terms of lowest maximum drawing force for both of A1050 and A1070. The final diameters of the drawn wires were very close to the specification 4.0 mm. The CAE simulations were able to find the optimum design of die. The designed drawing machine was capable of carrying out the experiments and obtaining the consistent results.

Acknowledgments
The authors would like to thank the MIRDC for the financial support of the granted project 103-EC-17-A25-1164.

7. References

[1] Q.C. Hsu, Y.D. Jhan, Y.C. Huang, Design and analysis of the circular metal wire drawing process, Forging, Vol. 19, No. 1, pp. 46-53, (2010)

[2] K.C. Tai, M.H. Caleb Li, H.P. Lin, C.F. Chuang, Improving productivity of SUS 304 stainless steel using Taguchi method, Journal of Technology, Vo. 120, No. 4, pp. 333-338, (2005)

[3] S.C. Yeh, Process simulation of cold tube drawing, Master Thesis, Department of Mold and Die Engineering, National Kaohsiung University of Applied Sciences, (2012)

[4] D.T. P., Process and die design of multi-pass cold steel tube drawing, Master Thesis, Department of Mold and Die Engineering, National Kaohsiung University of Applied Sciences, (2011)

[5] S.Y Lin, Process analysis and die design optimization of single-pass cold steel tube drawing, Master Thesis, Department of Mold and Die Engineering, National Kaohsiung University of Applied Sciences, (2011)

[6] S.Y. Shao, C.T. Kwan, W.C. Lee, H.S. Lin, G.Z. Ho, The Process Research of Titanium wire drawing, Forging, Vol. 16, No 4, pp.12-21, (2007)

[7] W.C. Lee, C.T. Kwan, H.R. Shih, An analysis of theory and experiment for stainless wire drawing with zirconia ceramic die, Proceedings of the 24th National Conference on Mechanical Engineering of CSME, No. D01-0012, CSME-P3196-3200, (2007)

[8] C.F. Fang, In Situ observation of grease lubrication in aluminum forming, Master Thesis, Department of Mechanical Engineering National Yunlin University of Science and Technology, (2004)

[9] S.M. Byon, S.J. Lee, D.W. Lee, Y.H. Lee, Y. Lee, Effect of coating material and lubricant on forming force and surface defects in wire drawing process. Transactions of Nonferrous Metals Society of China, Vol. 21, pp. 104-110, Supplement 1, (2011)

Gating System Design of Investment Casting for SUS316 Spray Nozzle

Pei-Hsing Huang[1], Hung-Mao Lin[2], and Wei-Jen Wu[3]

1Department of Mechanical Engineering, National Yunlin University of Science and Technology, Yunlin 640, Taiwan
2 Department of Chemical and Materials Engineering, Southern Taiwan University of Science and Technology, Tainan 710, Taiwan
3Department of Mechanical Engineering, National Pingtung University of Science and Technology, Pingtung 912, Taiwan
Email: huangph@yuntech.edu.tw

Abstract. Spray nozzles operate in harsh conditions with heat, corrosive substances, and high liquid pressure, which can easily cause corrosion and cavitation in the nozzle structure and threaten the safety of users. To reduce defects occurring during the casting process of spray nozzle and increase their service life, this study applied mold flow analysis to the casting process of a SUS316 stainless steel spray nozzle. We improved the configuration of the gating system, added water-cooling to the improved scheme, and examined the resulting filling and solidification conditions of the molten metal in the mold cavity. We employed the Niyama criterion to predict the probabilities of shrinkage defect formation in the spray nozzle and improve defect formation, thereby effectively lengthening the service life of the spray nozzle and increasing product quality.

1. Introduction

Casting dates back thousands of years. New developments occur every day in industrial technology, and with them are rising standards in workpiece quality. Correcting defects in workpieces to reach specifications has become a common goal in the entire casting industry [1-4]. To improve workpiece quality and yield while reducing personnel costs, a number of researchers have developed computer-aided engineering (CAE) software technology to increase the accuracy of defect predictions in mold flow analysis with the aid of software [5-8]. Most of the casting manufacturers in Taiwan search for optimal casting schemes through trial and error. In the event of defects, they use welding to repair the defects. Figure 1 displays a spray nozzle workpiece with severe shrinkage defects that welding cannot repair. Thus, finding a casting solution for spray nozzles is imperative. Wang et al. [9] employed ProCAST, a software program for casting process simulations, to simulate and optimize the investment-casting scheme for a thin-walled cylinder comprising aluminum alloy. They chose a bottom gating scheme, improved the gating system, and optimized the pouring parameters so as to prevent shrinkage defects and enhance workpiece quality. Zhou et al. [10] used the software Magmasoft to simulate the solidification processes of workpieces and determine where shrinkage pores and cavities are likely to appear in the workpiece. They added a cooling mechanism and risers and redesigned the casting parameters. Huang et al. [11, 12] used AnyCasting software to optimize various processing parameters for A356 aluminum alloy castings with the aim of minimizing shrinkage and porosity defects. Xie et al. [13] used ProCAST to analyze filling and solidification processes, predict shrinkage defects, and optimize the parameters of the pouring process. Their results indicated that the improved scheme could indeed reduce shrinkage defects. Zhen et al. [14] conducted numerical simulations of a wheel casting process and analyzed the temperature field of the solidification process and the causes of defect formation. Based on their

IOP Conf. Series: Materials Science and Engineering **644** (2019) 012020 doi:10.1088/1757-899X/644/1/012020

simulation results, they optimized the pouring system. Using AnyCasting, Zhang et al. [15] simulated the shrinkage defects that appear in the workpieces of steam generator supports in nuclear power plants with pressurized water reactors and optimized the process parameters. Huang et al. [16] used numerical software to simulate transient turbulence in a gating system in which the cross-sections were constricted using a V-shaped notch. Tian et al. [17] used Anycasting to analyze the filling and solidification processes of ductile iron crankshafts and predict where shrinkage defects may appear. This study used metal mold flow analysis to analyze the design of a gating system for spray nozzle and revise process parameters. We also applied water-cooling to reduce the defect formation probability and distribution and improve the output value and yield.

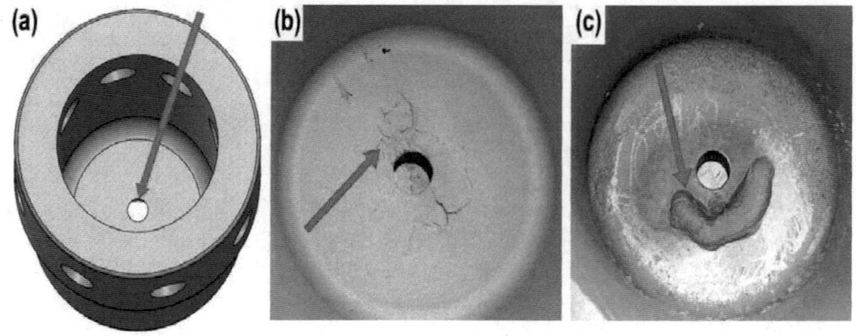

Figure. 1 Defects in spray nozzle casting: (a) location of defects in spray nozzle; (b) red arrow indicating surface at the bottom of the spray nozzle; (c) red arrow indicating the shrinkage defects after grinding.

2. Experimental and numerical methods

2.1. Parameter settings of investment casting process.

We used SUS316 stainless steel for the spray nozzle due to its resistance to heat, corrosion, acidity, and alkalinity. We input the physical properties of SUS316 into the formulas for mold flow analysis, including density (ρ) 7,930 kg/m3, specific heat (S) 494 J/kg·K, latent heat (L) 290 KJ/kg, thermal conductivity (K) 14.39 W/m·K, liquidus temperature (T_l) 1,454 °C, and solidus temperature (T_s) 1,392 °C, as shown in Table 1. We designed the process parameters of the initial scheme and the improved scheme based on the physical properties of the stainless steel spray nozzle. These process parameters included casting material, shell mold material, mesh number, ceramic shell temperature $T_{ceramic}$ (°C), casting temperature $T_{casting}$ (°C), pouring time $T_{pouring}$ (sec), ceramic shell thickness δ(mm), soaking depth $W_{cooling}$ (mm), gravity casting, and water soaking, as shown in Table 2.

Table 1. Thermophysical properties of SUS316 stainless steel [18].

ρ(kg/m^3)	S(J/kg·K)	L(kJ/kg)	K(W/m·K)	T_l (°C)	T_s (°C)
7930	494	290	14.39	1454	1392

IOP Conf. Series: Materials Science and Engineering **644** (2019) 012020 doi:10.1088/1757-899X/644/1/012020

Table 2. Process parameter design of gating system.

	Initial scheme	Improved scheme
Casting material	SUS 316	
Shell mold material	Zircon sand	
Mesh number	3,000,000	
$T_{ceramic}$ (°C)	950	950
$T_{casting}$ (°C)	1,580	1,580
$t_{pouring}$ (sec)	6	6
$\delta(mm)$	6.5	
$W_{cooling}$ (mm)	None	60

3. Results and discussion

3.1. Mold flow analysis of initial scheme

The filling conditions as the molten metal is poured into the mold cavity and the later solidification directions are major factors of workpiece quality. We therefore used AnyCasting to analyze the filling, solidification, poured short, and turbulent flow phenomenon, as displayed in Fig. 2. Figure 2(a) shows that the molten metal enters the ingate into the nozzle mold and quickly fills the bottom of the cavity. The molten metal collides with round opening of the nozzle, which creates splashes and turbulent flows during the filling process. Some air is trapped within the molten metal under the round opening, which disrupts the filling process. We can thus determine that the ingate is not positioned appropriately. The red circles in Fig. 2(b) indicate large portions of isolated residual melt at the bottom of the spray nozzle and below the round nozzle. Furthermore, no effective feeding channel is present near the ingate, so the feeding channel is broken during the solidification process, which leads to casting defects in the interior of the nozzle. Figure 3 presents the probability distributions of shrinkage defects resulting from the initial scheme. The color code shows that colors closer to dark red indicate higher probabilities of shrinkage defects, whereas colors closer to dark blue represent lower probabilities of shrinkage defects. Figure 3(a) displays the locations of the cross-sections in Figs. 3(b) through 3(d), which present the casting defects resulting from the initial scheme. As can be seen, the shrinkage defects all appear where the spray nozzle is thicker or where its geometric shape changes significantly. Due to the lack of consideration for the flow direction of the molten metal and feeding issues, the design of the initial scheme produced shrinkage defects in the casting structure.

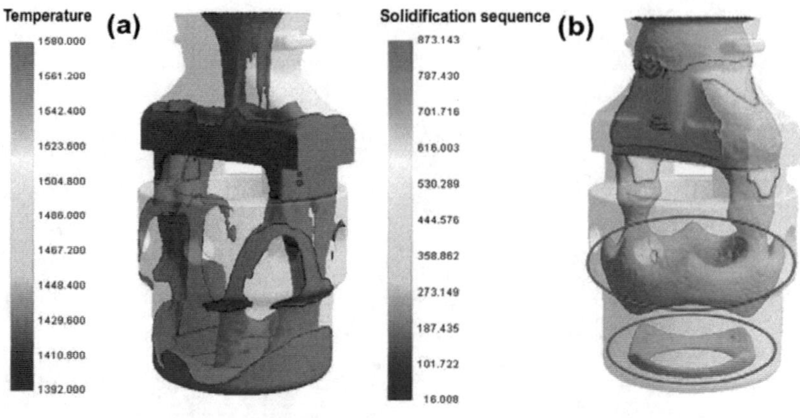

Figure 2. Analysis of filling and solidification sequence in initial scheme of spray nozzle: (a) filling sequence (t=1.80 s); (b) solidification sequence (t=316.0 s).

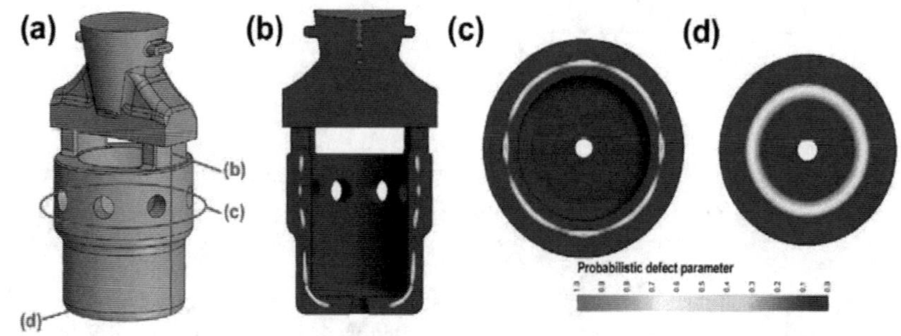

Figure 3. Probability distributions of shrinkage defects resulting from initial scheme of spray nozzle: (a) locations of cross-sections displaying casting defects; (b) side sectional view of pattern tree; (c)-(d) top sectional view of spray nozzle.

3.2. Gating system design and mold flow analysis of improved scheme

After examining the simulation results of the initial scheme, we moved the spray nozzle into a horizontal position in the improved scheme so that the molten metal flows in from the side. To improve the flow of the molten metal and achieve good feeding at the sprue and the ingate, the ingate was also slanted along the tangent of round opening to prevent the molten metal from directly hitting the round opening, dividing into two flows, and entrapping air at the bottom. Finally, we hoped to use water-cooling to reduce the casting defects during solidification and attain a good temperature gradient. Figure 4 displays the design of the water-cooling method applied to the improved casting scheme. As shown in Fig. 5(a), the molten metal slowly flows upward and can more effectively avoid the round opening than in the initial scheme so that there are no turbulent flows. As shown in Fig. 5(b), the soaking the bottom of the spray nozzle can accelerate the cooling rate, improve the solidification directions so that they are steadier and smoother than those in the initial scheme, and reduce the probabilities of defect formation. As shown in Figs. 6(b) through 6(d), an appropriate soaking depth can effectively improve the temperature gradient during solidification and narrow down the shrinkage defect distributions. The results were significantly improved defect distributions and probabilities when compared to those in Fig. 3.

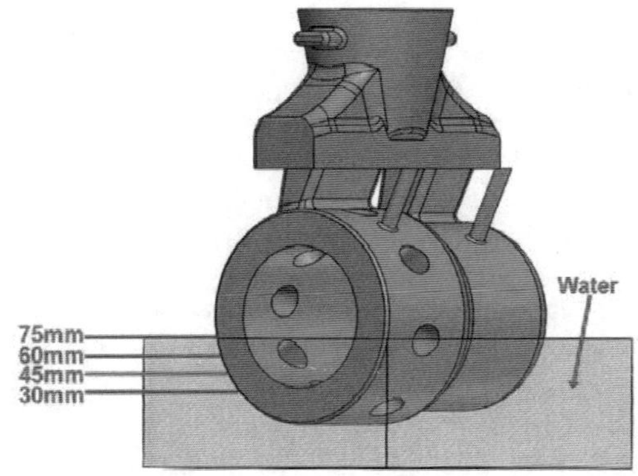

Figure 4. Design of water-cooling method in improved casting scheme.

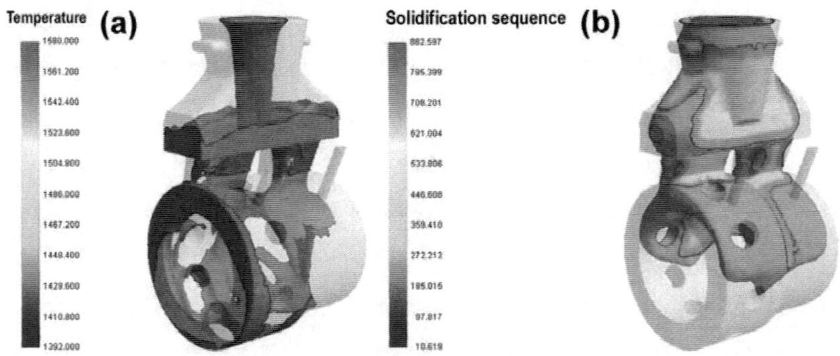

Figure 5. Analysis of filling and solidification sequence in improved scheme of spray nozzle: (a) filling sequence (t=1.19 s); (b) solidification sequence (t=246.0 s).

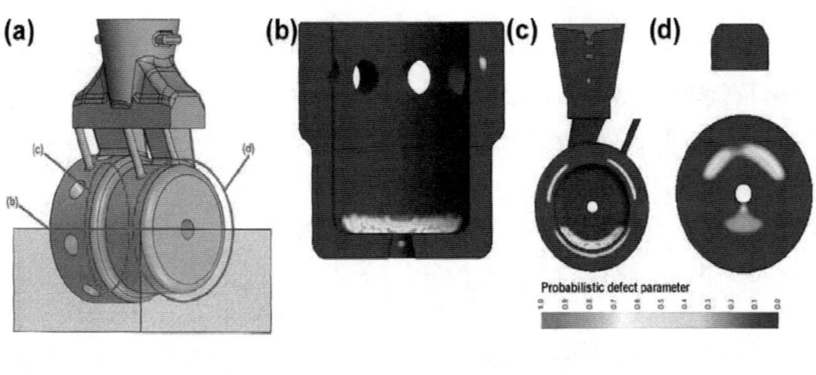

Figure 6. Probability distributions of shrinkage defects resulting from improved scheme of spray nozzle: (a) locations of cross-sections displaying casting defects; (b) top sectional view of spray nozzle; (c)-(d) side sectional view of spray nozzle.

4. Conclusions

Moving the spray nozzle into a horizontal position so that the molten metal flows in from the side and slanting the ingate along the tangent of round opening in the improved scheme effectively prevent the molten metal from directly hitting the round opening and creating turbulent flows in the mold cavity as it did in the initial scheme. The thicker structure of the spray nozzle caused inadequate feeding during the solidification process, which led to defect formation. We thus added a water-cooling method with a suitable soaking depth to the improved scheme to solve the solidification defect problems, thereby producing a good temperature gradient and better solidification directions and process. The simulation results indicated significant improvements in the probability distributions of shrinkage defects in the improved scheme compared to those in the initial scheme. Water-cooling with a soaking depth of 60 mm was especially effective in improving workpiece quality and reducing defects in the cross-section and bottom of the spray nozzle.

Acknowledgments
The authors gratefully acknowledge the support provided for this research by the Ministry of Science and Technology, R.O.C. under grants MOST 107-2221-E-224-060 and MOST 107-2622-E-224 -013 -CC3.

5. References

[1] P.H. Huang, C.J. Lin, Int. J. Adv. Manuf. Technol. 79, 997 (2015)

[2] J.K. Kuo, P.H. Huang, H.Y. Lai, J.R. Chen, Int. J. Adv. Manuf. Technol. 92, 1093 (2017).

[3] P.H. Huang, J.K. Kuo, T.H. Fang, W.R. Wu, MATEC Web Conf. 185, 00008 (2018)

[4] J.K. Kuo, P.H. Huang, M.J. Guo, Int. J. Adv. Manuf. Technol. 89, 101 (2017)

[5] P.H. Huang, M.J. Guo, Mater. Res. Innovations. 19, 77 (2015)

[6] P.H. Huang, J.Y. Luo, S.C. Hung, C.J. Lin, H.H. Cheng, Appl. Mech. Mater. 627, 46 (2014)

[7] J.K. Kuo, P.H. Huang, H.Y. Lai, W.J. Wu, Int. J. Adv. Manuf. Technol. 100, 529 (2019)

[8] P.H. Huang, W.J. Huang, IOP Conf. Ser.: Earth Environ. Sci. 186, 012001 (2018)

[9] Y.M Wang, K.R Zhang, J Zhang, FDY. 65, 342 (2016)

[10] X. Zhou, Z.X. Zhao, C.W. Feng, Q.X. Xu, FDY. 65, 1016 (2016)

[11] P.H. Huang, W.J. Wu, C.H. Shieh, Mater. Sci. Forum. 893, 276 (2017)

[12] P.H. Huang, W.J. Wu, C.H. Shieh, Appl. Mech. Mater. 864, 173 (2017)

[13] Q.F. Xie, X. Xue, Y.F. Wu, X.F. Chen, J.T. Wu, W. Li, J.T. Li, FDY. 64, 647 (2015)

[14] Z.W. Zhen, R.F. Chen, Z.L. Li, H.F. Lu, J. Zhao, G.S. Cai, FDY. 64, 430 (2015)

[15] Z.X. Zhang, F. Huang, W.N. Meng, X.Y. Liao, FDY. 64, 1239 (2015)

[16] P.H. Huang, Y.T. Chen, B.T. Wang, Int. J. Adv. Manuf. Technol. 74, 1275 (2014)

[17] Y.X. Tian, W.H. Zeng, X.Q. Zeng, D.J. Li, FDY. 64, 1102 (2015)

[18] AnyCasting, tutorial manual. Ver. 6.0., AnyCasting, Co., Ltd. Gangseo-gu, Seoul City. (2001)

The Structure Design and Analysis of a Parallel Kinematic Mechanism Based on Abbe's Principle

Chih-Liang Chu and Hung-Chi Chen

Department of Mechanical Engineering, Southern Taiwan University of Science and Technology, No. 1, Nan-Tai Street, Yungkang Dist., Tainan City 710, Taiwan, Taiwan
Email: cliang@stust.edu.tw

Abstract. The purpose of this study is to design and analysis of a parallel kinematic mechanism based on Abbe's principle can be applied to the 3D-coordinate measuring machine. In this study is the design and analysis of the parallel kinematic mechanism, which includes the mechanism design of linear drive rod mechanism and displacement sensing position coaxial, and the intersection of the three linear drive rod mechanisms are located at the contact probe. Therefore, the design of this study improves the displacement of the linear drive rod and Abbe error resulted from the angle variation of the displacement sensor. Furthermore, the mechanism performance has been verified experimentally. Through precision tests of the machine by a laser Doppler interferometer, the linear drive rod mechanism positioning precisions are within 1μm.

1. Introduction

Coordinate Measuring Machines (CMMs) are extensively used throughout the industry to measure the geometries, dimensions and profiles of a wide range of components. With advances in micro-electro-mechanical systems (MEMS) and ultra-precision manufacturing technologies in recent decades, the characteristic scale of many devices has reduced to the micro- or even nanometer level. Consequently, a requirement exists for CMMs with commensurate measurement capabilities. Leach et al. [1] described three instruments developed at the National Physical Laboratory in the UK for facilitating accurate and traceable measurements at the nanometer scale, namely (1) an atomic force microscope (AFM) / surface texture measuring instrument; (2) an optical / X-ray interferometer calibration system, and (3) a CMM with a working volume of $(50\ \mathrm{mm})^3$ and a volumetric accuracy of 50 nm. Fan et al. [2] presented a high-precision, low-cost micro-CMM featuring an arch-shaped bridge designed to provide an improved stiffness and thermal accuracy and a co-planar stage designed to reduce the Abbe error. The experimental results indicated that each axis had a positioning resolution of 1 nm, while the overall working area had dimensions of $25 \times 25 \times 10\ \mathrm{mm}^3$.

Fan et al. [3] presented design considerations of a precision micro-CMM system and its mechatronic modules. The basic design concept is to meet the requirements of high stiffness, force balance, thermal balance, Abbe principle, metrology frame and vibration-free. Based on these criteria, a novel bridge of pagoda shape was designed and analyzed by optimization to verify its superior stiffness with force balance and thermal balance structure due to its symmetrical geometry, moreover, the co-planar XY stage, the ram and the nano-scale encoder. The experimental results indicated that each axis had a positioning accuracy of ±10 nm, while the overall working area had dimensions of $20 \times 20 \times 10\ \mathrm{mm}^3$. Brand et al. [4] modified a commercial CMM through the addition of high resolution scales and optimized air bearings to enable the dimensional metrology of micro system components with a measurement uncertainty of less than 1 nm. The proposed system incorporated two sensing systems, namely an opto-tactile 3D-sensor and a tactile touch trigger sensor, respectively, and incorporated a metrology frame comprising three miniaturized plane mirror laser interferometers which enabled the simultaneous measurement of both the displacement and the angle of the object of interest. Looking at the literature of many research units in the market or abroad, the results of CMM [5-12], most of their designs are serial-type CMM, so this study will design a parallel kinematic mechanism based on Abbe's principle can be applied to the 3D-coordinate measuring machine.

IOP Conf. Series: Materials Science and Engineering **644** (2019) 012021 doi:10.1088/1757-899X/644/1/012021

2. Design of the parallel kinematic mechanism

The mechanism is designed with three linear drive rods mounted on the three universal joints of the top casting. The other end of each drive rods is directly connected to the probe stage to form a three-degree-of-freedom parallel kinematic mechanism, as shown in Figure 1. The optical linear encoder provides instant feedback to the system when the drive rod moves. This design method creates to the driveshaft and the optical linear encoder to be positioned coaxially, and the intersection point of the three drive shafts is at the contact probe sphere, as shown in Figure 2. This design improves the Abbe error caused by the difference between the displacement of the drive rod and the angle of the optical linear encoder, improve the accuracy of the machine. In addition, the mechanism has a working volume of XY axes is Ø200 mm and Z-axis is 200 mm.

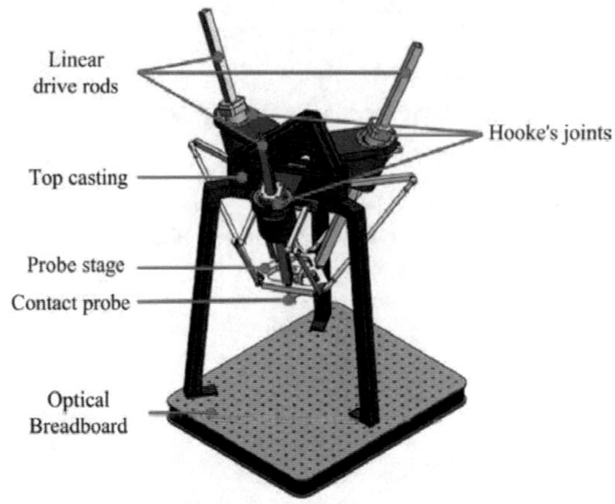

Figure 1. Schematic diagram of the parallel kinematic mechanism.

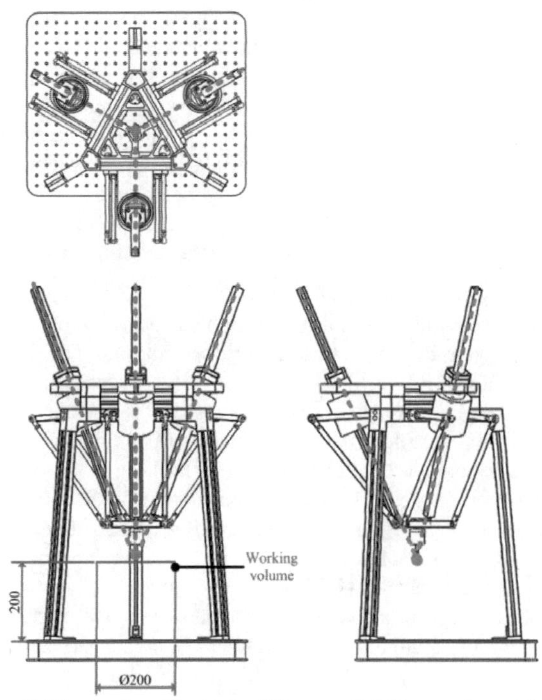

Figure 2. Three-view diagram of the parallel kinematic mechanism.

3. Design of the linear drive rod mechanism

The linear drive rod mechanism has been adopted for the actuator of the parallel kinematic mechanism. Figure 3 displays the basic construction of the linear drive rod mechanism of the parallel kinematic mechanism. A capstan roller is rotated by a servo motor with a Harmonic Drive. And the pre-load is obtained to the capstan roller by two idle rollers with a pre-load spring.

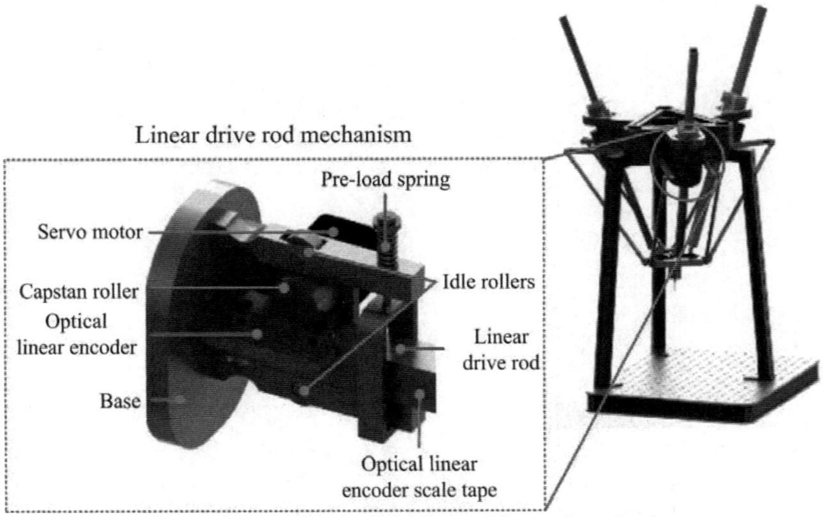

Figure 3. Photograph of the linear drive rod mechanism.

4. Experimental set-up and measurement results

The optical linear encoders in the parallel kinematic mechanism were to measure the displacements of the micro-positioning stages during the mechanism motion procedure. To confirm the precision of the optical encoders, a series of positioning tests were performed in which the displacement measurements obtained from the optical encoders were compared with those obtained using a laser Doppler interferometer. The experimental configuration is shown in Figure 4.

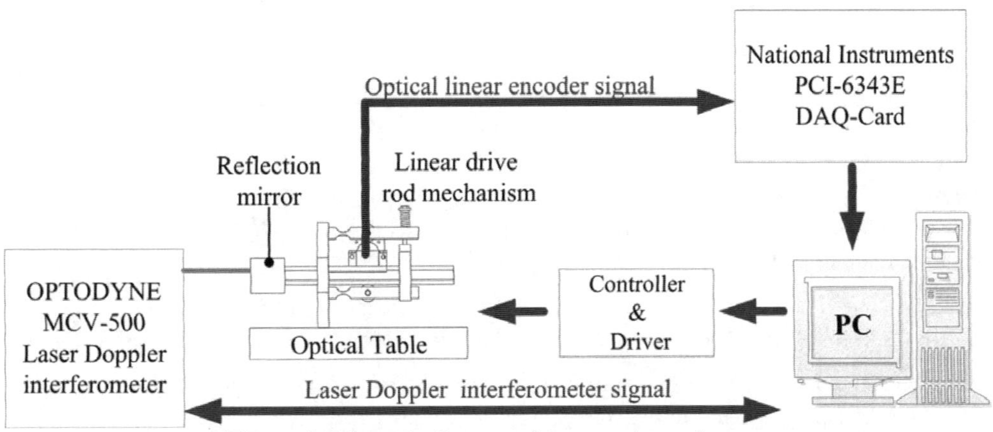

Figure 4. Schematic diagram of the experimental set-up.

In the experiments, the linear drive rod was driven through a total distance of 200mm in incremental steps of 10mm and was then returned to its starting point again in the same incremental steps. The experiment was repeated seven times for each stationary position of the mechanism. Figure 4 plot the differences between the seven times positioning precision measurements and the corresponding Doppler measurement at each stationary position. From inspection, it was determined that the displacement measurements of the encoder deviated from those of the laser interferometer by no more than 1μm. If the influence of the environment was isolated, the positioning precision of the linear drive rod could be further improved.

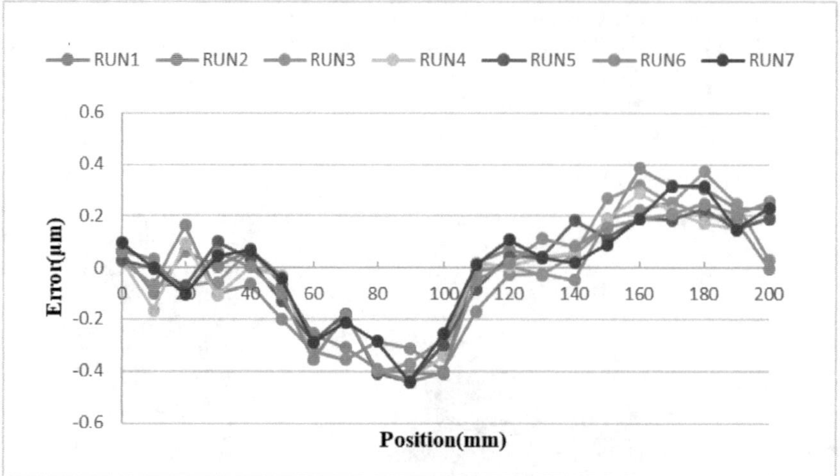

Figure 5. Measurement error of the mechanism.

5. Conclusion

This study successfully developed a parallel kinematic mechanism that can be applied to CMM, and its characteristics are as follows:

1. In order to reduce the influence of Abbe error on the parallel kinematic mechanism. Therefore, this study uses the design of the linear drive rod and displacement sensing position coaxial, and the intersection of the three linear drive rod mechanisms are located at the contact probe. The design of this study improves the displacement of the linear drive rod and Abbe error resulted from the angle variation of the optical linear encoder, and improves the accuracy of the machine. In addition, the mechanism has a working volume of XY axes is Ø200 mm and Z-axis is 200 mm.

2. In order to improve the positioning accuracy of the CMM machine, this study designed a linear drive rod mechanism to drive the parallel kinematic mechanism to achieve the purpose of 3-DOF motion. The experimental results have shown that the mechanism had a positioning accuracy of 1μm.

Acknowledgments

The present authors gratefully acknowledge the support provided to this project by the Ministry of Science and Technology, Taiwan, ROC under contract no. MOST 107-2221-E-218-028.

6. References

[1] R.K. Leach, J. Haycocks, K. Jackson, A. Lewis, A. Oldfield, and A. Yacoot, Advances in traceable nanometrology at the National Physical Laboratory, Nanotechnology, Vol. 12, pp. R1-R6, 2001

[2] K.C. Fan, Y.T. Fei, X.F. Yu, Y.J Chen, W.L. Wang, F. Chen, and Y.S. Liu, Development of a low-cost micro-CMM for 3D micro/nano measurements, Meas. Sci. Technol, Vol. 17, pp. 524-532, 2006

[3] K.C. Fan, F. Cheng, H.Y. Wang, and J.K. Ye, The system and the mechatronics of a pagoda type micro-CMM, J. Nanomanufacturing, Vol. 8, Nos. 1/2, pp. 67-86, 2012

[4] U. Brand, T. Kleine-Besten, and H. Schwenke, Development of a special CMM for dimensional metrology on microsystem components, Proc. of the 2000 Annual Meeting of the ASPE, Scottsdale, USA, pp. 361-364, 2000

[5] G.N. Peggs, A. J. Lewis, S. Oldfield, Design for a compact high-accuracy CMM, Annals of the CIRP, Vol. 48, No. 1, pp. 417-420, 1999

[6] M.M.P.A. Vermenulen, et. al, Design of a High-Precision 3D-Coordinate Measuring Machine, Annals of the CIRP, Vol. 47, No. 1, pp. 447-450, 1998

[7] W.O. Pril, et. al, Development of 2D probing system with nanometer resolution, Proc. of 16th ASPE, pp. 438-442, 1997

[8] H. Haitjema, et. al, Development of a silicon-based nanoprobe system for 3-D measurements, annals of the CIRP, Vol. 50, No. 1, 365-368, 2001

[9] H. Schwenke, et. al, Opto-tactile sensor for 2D and 3D measurement of small structures on coordinate measuring machines, Annals of the CIRP, Vol. 50, No. 1, pp. 361-364, 2001

[10] H. Schwenke, et. al, Future challenges in coordinate metrology: addressing metrological problems for very small and very large parts, Proc. of IDW Workshop 2001, Knoxville, pp. 2-12, 2001

[11] S.W. Kim, New design of precision CMM based upon volumetric phase-measuring interferometry, Annals of the CIRP, Vol. 50, No. 1, pp. 357-360, 2001

[12] K.C. Fan, Y.T. Fei, and X.F. Yu, Development of a micro-CMM, in International Manufacturing Leaders Forum on Global Competitive Manufacturing, Adelaide, Australia, pp. 1-7, 2005

IOP Conf. Series: Materials Science and Engineering **644** (2019) 012022 doi:10.1088/1757-899X/644/1/012022

Optimization Design of Wind Turbine Tunnel Based On Fluid Analysis Results

Po-Chuan Chen[1] and Jiunn Fang[2]

1 Ph.D. Program of Mechanical and Aeronautical Engineering, Feng Chia University, No.100, Rd Wenhua, Taichung, Taiwan
2 Department of Aerospace and Engineering, Feng Chia University, No.100, Rd Wenhua, Taichung, Taiwan
Email: chenname0317@gmail.com

Abstract. At present, the problem of wind turbines affecting flying organisms has been highly valued. Wind turbines will not only attack flying creatures through rotating blades, but also the flying organisms will affect the habitat of ground creatures because of the reduction of flying height. In this study, a fan with protection function is designed by designing a wind turbine system with a wind tunnel design that can be smoothly adopted by birds. Because the blade is not in direct contact with the wind, it is introduced into the blade through the wind tunnel. Therefore, in the design of the wind tunnel, not only the wind speed of the introduction is not weakened, but also the wind speed can be fastened through the structural angle. Therefore, the blades can be rotated as scheduled to achieve the purpose of wind power generation. In the design of the wind tunnel, the study conducted the wind speed test of the structure through continuous fluid analysis. In the results obtained, correct all structures that may affect wind speed and re-analyze to ensure that the wind speed can successfully to achieve the goal. At the same time, the maximum wind speed obtained will be used to test the torque and power generation by the formula to see if the design is able to reach the goal of power generating.

1. Introduction

Wind power can be hailed as the world's best clean energy, and the reason why wind power can be so high credited is because of its power generating efficiency and the source of power. Although wind power equipment has excellent power generation benefits, few people realize that wind power equipment has a slight impact on the ecology. Wind power plants are often built in high mountains and coastal areas where to find strong wind speeds, and these areas are also easy to inhabit and migrate. In addition to the damage caused by human forced land acquisition, the noise generated by wind turbines, as well as bird strikes and even forced flying creatures to reduce flight altitude, will have an impact on the ecology. Therefore, this study developed a whole new wind turbine type which has both benefits of eco-friendly accelerated wind speed design.

1.1. Comparison of vertical axis wind turbine and horizontal axis wind turbine.

The blade is the most important part of the wind turbine and is also the key structure for converting kinetic energy into mechanical energy. However, because the source and speed of the world wind will cause different wind conditions depending on the terrain, the seasons and other factors. In order to capture the best performance, the blades are designed to adjust to different situations. At present, wind turbines can be divided into two categories in the world, horizontal axis wind turbines and vertical axis wind turbines. The biggest difference between the horizontal axis wind turbine and the vertical axis wind turbine is the blade device. The horizontal axis fan chamber and the rotating shaft are perpendicular to each other, and the reason why it is called the horizontal axis wind turbine is because the wind direction and the rotating axis are presented. The parallel state is called a horizontal axis wind turbine. The difference between the horizontal axis wind turbine and the vertical axis wind turbine is very large. The advantage of the horizontal axis fan is that the height of the windmill tower and the hub can be set higher than that of the vertical axis unit, and it can be obtained because of the high altitude relationship. The wind speed is also faster. Generally, the number of horizontal axis wind turbine blades mostly falls between 2 and 3 pieces. This number of wind turbines can achieve a higher wind energy utilization coefficient during operation. Although the starting wind speed is relatively higher than that of the vertical axis fan, however, because of the small number of blades and the light weight, the power generation efficiency is relatively high. This advantage is also the biggest difference between a horizontal axis fan and a vertical axis fan. In addition, the horizontal axis wind turbine will also fix the aerodynamic load when the windward angle is fixed, and the damage caused to the unit will be less. However, compared to the vertical axis fan, the rotor of the horizontal axis fan must have a roll control facing the wind direction, and when the direction is controlled, the fan body will vibrate due to the force to withstand the wind direction. At the same time, not only the pressure caused by the wind direction, the large horizontal axis blades will bear the bending moment load due to the wind cut relationship during the movement, and the force will affect the life of the fan, etc., so the fan design The upper horizontal axis unit must consider these factors. The vertical axis wind turbine is a structure in which the blades are rotated around the axis of the vertical axis, and also because the direction is called a vertical axis wind turbine. The vertical axis wind turbine is a bit so that it does not require a special windward surface. When the wind comes from any direction, it can be plucked and converted into mechanical energy without the need to adjust the setting of the device. Vertical axis fans are also less susceptible to wind cuts and therefore generate less load. At the same time, the disadvantage is that because the height of the rotor is relatively low, the wind speed is usually low, so the performance that cannot be achieved under long-term high-speed operation is low. The structure of the H-type fan is relatively simple, but the centrifugal force caused by this structure will cause severe bending stress at the reconnection point of the blade. In addition, linear blades also need to use the crossbar as a connection, which also produces some slight resistance and thus reduces efficiency. The table shows the change and comparison of the type ratio of vertical axis fan and horizontal axis fan.

Table 1. Vertical axis fan and horizontal axis fan type change ratio.

	Diameter (m)	Height (m)	Frontal Area (m^2)	Rotating Speed (RPM)	Output (kW)	Torque (Nm)
Vertical Axis Wind Turbine (VAWT) 1:1	10	10	67	95	23	2300
VAWT 1.5:1	10	15	100	95	34	3400
VAWT 1.8:1	10	18	120	95	41	4100
VAWT 3:1	10	30	200	95	68	6800
Horizontal Axis Wind Turbine (HAWT)	14.7	-	170	85	68	9960

1.2. The threat wind energy poses to flying creatures.

According to a journal published in the Mammal Study, researchers conducted bat death surveys from three different wind turbine wind farms in Taiwan between 2007 and 2010 to verify that wind turbines are creating clean energy. At the same time, it also hurts the flying creatures living in this environment.

According to the survey, five different types of bat corpses were found under the fan. The five bat species are: Pipistrellus Abrades, Scotophilus Kuhlii, Eptesicus Serotinus Horikawai), Myotis Formosus and Nyctalus Plancyi Velutinus. Among these different types of bats, most of the bats from Southeast Asian species are ridiculous, but the Chongchuan's brown bat and the golden rat ear bat are endemic to Taiwan. And although the bat is a night animal, the reason why the bat will die due to the wind turbine is that the bats are used to flying in the wind direction, just because the wind turbines are going to take better wind energy. After setting it on the best windward side. When the bat moves around the fan, the different sounds, magnetic fields, and changes in air pressure generated by the wind turbine may cause confusion to the bat itself, thus causing death to the blade.

At the same time, in the group of birds that are generally slightly known to attack birds, water birds have the greatest impact. As mentioned earlier, wind turbines are set in the optimal wind field in order to achieve the best performance, and the migration routes of these wind farms are usually the breeding and habitat of waterfowl. According to the research, wind turbines are pointed out. After the erection, the waterfowl changed the path because of the continuous impact of the impact. Whether it is changing the original path or adding a new flight path, it will leave the distance of the fan more than 100 meters, but not all birds can have such consciousness, so temporarily changing the direction also caused a lot of damage to them.

2. Wind tunnel design

2.1. The threat wind energy poses to flying creatures.

The main purposes of this study are to provide a safe passage space for flying animals, so there is no direct exposure setting for wind blades. Under such a concept, how to collect and introduce the flowing wind into the air duct is a very difficult problem in design.

This study takes advantage of a very well-known and very efficient physical phenomenon – the Venturi effect, also known as the Venturi effect, as a major conceptual basis for design. The Venturi effect is that a gas or liquid in a high-speed flow generates a low pressure in the vicinity and thus generates an adsorption. At the same time, when the restricted flow of gas or liquid passes through the greatly restricted flow cross section, the flow velocity and the flow increase phenomenon will occur instantaneously, and the flow rate will be inversely opposite to the flow cross section. This phenomenon can be called the Venturi effect and this effect is also very suitable for the design concept of this study.

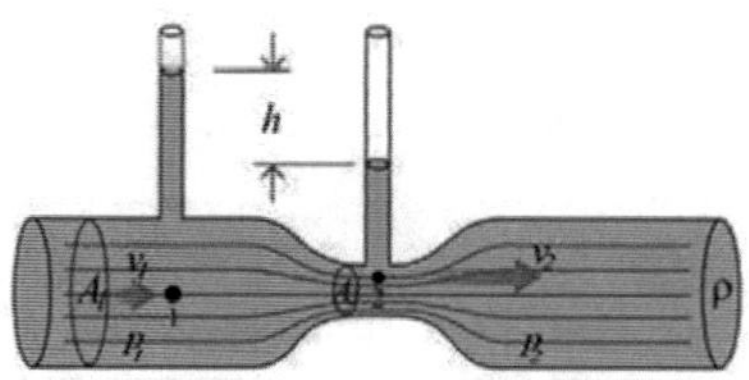

Figure 1. Venturi effect.

In this study, the Venturi effect was used to design the channel with curvature at the air inlet, which greatly reduced the size of the air inlet to enhance the wind speed. In addition to the channel design of the wind turbine blade reinforcement zone.

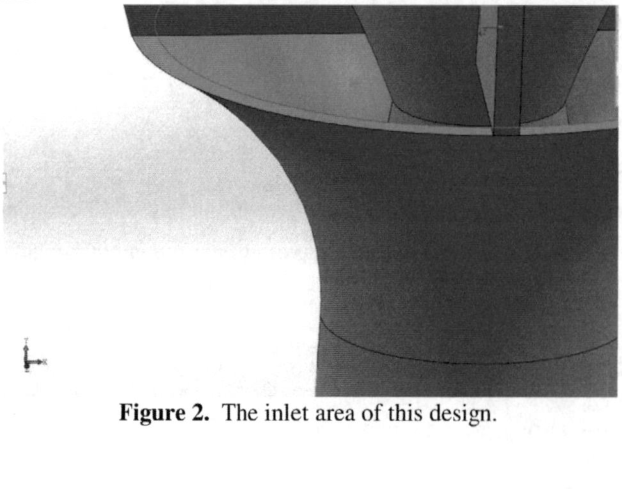

Figure 2. The inlet area of this design.

Figure 3. Venturi effect used on this design.

3. Fluid analysis of wind tunnel

The general mathematical model of solid-force analysis is mainly FIA (Finite Element Analysis) and can also be called Finite Element Analysis (FEM). The fluid analysis architecture is not using these mathematical models.

The operation is performed using the FVM (Finite Volume Method). The finite volume method does not use the explicit pressure equation, but uses the velocity equation to do the calculation. In the calculation, the three Momentum Equations momentum equations plus the Continuity Equation continuous equation form four equations to solve four different variable values.

The finite volume method can evaluate the mesh of the inlet and outlet momentum for each grid. When the momentum is unbalanced, the pressure will change and it will be calculated until the equilibrium is reached.

In order to find the best air duct design, the study is corrected by constant airway angle, and at the same time, the same preset wind speed is added to the same Domain size with 15 m/s for flow field analysis. When the direction flows to this air duct, whether the wind enters the air duct can smoothly introduce the wind into the air duct, or even has an acceleration function. The following figure shows the flow field analysis results of each air duct design.

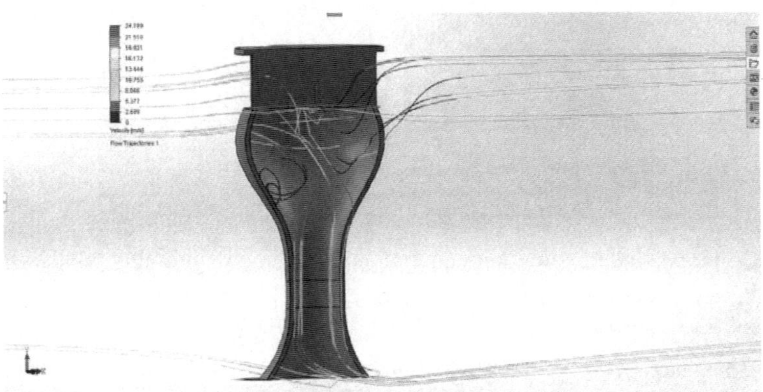

Figure 4. First design fluid analysis result.

This model is the first wind turbine of the wind turbine designed by the Study. At the beginning, this study designed the design as a torch in the city. The reason why this image is used is because modern people often feel helpless and helpless in the life of the city, and the torch torch gives the impression of a burning power. Such a concept stands in the city. When people feel overwhelmed by the pressure of life, looking up at the city's functional landscaping in the city can remind them to stick to it.

However, such a design seems to still have to be adjusted in actual use. According to the results of the flow field analysis, although the wind received by the air duct has flowed into the air duct, due to the structural design, the wind accumulated in the middle section and being introduced into the air is instantaneously dissipated and the wind speed is immediately weakened, representing the wind. Road needs to be improved through structural correction.

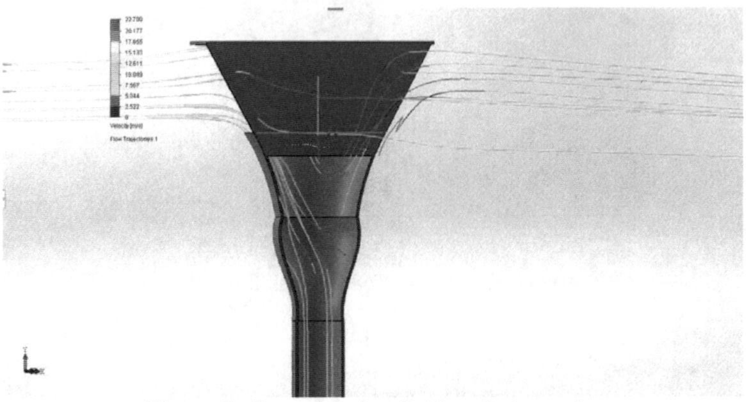

Figure 5. Second design fluid analysis result.

In the second edition of the wind tunnel design, in order to continue the original torch concept, it must be adjusted for efficiency. Therefore, in the middle part, the design of the torch hand is still left, but like the first version of the design, the wind speed at the beginning, there was a certain wind speed at the time of inflow, but when the fluid was introduced into the middle handle, the wind speed was still dispersed and weakened instantaneously, so the research still needs to be adjusted.

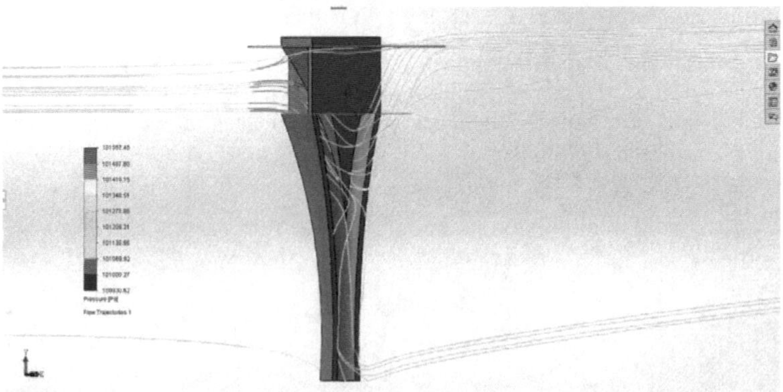

Figure 6. Third design fluid analysis result.

This model is the third edition of the wind tunnel design. This study still uses a relatively slender design concept to introduce the design, and also continues the concept of a torch like trying to bring wind turbines into inspirational life aesthetics.

According to the previous version of the duct design, it seems that the middle part of the road seems to be widened in an instant, so the introduced wind speed will be slowed down and unable to concentrate in an instant, so the design is reduced in the middle part, but it seems that the angle into the duct is too sharp, so Wind speed not only does not increase in the air duct, but also slows down the wind speed and reduces the flow. Therefore, the design of the wind turbine is continuously improved.

IOP Conf. Series: Materials Science and Engineering **644** (2019) 012022 doi:10.1088/1757-899X/644/1/012022

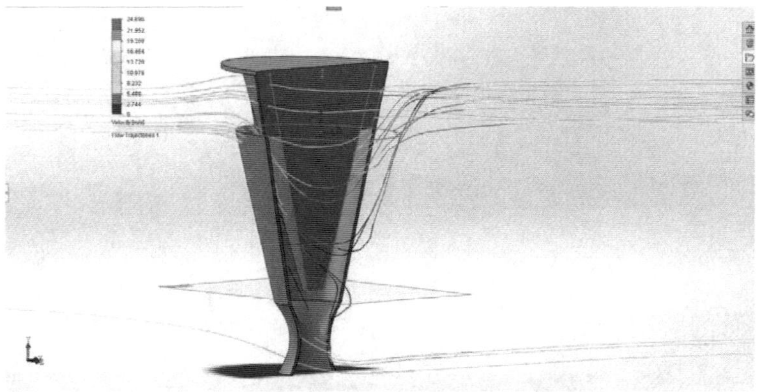

Figure 7. Fourth design fluid analysis result.

Since the windshield design of the first three editions has not been able to achieve the expected wind speed progress, in the fourth edition, the concept of life aesthetics that wants to be included in the torch was broken, and the practical aspect was mainly changed. Since the first three editions have been found in the flow field analysis, it has been found that the wind cannot be gathered and therein, so in particular, try to lengthen the partition at the center point to see if the wind can be gathered and accelerated. However, according to the results of the flow field analysis, it is shown that such a structure can not only concentrate the wind, but the wind speed has completely dropped to 0 m/s when the wind flows to the tail end, indicating that there is no wind speed at all, so this structure is still not feasible. Previously, in order to find the best design, it took too much time to design the wind turbine's air duct. Therefore, when the fifth edition design was carried out, it was considered that it should be integrated into the theoretical research. Therefore, in the theoretical research, the venturi theory mentioned in the literature was found. This study immediately considers that if venturi theory can be incorporated into this airway design, it is bound to achieve the goal of concentrating and accelerating wind speed at the beginning of this study.

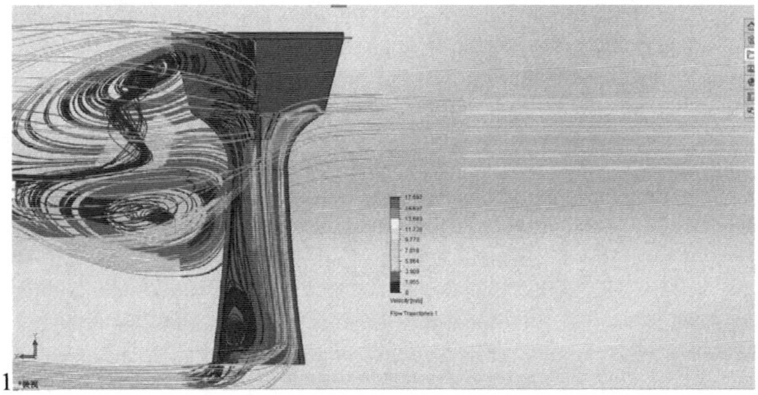

Figure 8. Best design fluid analysis result.

By using venturi theory, the angle of the air inlet is still reserved for the design of the wide mouth, but the venturi theory is applied to the trend of the venturi and narrowly narrowed. After establishing this design and analysing the flow field through computer 3D modelling, it is found that the wind speed starts to slow down when the wind flows into the entrance, but it is accelerated because of the narrowing of the wind channel. In this analysis, The wind speed in these areas even maintained the original set wind speed of 15 m/s, which was not weakened at all. Therefore, the design of the air duct was designed as the best design, and the air duct design to be used for this study was also determined.

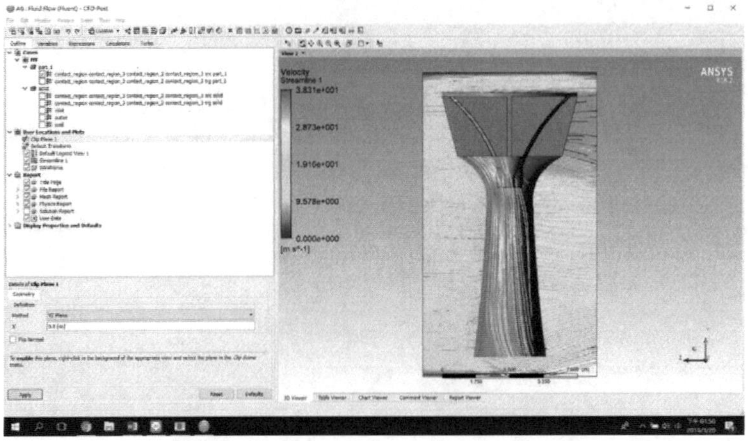

Figure 9. Best design fluid analysis result by using Ansys Software.

In the previous analysis, the flow field of 15 m/s through Flow Simulation of the modelling software Solid works, the result shown in the previous section is that the wind speed is maintained at a certain wind speed. Therefore, the wind speed of the inlet is set to 15 m/s as in the Ansys Fluent analysis, but the Domain is pulled down because the wind is an unstable fluid phenomenon. The Fluent analysis of the small Domain shows that when the wind enters the wind turbine, the wind speed is sharply enhanced. The strongest wind speed reaches 38.31 m/s, which is more than twice the present wind speed of 15 m/s. However, because this Domain's inlet distance is relatively close, the wind speed is likely to be so strong, but it can be understood that the design can maintain or even accelerate the design regardless of the analysis software.

4. Conclusions

According to the analysis, different angles of the wind tunnel will affect the result and the efficiency of the wind flow. In the previous designs, the inlet was designed too large and also there is no area to speed up the wind flow. And because of the analysis results, allows the design to be modified immediately and made it better. By using Venturi Effect into the design to solve the problems of wind speed slowed down when the wind blew into the tunnel. This effect not only maintains the same speed of the wind but also speed up the wind flow. When designing a new engineering design, analysis on variety of functions is very important. By observing every result came from analysis software allows designers and engineers to modify the structure immediately. This helps cooperation or any facility to cost down the fee of development new items. In this study, the main concept is to design a brand-new wind turbine system which has the function of eco-preservation. By putting eco-friendly idea into wind turbine is hard enough, but the harder thing is how to overcome the fact that it is very hard to balance the efficiency and the environment preservation. Therefore, by using analysis software helps this study rapidly modify the structure and test again to reach the optimization design. And by non-stop modifying and testing, finally find the best design for this study.

5. References

[1] Damila Rekioua, Wind Power Electric Systems:Modeling, Simulation and Control(Berlin: Verlag), 2014

[2] R.K. Kavade and P.M. Ghanegaonkar, Journal of Clean Technologies. **5** 353-358, 2017

[3] V. Lyatkher, Wind Power: Turbine Design, Selection and Optimization (Berlin : Wiley), 2014

[4] K. Seki and I. Ushitama, Vertical Axis Wind Turbine (Taipei : National Taiwan University) , 2011

[5] The Guardian Is the Wind Turbine Really a New Apex Predator? http://www.theguardian.com , 2018

[6] KKNEWS, The Effect That Changes the World – Venturi Effect. Retrieved February 20, 2019 from www.kknew.com, 2015

[7] P.J. Tang, 2016 ANSYS Fluent 16.0 Super Learning Manual (Beijing: Posts and Telecom Press)

[8] Y.H. Lin, C.I. Chen, Tsung Hseik Tu, Application of ANSYS CFX in Wing Turbine Fluid Structure Interaction Simulation. Journal of Applied Sound and Vibration, **9**, 1&2, 39-47, 2017

[9] Z.G. Tsai, 2013 ANSYS/Workbench Finate Element Analysis and Applied Engineering (New Taipei City : Bookcity)

[10] M. Kulak, M. Karczewski, K. Olasek, K. Jozwik. CFD Analysis of Diffuser Augmented Wind Turbine Model for Wind Tunnel Investigation. IECON, IEEE(2016)

IOP Conf. Series: Materials Science and Engineering **644** (2019) 012023 doi:10.1088/1757-899X/644/1/012023

The Invention of Cam-Linkage Grinding Device with Double-Circle Grinding Traces

Tzu-Hsia Chen[1] and Long-Chang Hsieh[2]

1Department of Mechanical and Computer-Aided Engineering, Feng Chia University, Taiwan
2Department of Power Mechanical Engineering, National Formosa University, Taiwan
Email: summer34134@gmail.com

Abstract. The double-circle grinding traces has 0% velocity deviation percentage (V_{dp}=0%), which is the optimum polishing trace. This paper focuses on the invention of cam-linkage grinding devices with double-circle grinding traces. First, based on the modified Yan's creative design methodology, 1st cam-linkage grinding device is synthesized. Then, according to Osborn's Checklist Method, 2nd cam-linkage grinding device is synthesized. These two cam-linkage grinding devices can be designed to have optimum polishing traces and 0% velocity deviation percentage (V_{dp}=0%). This paper provides a new creative design methodology for creating some or all possible design concepts that have the different topological characteristics as existing designs, e.g. with the different degrees of freedom. It can be said to create some or all possible design concepts from nothing (no existing design). The results of this paper can be used as the teaching material of innovative design course.

1. Introduction

In the past researches, the mechanisms of grinding devices can be planetary-type [1-6] and the link-type [7-9]. The link-type grinding devices, shown in Fig. 1(a), provide 8-shape grinding traces, shown in Fig. 1(b), and have velocity deviation percentages (V_{dp}) between 36.5520%~76.8652%. The mechanism, shown in Fig. 1(a), has two degrees of freedom and its links 2 and 6 are the two inputs (having the kinematic constraint "$\Delta\theta_6 = 2\Delta\theta_2$") to get 8-shape grinding g trace. Fig. 1(b) shows the 8-shape grinding traces with 3 straight line segments, so the grinding qualities are not good. The link-type grinding devices cannot be designed to have zero velocity deviation percentages (V_{dp} =0%). The ideal grinding traces, shown in Fig. 2, is grinding traces with double circles (OO-shape grinding traces). In 1997, Buzzetti [10] used the relatively high-cost PLC to control the link-type polishing device to get grinding traces with double circles. In order to reduce the manufacturing cost, other patents [11-13] of grinding devices are proposed.

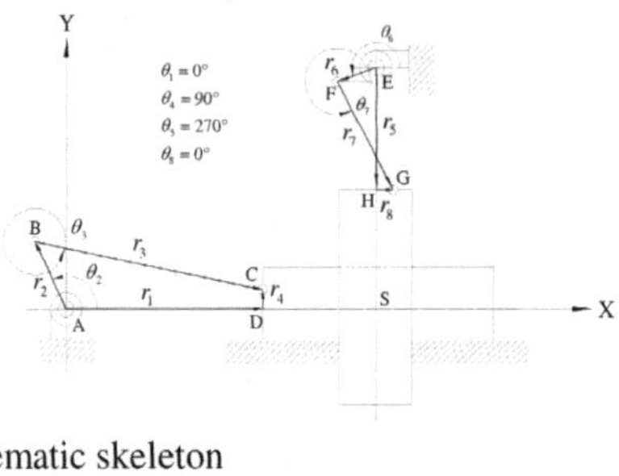

(a) Kinematic skeleton

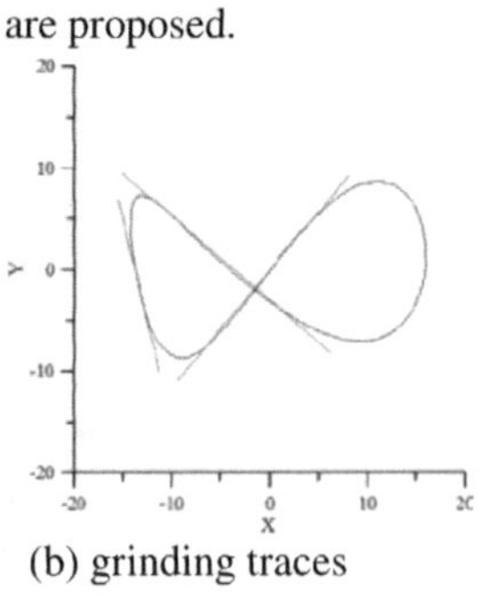

(b) grinding traces

Figure 1. Link-type grinding device.

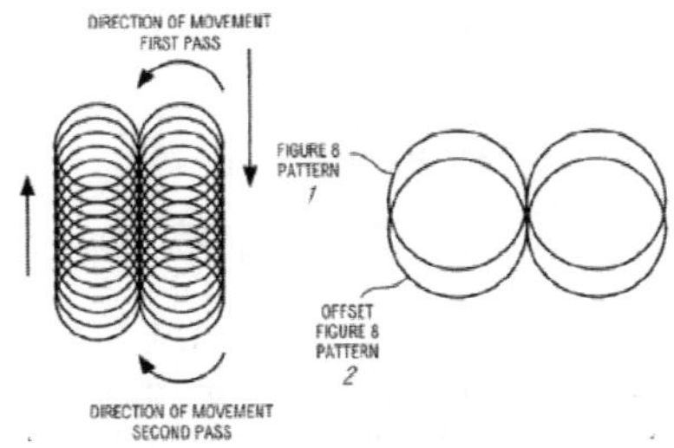

Figure 2. The ideal grinding traces with double circles [11].

This paper focuses on the invention of cam-linkage grinding devices with ideal grinding traces (double-circle grinding traces). Yan's creative design methodology is the method for creating some or all possible design concepts that have the same topological characteristics as existing designs, e.g. with the same degrees of freedom. First, based on the creative design of mechanical devices [14-17], the modified Yan's creative design methodology is proposed to create some design concepts that have the different topological characteristics as existing designs, e.g. with the different degrees of freedom. Then, according to the modified Yan's creative design methodology, 1st cam-linkage grinding device is generated. Finally, according to Osborn's Checklist Method [18-23], 2nd cam-linkage grinding device is synthesized. These two cam-linkage polishing devices are designed to have ideal grinding traces (double-circle grinding traces).

2. First invention by modified Yan's creative design methodology

Conceptual design is the initial stage of the engineering design process, which is the most difficult part. There is no existing mechanism (F=1) which can provides double-circle grinding traces, the Yan's creative design methodology [15-18] must be modified. Fig. 3 shows the modified Yan's creative design methodology, and its steps are:

Step 1. Based on the existing designs (F=2), conclude the topological characteristics and design requirements and constraints.

Step 2. According to the topological characteristics, synthesize the atlas of generalized chains (F=1) that can be used to design cam-linkage polishing devices.

Step 3. Based on the design requirements and constraints, synthesize the atlas of feasible specialized chains by assigning link types and joint types to each generalized chain.

Step 4. Particularize each feasible specialized chain into its corresponding kinematic skeleton, the atlas of cam-linkage polishing devices can be obtained.

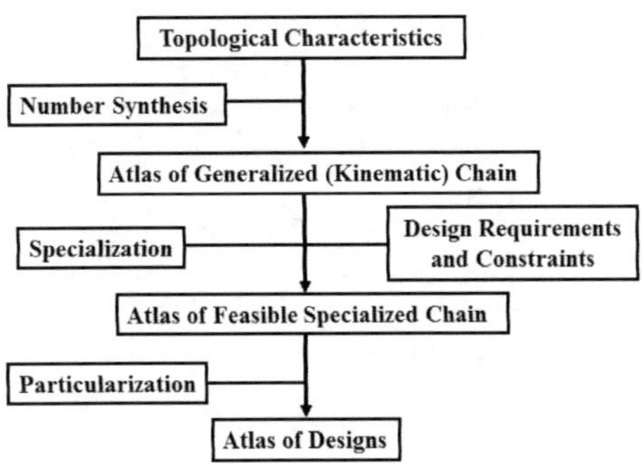

Figure 3. Modified Yan's creative design methodology.

2.1. *Topological characteristics.*

The 1st step of this invention is to conclude the topological characteristics of existing grinding devices. The link-type grinding devices are the mechanisms with 2 degrees of freedom. The purpose of this paper is to invent mechanism with 1 degree of freedom (DOF) to provide double-circle grinding traces. For the mechanism having double-circle grinding traces, it must have cam pairs. Eq. (1) shows the degrees of freedom of planar mechanical devices.

$$F = 3(N-1) - 2J_1 - J_2 \tag{1}$$

Where: N is the number of links, J_1 is the number of joints with 1 degree of freedom, and J_2 is the number of joints with 2 degrees of freedom.
According to Eq. (1), if N=4, J_1=3, and J_2=2, the corresponding generalized chain (4, 5) has 1 degree of freedom. Hence, the grinding device must have 2 cam pairs. The topological characteristics of grinding device with double-circle grinding traces are concluded as belows:

1. It is a mechanism with 4 links and 5 joints.
2. It has 4 links including a ground link (Gr), cam link (C_A), 1st slider moving in X direction (S_1), 2nd slider moving in Y direction (S_2). The 2nd slider of grinding device can be also named as grinding disk.
3. It has 5 joints including 1 revolute joint (J_R), 2 prismatic joints (J_P), and 2 cam joints (J_A).

2.2. Atlas generalized chain.
The 2nd step of this invention is to synthesize all possible generalized chains. According to above topological characteristics, Fig. 4 shows the generalized chain with 4 links and 5 joints.

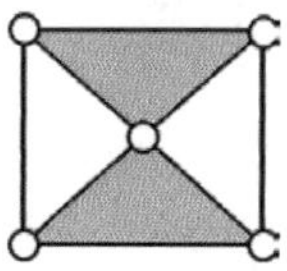

Figure 4. Generalized chain (4, 5).

2.3. Design requirements and constrains.
Based on the topological characteristics, the design requirements and constraints of the cam-linkage grinding devices are:

1. There must be a ground link (G_R), cam link (C_A), 1st slider moving in X direction (S_1), 2nd slider moving in Y direction (S_2).
2. The ground link (G_R) must be adjacent to cam link (C_A) with revolute pair and adjacent to first slider (S_1) with prismatic pair.
3. The ground link (G_R) cannot be incident to cam pair.
4. The cam link (C_A) must at least have 1revolute pair and 1 cam pair.
5. 1st slider (S_1) must be adjacent to the ground link (G_R) with prismatic pair.
6. 1st slider (S_1) must be adjacent to the second slider (S_2) with prismatic pair.
7. 2nd slider (S_2) can't be adjacent to the frame.
8. There are 5 joints including 1 revolute pair (J_R), 2 prismatic pairs (J_P) and 2 cam pairs (J_A)
9. There is at least 1cam pair in 3-bar loop.

2.4. Specialization.
The 3rd step of this invention is to assign specific types of members and joints to each generalized chain, subject to design requirements and constraints to have specialized chains. The specializing steps for polishing devices are:

1. Identify the ground link (G_R) to each generalized chain for all possible cases. There are 2 possible identifications as shown in Figs. 5(a) and 5(b).
2. Identify the 2nd slider (S_2) to each case obtained in Step 1. For the specialized chain shown in Fig. 5(b), based on the design requirements and constraints, there is no possible identification of 2nd slider (S_2). So, there is only 1 possible identification, shown in Fig. 6.
3. Identify the cam link (C_A) and 1st slider (S_1) to each case obtained in Step2. There is only 1 possible identification, shown in Fig. 7.
4. Identify the corresponding revolute pairs (denoted by ○), prismatic pairs (denoted by ●), and cam pairs (denoted by ●) to each case obtained in Step 3. There is only 1 feasible specialized chain, shown in Fig. 8.

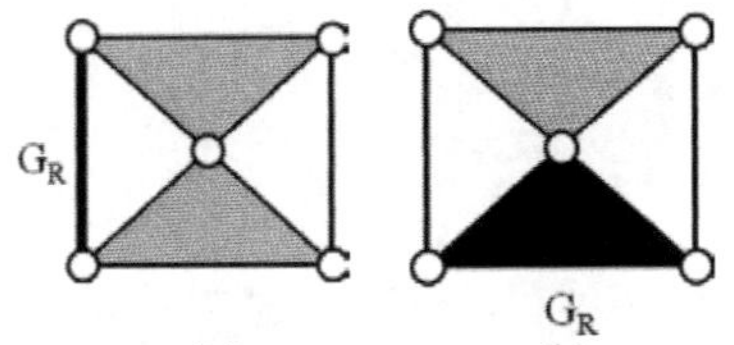

Figure 5. Identify ground link (G_R).

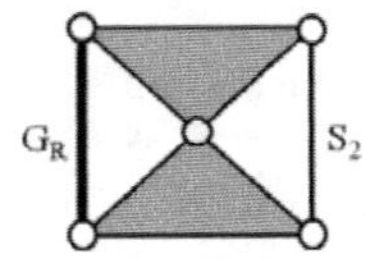

Figure 6. Identify 2nd slider (S_2).

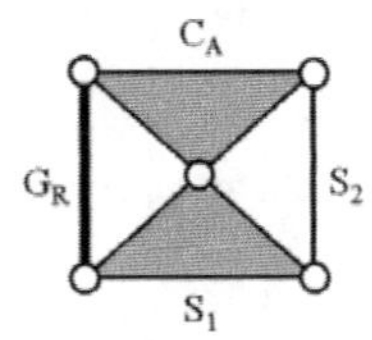

Figure 7. Identify cam link (C_A) and 1st slider (S_1).

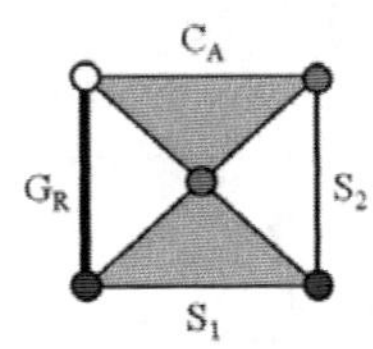

Figure 8. Feasible specialized chain.

2.5. Particularization.

The last step of this invention is to particularize the feasible specialized chain into its corresponding kinematic skeleton. Particularization is the reverse process of generalization, and can be done by applying the generalizing rules in reverse order. Fig. 9 shows the corresponding kinematic skeleton of the feasible specialized chain as shown in Fig. 8. The mechanism, shown in Fig. 9, has 4 links and 5 joints (J_1=3 and J_1=2), its' degrees of freedom can be calculated as:

$$F = 3(4-1) - 2\times 3 - 1\times 2 = 9 - 8 = +1 \tag{2}$$

It is a grinding device with 1 degree of freedom, and can be designed to have double-circle grinding traces (ideal grinding traces).

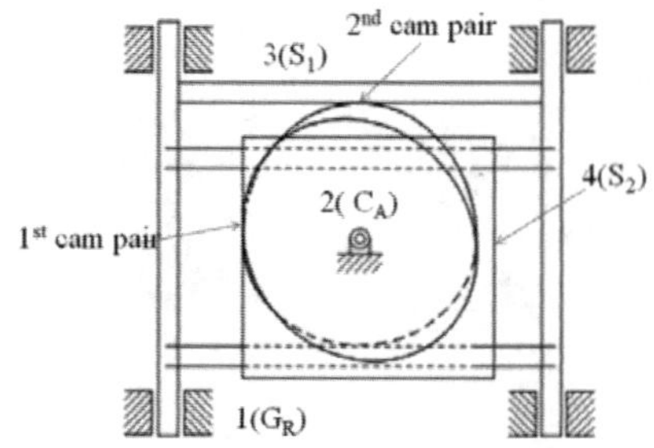

Figure 9. 1st new design concept of the cam-linkage grinding device.

3. Second invention by checklist method

The checklist is the simplest method for problem solving. It is a method that provides hints by checking the items on a prepared list against the problem. In 1953, Alex Faickney Osborn published his book *"Applied Imagination: Principles and Procedures of Creative Problem Solving*" in which Osborn presented the technique of *checklist*. Professor Dior of Harvard University summarized Osborn's 83 questions to 9 items. The 9 items are : 1. Reverse, 2. Transfer, 3. Combine, 4. Change, 5. Extend, 6. Enlarge, 7. Reduce, 8. Substitute, 9. Rearrange.

Here. based on the "change method", the new invention could be realized. For the mechanism shown in Fig. 9, the 2nd cam pair is incident to cam link (C_A) and first slider (S_1). Here, we make a minor change "2nd cam pair is changed to be incident to cam link (C_A) and 2nd slider (S_2)". Based on this minor change, the corresponding kinematic skeleton is shown in Fig. 10. The mechanism, shown in Fig. 10, has 4 links and 5 joints (J_l=3 and J_l=2), its' degrees of freedom can be calculated as:

$$F = 3(4-1) - 2\times 3 - 1\times 2 = 9 - 8 = +1 \tag{3}$$

It is a grinding device with 1 degree of freedom, and can be designed to have double-circle grinding traces (ideal grinding traces).

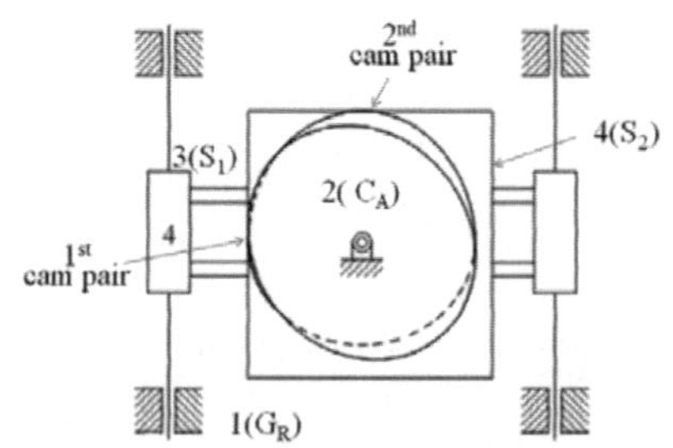

Figure 10. 2nd new design concept of the cam-linkage grinding device.

4. Cam profile design and dynamic simulation

The cam is a mechanical element with a special profile, and the follower will be driven to have the desired motion. It has the advantage of small size, compact construction, good dynamic performance, and easy designed to meet motion requirements. It is widely used in automated machinery, and plays an irreplaceable role. According to Fig.10, the schematic diagram of cam pairs can be redrawn as Fig.11. The cam link is a dual cam which is coupled by two disc cams.

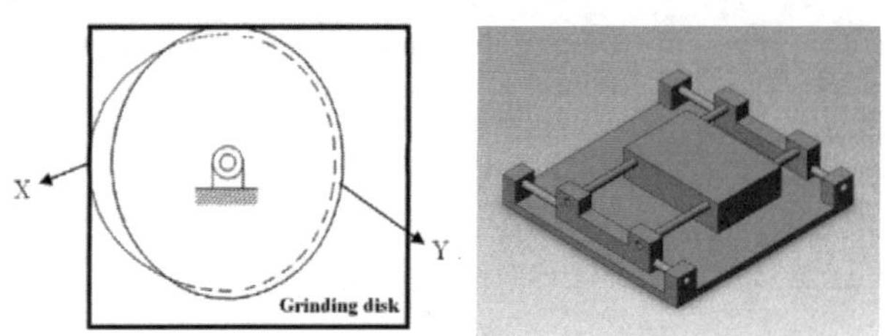

Figure 11. Schematic diagram of cam pairs.

4.1. Cam profiles of x-cam and y-cam.

X-cam controls the displacement of grinding disk in X direction, and Y-cam controls the displacement of grinding disk in Y direction. By the special profiles of X-cam and Y-cam, grinding disk (S_2) can provides the optimum grinding traces (double-circle grinding traces) as shown in Fig. 12.

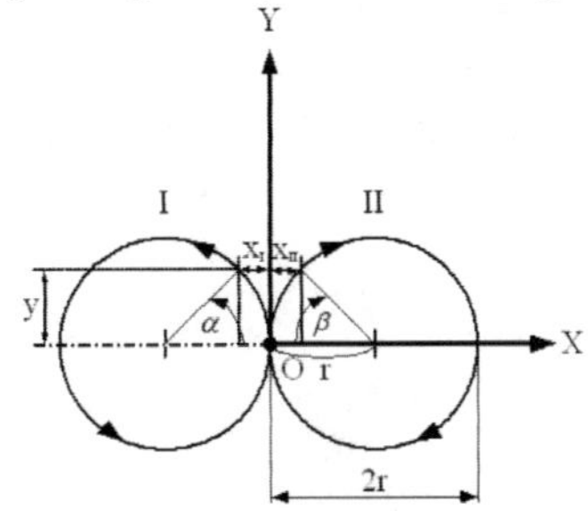

Figure 12. Optimum grinding traces(double-circle grinding traces).

By analytical method, Chen [24] synthesized the cam profiles of X-cam and Y-cam, shown in Figs. 13(a) and 13(b). If the rotation speed of cam is 30rpm, Fig. 14(a), 14(b), and 14(c) show the corresponding speed in X-direction, Y-direction, and linear velocity of grinding disk. The linear velocities of grinding disk are constant value "V= 47.12 mm/sec". According to Fig. 14(c), it is concluded that this invention "Cam-Linkage Grinding Device" has zero velocity deviation percentages (V_{dp})=0%. And, Fig. 15 shows its theoretical grinding traces.

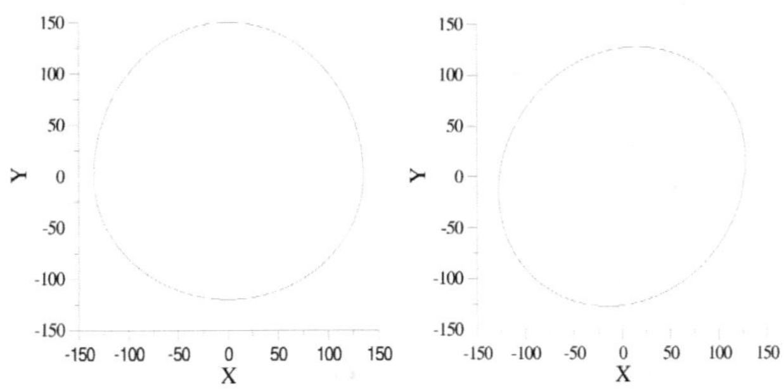

(a) X-cam (b) Y-cam

Figure 13. Cam profiles of X-disk and Y-disk (with conditions r =7.5mm and r_b =240mm).

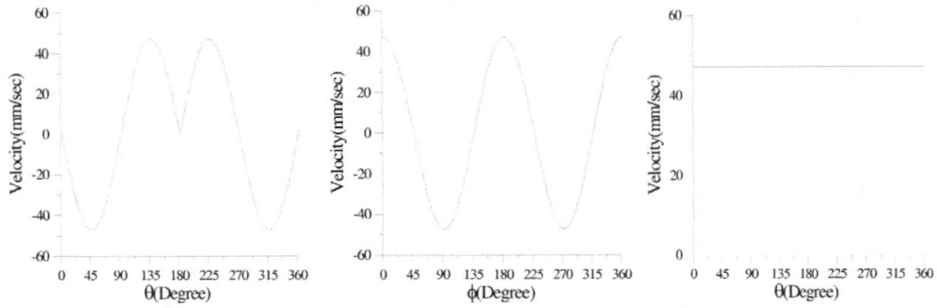

(a) X-direction (b) Y-direction (c) Linear velocity

Figure 14. Grinding velocity of grinding disk.

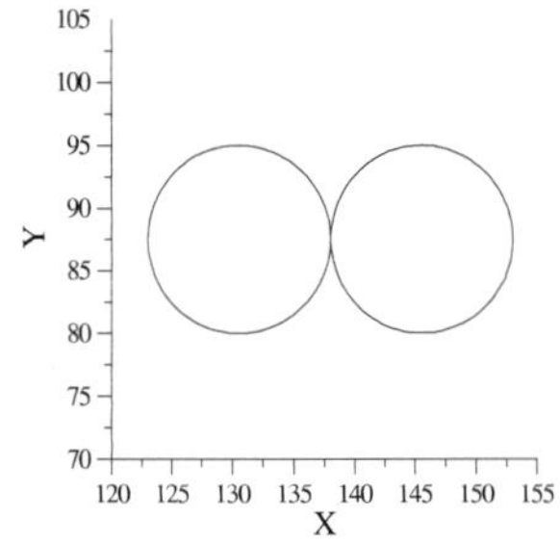

Figure 15. Grinding traces of the invention "Cam-Linkage Polishing Device".

5. Conclusions

Traditionally, for link-type polishers, the velocity deviation percentages (V_{dp}) are between 36.5520%~76.8652%. It can't be designed to have zero velocity deviation percentages (V_{dp})=0%. This paper provides a new creative design methodology of for creating some or all possible design concepts that have the different topological characteristics as existing designs, e.g. with the different degrees of freedom. First, based on the modified Yan's creative design methodology, 1st cam-linkage grinding device is synthesized. Then, according to Osborn's Check-List Method, 2nd cam-linkage grinding device is synthesized. These two cam-linkage grinding devices can be designed to have optimum polishing traces and 0% velocity deviation percentage (V_{dp}=0%). The results of this paper can be used as the teaching material of innovative design course.

Acknowledgements

The authors are grateful to Chumpower Machinery Co., Ltd. for the support of this research under 105AF71.

6. References

[1] M. Takahashi, Elastic polishing plate method and conditions for forming angled convex surface on ferrule endface, Journal of Lightwave Technology, Vol. 15, No. 9, pp.1675-1680, 1997

[2] H.X. Zhang, M.G.Wu, and H.G Gao, Simulation of even material removal during eccentric plane polishing by a tin polisher, Optics and Precision Engineering, Vol. 6, pp. 77-82, 1998

[3] C.Y. Huang, Polishing analysis and simulation of optical fiber endface, National Taiwan University, Department of Mechanical Engineering, master's thesis, 2000

[4] L.Yin, H. Huang, W.K. Chen, Z. Xiong, Y.C. Liu, and P.L.Teo, Polishing of fiber optic connectors, International Journal of Machine Too1s & Manufacture, Vol. 44, pp. 659-668, 2004

[5] L.C. Hsieh, H.S. Lee, and Z.Y. Wang, The design of planetary gear trains–application of optical fiber polishers, International Journal of Mechanical Engineering Education, Vol. 36, No .1, pp. 16-36, 2008

[6] L.C. Hsieh and T.H. Chen,The Systematic Design of Planetary-Type Grinding devices for Optical fiber ferrules and wafers, Transactions of the Canadian Society for Mechanical Engineering, Vol. 40, No. 4, pp. 619-630, 2016

[7] L.C. Hsieh, H.S. Lee, and Z.Y. Wang, Polishing trace analysis of optical fiber end-face polisher, 2nd Conference on Precision Machinery and Manufacturing Technology, pp. 171-181, Pingtung, Taiwan, 4-6 June, 2004

[8] L.C. Hsieh, and T.H. Chen, Optimal design of polishing trace for link-type optical fiber polisher, Journal of Advanced Materials Research, Vol. 328-330, pp. 900-905, 2011

[9] L.C. Hsieh, and T.H. Chen, The systematic design of link-type optical fiber polisher with single flat, Journal of Advanced Science Letters, Vol. 9, pp. 318-324, 2012

[10] M. Buzzetti, Computer-Controlled Method For Polishing, U.S. Patent, No. 5947797, 1997

[11] M. Buzzetti, Method and Apparatus for Polishing, U.S. Patent, No. 6428391, 1999

[12] L.C. Hsieh, S.L.Chang, J.U. Liu, and S.T. Zhou, Optical Fiber End-Surface Polisher, ROC patent, M175358, 2001

[13] T. Boyer, Polishing Pad Assembly for Fiber Optic Cable Connector Polishing Apparatus, U.S. Patent, No. 6641472, 2003

[14] H.S.Yan, Creative Design of Mechanical Devices, Springer-Verlag, Singapore, 1998

[15] H.S.Yan & L.C. Hsieh, Conceptual Design of Gear Differentials for Automotive Vehicles, ASME Transactions, Journal of Mechanical Design, Vol. 116, pp. 565-570, 1994

[16] L.C. Hsieh, H.S.Lee, and B. Fu, Innovative Design and System Development of Active Orthosis,Journal of machine design and research, Vol. 23, No. 4, pp. 163-167, 2007

[17] L.C. Hsieh & T.Z. Chen, The Systematic Design of Link-Type Optical Fiber Polisher with Single Flat, Journal of Advanced Science Letters, Vol. 9, pp. 318-324, 2012

[18] A.F. Osborn, Wake Up Your Mind: 101 ways to develop creativeness, New York, London: Charles Scribner's Sons, 1952

[19] A.F. Osborn, Applied Imagination: Principles and Procedures of Creative Problem Solving, New York: Charles Scribner's Sons, 1953

[20] L.C. Hsieh, and H.S. Lee, Creative Design of Wafer Transport Mechanism, The TRIZ Journal, 2004/07, pp. 1-30, 2004

[21] L.C. Hsieh, H.S. Lee, and Z.Y Wang, Innovative Design for Link-Type Optical Fiber Polisher, The TRIZ Journal, 2006/07

[22] L.C. Hsieh & T.Z. Chen, The Innovative Design of Quick Folding Bicycle with High Rigidity, Journal of Advanced Science Letters, Vol. 9, pp. 24-29, 2012

[23] L.C. Hsieh, & H.C. Tang, The Innovative Design of Automatic Transmissions for Electric Motorcycles,Transactions of the Canadian Society for Mechanical Engineering, Vol. 37, No. 3, pp. 741-753, 2013

[24] T.H. Chen, The Systematic Design and Prototype Verification of Cam-Linkage Polishing Devices With Double-Circle Polishing Traces, Journal of Advances in Mechanical Engineering, Vol. 10, No. 12, pp. 1-8, 2018

Investigation on an Active Thermoelectric Vacuum Sensor with Low Frequency Modulation

Shu-Jung Chen and Ruei-Ting Ding

Department of Mechatronics Engineering, National Changhua University of Education, No. 2, Shi-Da Road, Changhua, Taiwan
Email: sjchen@cc.ncue.edu.tw

Abstract. In this research, we propose and investigate a thermoelectric-type vacuum sensor with an active heater in the center of sensor under low frequency modulation. Thermal type sensor for vacuum measurement with active-heating is proposed for the suspended thermoelectric element to reach a better performance. Using the TSMC 0.35 μm CMOS-MEMS process, the proposed thermoelectric sensor with active heater is designed and fabricated with central-symmetrical thermocouples positioned. There are 64 pairs of thermocouples placed around the center of membrane and each thermocouple is designed with two kinds of material, n+ Poly and metal. The material of Metal 1 ~ Metal 4 layers is aluminum, and titanium is used as contact layers to connect these metal layers. The total resistance of the structure with four units is measured to be about 14.8 kΩ and the resistance of each unit is about 3.7 kΩ. Four quarters of sensing units of thermopile were connected and the sensing area at the center will give a temperature difference. According to the Seebeck effect, serial connected thermocouples will produce a weak voltage difference between the hot and cold junctions after heat exchange under air convection related to vacuum pressure. The heating of thermopile by active heater is proceeded under different frequencies and the output signal is acquired with a Phase-Lock-Loop (PLL) amplifier. When the periodic heat exchange will take away the heat to cause a temperature drop of sensing area gives a corresponding periodic weak voltage between the cold and hot end of the thermocouples. According to a careful investigation of the PLL measurement with a wide range of 10 m~100 Torr, our proposed sensing scheme based on a thermoelectric type sensor is proved for practical vacuum detection. Most of all it is proved as a new approach to use a commercial thermopile with heater, which is easier to include than a special custom design.

1. Introduction

With the rapid advancement of MEMS process technology, the size of the vacuum gauge has gradually become smaller. From the research literature of the vacuum gauge from 2005 to 2017, it can be found that the development of the vacuum sensor focused on the microelectromechanical processes and CMOS processes. Some papers propose different approaches to improve vacuum response measurement, or to extend the sensing range of the vacuum gauge in a special process. For these researches, the heat sensing for micromachined transducers are popular for applications that are of interest for vacuum sensing because the microelectromechanical industry with outstanding advances in semiconductor process develops rapidly. Application of this new technology for vacuum sensing enables products miniaturized in nanometer, at the same time, improves its performance, as well as reliability and reduces the cost of manufacture [1].

Especially, vacuum sensors intended for measurements of pressures below atmospheric are developed by using well-known and mature CMOS-MEMS devices which explore many important applications [2]. Thermopile-based MEMS vacuum sensors [3] give a new approach with mature integration of MEMS sensors and circuits and bring advantages with wide pressure range. The thermopile devices based on Seebeck effect are also utilized for gas flow sensors, vacuum detectors, thermal converters, etc. Some researches reveal a broad range of commercial applications [4-7].

2. Sensors and sensing scheme

2.1. Design and fabrication of proposed Metal-N-Poly thermoelectric-type sensor.

Using the standard CMOS TSMC 0.35 μm process and CMOS materials the proposed vacuum sensor is fabricated in Fig. 1. A series of post-processes including an isotropic RIE and wet etching processes are proceeded after the CMOS processes to remove the silicon under the sensing structures. Each thermocouple is fabricated with standard CMOS materials of the Metal 4 layers and N-polysilicon which is serial connected to each other. There are 64 pairs of thermocouples around the centre of sensing membrane. To remove the silicon substrate beneath the membrane efficiently, we design the etching windows of slim-long type situated between each two thermocouples which allow the etchant to flow through. The active area of thermopile is around 800x800 μm^2.

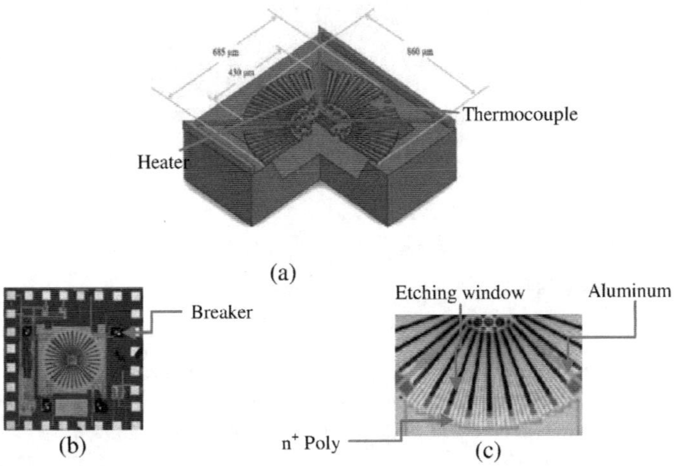

Figure 1. (a) Schematic drawing of thermopile structure; microscopic picture of (b) full chip after fabrication and (c) a quarter of the thermopile.

2.2. Proposed sensing scheme for thermoelectric-type sensor.

Installed in a vacuum chamber, the active heater of thermopile sensor is operated under periodic modulation of heating which acts as a heat source for the sensing membrane. Besides of the parameters of operation in our experiment structure, the sensitivity of vacuum sensor also depends on thermal properties including the heat capacity and heat conductance which is governed by the heat equation.

3. Measurement for proposed thermoelectric type sensor

The experimental structure includes a vacuum sensor, an input signal source, a vacuum chamber and a PLL amplifier, which is shown in Fig. 2. The input signal resource for the modulation is heating and cooling, and the vacuum sensor for modulation is vacuum level. After amplification and low-pass filtering, the signal is sent to the PLL amplifier. A time constant is setup at 10 seconds to acquire a stable reading for low frequency measurement.

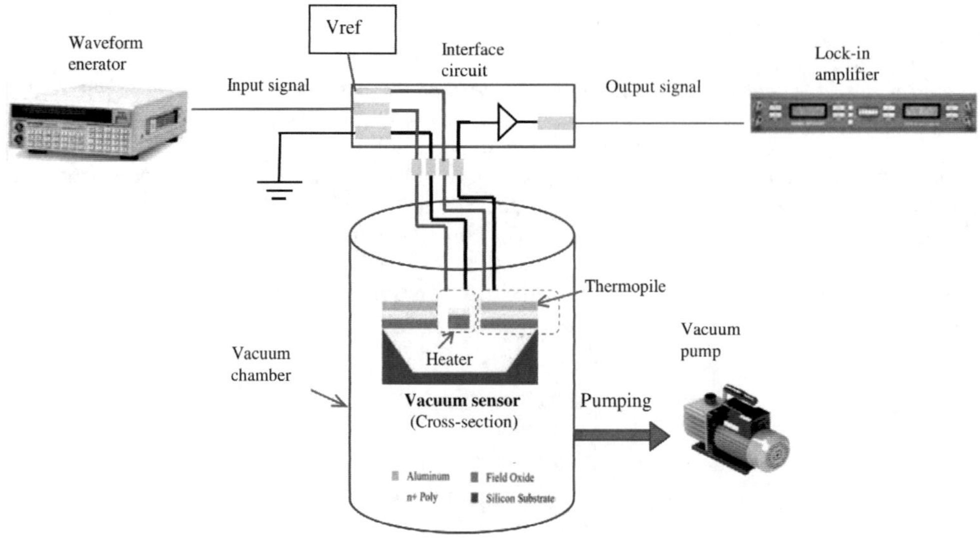

Figure 2. Setup of measurement for proposed thermoelectric type sensor.

4. Vacuum measurement and analysis

4.1. Voltage vs. frequency.

To investigate the frequency response, the proposed thermoelectric sensor is operated and sensing under the modulated heating for frequency 1~100 Hz at different vacuum conditions. In order to test thermopile output voltage at different vacuum levels, when heater input voltage at different frequencies from low to high, the voltage from the output of the thermopile will be changed, the data is collected and arranged which is shown in Fig.2. The modulation variable of this experiment is mainly the frequency of the heater input voltage. The experimental input voltage amplitude for heater is fixed at 0.5V, the offset voltage is 0.26V which gives a sinusoidal bias voltage to the heater at different vacuum.

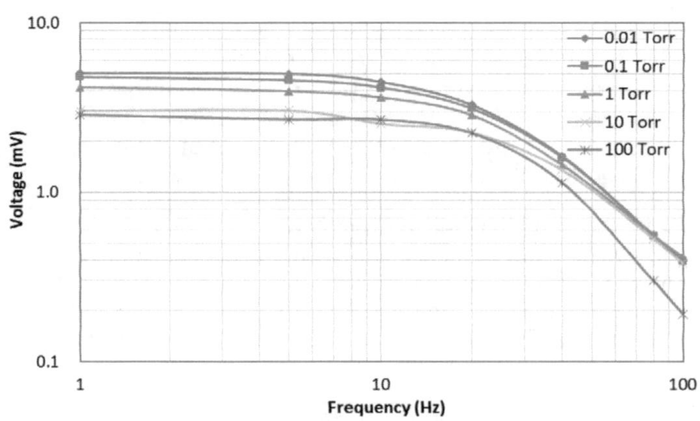

Figure 3. Thermopile voltage vs. frequency of input voltage for the heater.

At the same time, we can see that in the Fig. 3, when the input frequency exceeds 20 Hz, the output voltage begins to drop sharply, and when the frequency exceeds 60 Hz, the output voltages of different vacuum degrees are nearly equal.

4.2. Voltage vs. pressure.

Moreover, in order to observe that the vacuum gauge uses different degrees of vacuum at the different frequencies of input voltage for heater, the voltage output of the thermopile changes, Fig. 4 changes the X axis to vacuum and the line segment changes to the input frequency. The modulation variables of this experimental are mainly degrees of vacuum.

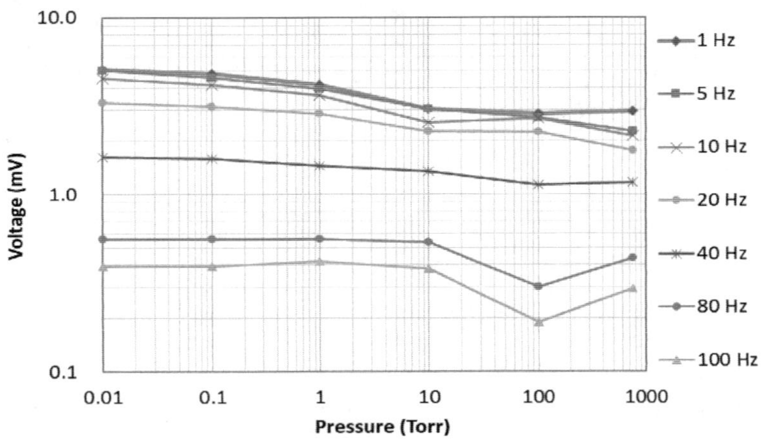

Figure 4. Thermopile voltage vs. pressure of chamber.

Fig. 4 changes the X-axis from frequency to vacuum. When the X-axis becomes vacuum, we can see the change of the thermopile output voltage under different vacuum degrees. In Fig. 4, it can be observed that when the degree of vacuum is lower than 10 Torr, the line segments of different frequencies can be clearly distinguished, but when the degree of vacuum is higher than 10 Torr, the output voltage at different input frequencies is irregular.

4.3. Voltage at different vacuum intervals.

When the heater input voltage with different frequencies and vacuum degree change experiments are completed, the data can be used to obtain trend line of output signal in the different vacuum intervals when the frequency of heater input voltage from low to high, and calculate the sensitivity of the chip under different vacuum degrees which is shown in Figure 5.

Fig. 5 takes frequency as the X axis, voltage as the Y axis, and vacuum as the fitting line of the vacuum range. In this figure, we can clearly see the voltage trend line graph at various frequencies. It shows the frequency of input voltage for the heater from low to high and the change of the sensitivity curve under different vacuum degrees. Between the different vacuum interval curves, it is found that the best sensitivity curve is obtained when the vacuum is 1 Torr to 10 Torr which is shown in Fig. 5. It means that we can find the best measurement sensitivity from these different vacuum intervals. And when we find the best sensitivity, we can confirm the best working vacuum of this vacuum gauge.

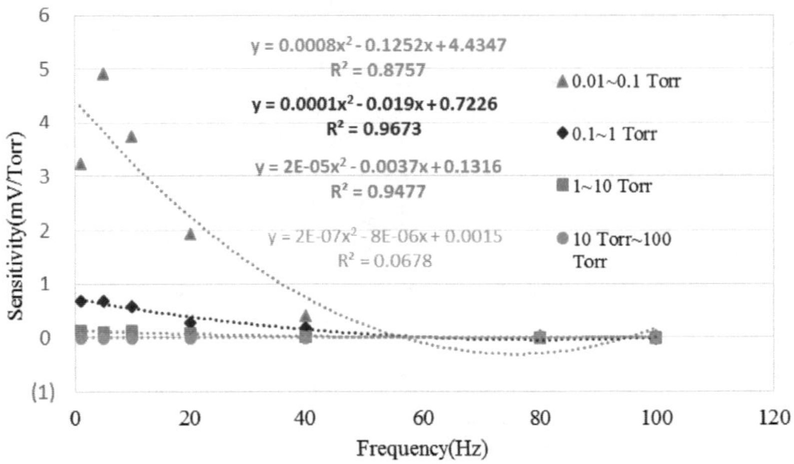

Figure 5. Thermopile voltage vs. frequency in different intervals.

5. Conclusions

In this research work, a thermoelectric-type vacuum sensor with heating modulation is proposed and the frequency response within 1 ~100 Hz is investigated. The periodic heating is proceeded based on the sinusoidal heating and PLL reading of the output signal from the amplified voltage of CMOS-MEMS thermoelectric sensor. The proposed thermopile with active heaters is designed and fabricated with 64 pairs by using the TSMC 0.35 μm CMOS-MEMS process and the post MEMS process. The results of vacuum measurement with a wide range of 10 m~10 Torr shows a good signal and better stability under low frequency modulation. A new sensing scheme based on a thermoelectric type sensor is proved for practical vacuum detection at low frequency modulation of heating and this research work is proved as a new approach to use an active thermopile with heater for vacuum measurement.

Acknowledgment

This research was supported in part by grant of making chips from National Chip Implementation Center (CIC). The authors would also like to thank the National Science Council of the Republic of china, Taiwan for financially supporting this research under Contract. MOST 106-2221-E-018-026.

6. References

[1]J.T.W. Kuo, L. Yu and E. Meng, Micromachined Thermal Flow Sensors—A Review, Micromachines, 3(3),550–573 (2012)

[2] A. Gorecka-Drzazga, Miniature and MEMS-type vacuum sensors and pumps, Vacuum, 83, 12 1419-1426 (2009)

[3] A.W. Van Herwaarden, P.M. Sarro and H.C. Meijer, Integrated Vacuum Sensor, Sens.Actuators, 8(3) (1985) 187-196

[4] D. Randjeloviu, A. Petropoulos, G. Kaltsas, M. Stojanoviua/DLu, Z. Djuriu, M. Matiu, Multipurpose MEMS thermal sensor based on thermopiles, Sensor Actuat A-Phys, 141, 404-413 (2008)

[5] D. Randjeloviu, Djuriu, Multipurpose MEMS thermal sensor based on thermopiles, Microelectron. Eng., 86, 1293-1296(2009)

[6] Xensor Integration, http://www.xensor.nl /txtfiles/ hfdfiles/ prodstan.html

[7]D. Randjeloviu, Z. Djuriu, M. Matiu, Vacuum MEMS sensor based on thermopiles-simple model and experimental results, Microelectronics, MIEL 2008. 26th International Conference on, 367-370 (2008)

IOP Conf. Series: Materials Science and Engineering **644** (2019) 012025 doi:10.1088/1757-899X/644/1/012025

40-GHz Millimeter-Wave Balanced Bandpass Filter Design in Standard 0.35-μm CMOS Process

Hsun-Hsiang Chen[1], Ching-Her Lee[2], Chung-I G. Hsu[3], and Shang-Wei Wang[2]

1 The Department of Electronic Engineering, National Changhua University of Education, Changhua 500, Taiwan.
2 The Graduate Institute of Communications Engineering, National Changhua University of Education, Changhua 500, Taiwan.
3 The Department of Electrical Engineering, NationalYunlin University of Science and Technology, Douliu 640, Taiwan.
E-mail: chenhh@cc.ncue.edu.tw

Abstract. In this work, a 40 GHz mm-wave balanced band-pass filter (BPF) is realized using coupled structure consisting of symmetric feed lines and stepped-impedance resonators (SIRs). For differential-mode operation, the BPF circuit exhibits band-pass characteristic; for common-mode operation it behaves as a band-stop unit. The designed balanced BPF circuit is realized by using TSMC 0.35-μm standard process. Good agreement was observed between simulated and measured results. The measured passband center frequency is at 38.8 GHz, with a minimum differential-mode insertion loss of 5.87 dB and a 3-dB bandwidth of 9.6 GHz.

1. Introduction

Recently, as the cost and the level of integration is concerned, silicon-based CMOS technology had become a good candidate as the solution of a highly integrated single-chip system, which incorporates digital, analog and radio frequency (RF) circuits. Filters are essential components in the RF front end of wireless communication systems [1-4]. While CMOS devices have been successfully applied to RF integrated circuit design. The on-chip filter is an essential component in RF or microwave circuits to achieve system-on- a-Chip (RF-SoC). However, it suffers from the high insertion loss due to the lossy substrate as well as the large occupied area in a standard CMOS process. Consequently an external off-chip filtering circuit is typically employed in a high performance system. CMOS based BPFs with a low insertion loss were earlier developed by [5-7] to overcome the high insertion loss due to the lossy substrate, so there are extensive research has been conducted in the area of on-chip filter design. Furthermore, millimeter-wave transceivers are presented with new design challenges such as increased phase noise, limited amplifier gain, and the need for transmission line modeling of circuit components [8-9]. So that , we designed such a balanced filter to reject the noise which is reduced from common mode terminal, and several balanced circuit topologies have been developed to connect with the monopole antenna for which a balun is needed to convert a balanced signal into unbalanced.

The on-chip integrated antenna usually takes the form of a dipole, which needs a balanced filter for differential operation. A differential-to-differential balanced bandpass filter usually has the desired bandpass frequency response in differential operation and has good common-mode signal rejection, so it can increase the signal-to-noise-ratio (SNR).

2. Filter design

The proposed bandpass filter was implemented in TSMC 0.35 μm CMOS technology 2P4M standard process as an immersed microstrip line (IMSL) structure, the CMOS wafer cross-sectional view is shown in Fig. 1.

The top metal layer (M4) is used to form the filter circuit and bottom metal layer (M1) is used as ground plane. The proposed balanced bandpass filter structure is shown in Fig. 2. which consists a typical coupled line structure combined with a $\lambda/4$ stepped-impedance resonator, a plan of symmetry exists in the circuit center plan. When operated in differential mode, the plan of symmetry is grounded and the equivalent half circuit is shown in Fig. 3(a). The circuit has a bandpass characteristic, for common-mode operation, the plan of symmetry is opened, the equivalent half circuit is shown in Fig. 3(b). This circuit has all stop characteristic.

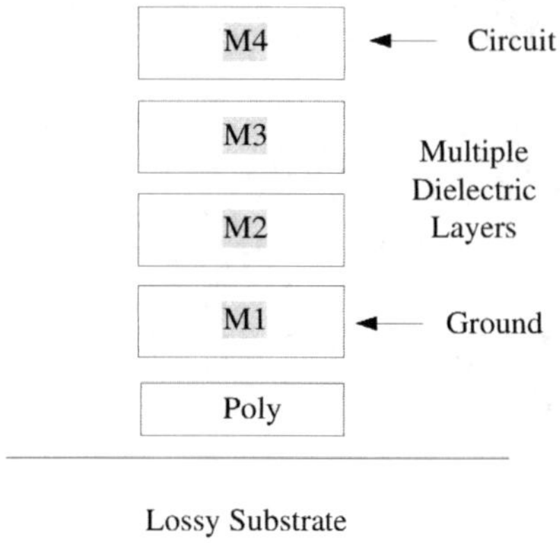

Figure 1. Stack up of metal layers in 0.35 CMOS process Cross-sectional view.

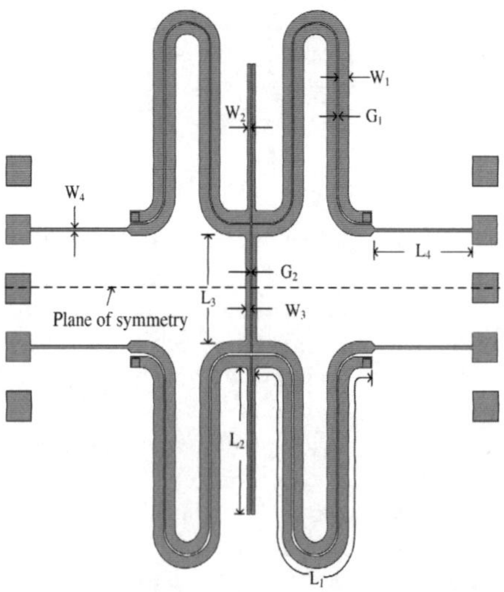

Figure 2. The proposed circuit, where L_1=702, L_2=348, L_3=200, L_4=47.17, W_1=11.66, W_2=3.15, W_3=13.15, W_4=47.17, G_1=1.18, G_2=0.85 (Unit: μm).

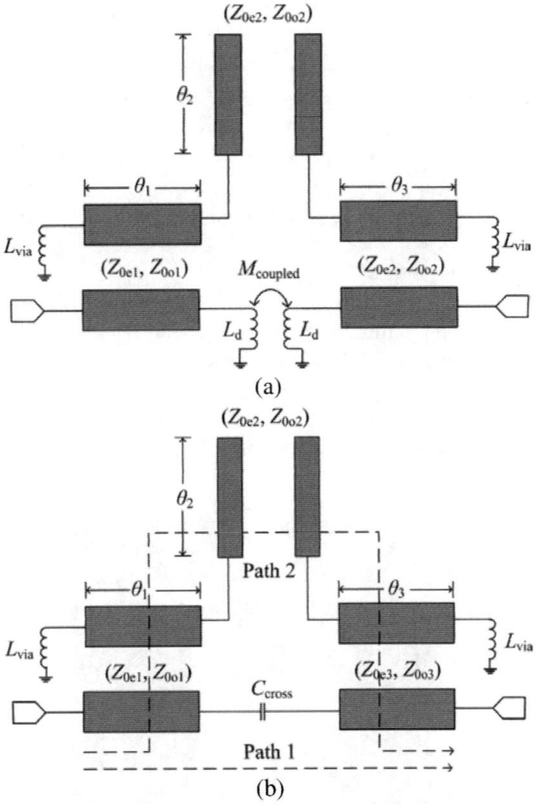

Figure 3. The BPF equivalent half circuit of (a) differential mode, (b) common mode.

We can analyze the filter characteristics of the coupled line by calculating the image impedance in terms of the Z-parameters, then the equations can be solved for even- and odd- mode line impedances to give

$$Z_0 J_1 = \sqrt{\frac{\pi \cdot \Delta}{2g_1}} \tag{1}$$

$$Z_0 J_2 = \frac{\pi \cdot \Delta}{2\sqrt{g_1 g_2}} \tag{2}$$

$$Z_0 J_3 = \sqrt{\frac{\pi \cdot \Delta}{2g_2 g_3}} \tag{3}$$

$$Z_{0e} = Z_0[1 + JZ_0 + (JZ_0)^2] \tag{4}$$

$$Z_{0o} = Z_0[1 - JZ_0 + (JZ_0)^2] \tag{5}$$

According to the two-stage 3 dB Chebyshev filter theory, the coefficients is given as, $g_1 = 3.1013$, $g_2 = 0.5339$, $g_3 = 5.8095$. In this work, the passband of differential mode is designed to be centered at 40 GHz with 11% fractional bandwidth. Note that (1), (2), (3), (4), (5) has been given in[10], coupled line is composed of a quarter-wavelength section with two ends shorted, the odd-mode equivalent circuit exhibits a bandpass frequency response with the center frequency in f_0, then the plane of symmetry looks like short, and as an odd-mode. In another hand, the even-mode equivalent circuit would present the all-stop frequency response,

IOP Conf. Series: Materials Science and Engineering **644** (2019) 012025 doi:10.1088/1757-899X/644/1/012025

because it consists of a coupled-line section with one end shorted, and the plane of symmetry looks like open [11-12].

The even and odd-mode characteristic impedances of this design are $Z_{0e1} = Z_{0e3} = 45.24$, $Z_{0o1} = Z_{0o3} = 24.15$, $Z_{0o1} = 107.2$, $Z_{0o2} = 36.5$, which can be used to decide the width and gap between coupled line. In order to decrease the loss of circuit, we choice $Z_0 = 33$ for reference impedance which has wider line width(W_1). The second couple ine width (W_2) is smaller the W_1 to increase coupling efficiency, the electrical length θ_1 at 40 GHz is near 90 degree, which corresponds to a length L_1 include L_2 is 1032 μm, and the BPF is simulated by EM simulation software of Agilent technology ADS momentum.

3. Simulation and measurement results

Fig. 4 shows the die photo of the proposed filter, and the measurement and simulation S-parameter results in differential mode and common mode of the filter with slot ground are shown in Fig. 5. From the Fig. 5, we can see that the simulation result and measured data is well matched. The measured passband centre frequency is at 38.8 GHz, with a minimum differential-mode insertion loss of 5.87 dB and a 3-dB bandwidth of 9.6 GHz. In common mode, the common mode rejection is very good in wide frequency range, from the measured data S_{21} < -18.02 dB within 20 ~ 50 GHz and S_{21} = -25.8dB at 38.8 GHz. The detail simulation and measured results are listed in table 1.

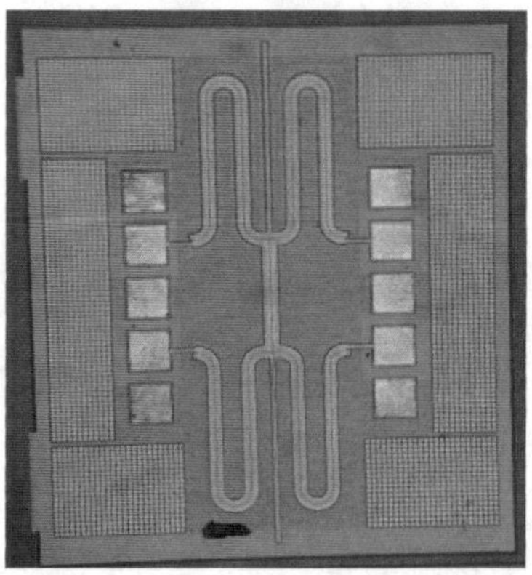

Figure 4. The die photo of the proposed filter.

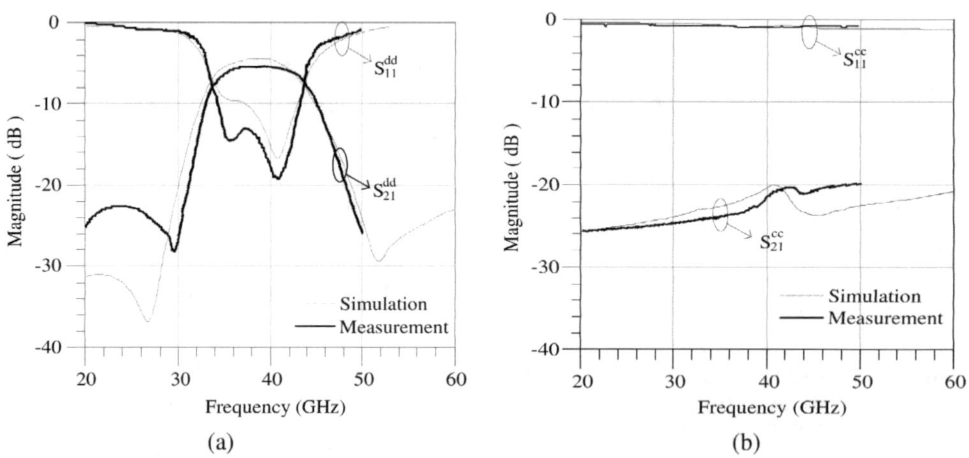

Figure 5. The simulation and measurement results (a) Differential-mode (b) Common-mode.

Table 1. The summary of Simulation and measurement.

Process		TSMC 0.35-μm CMOS	
Operation type		S_{11}	S_{21}
Simulation	Differential-mod e	-17.374 dB @ 40 GHz	-5.084 dB @ 40 GHz
	Common-mode	S_{21} < -20.92 dB within 20 ~ 60 GHz and -25.76 dB @40 GHz	
Operation type		S_{11}	S_{21}
Measurement	Differential-mod e	-21.92 dB @ 38.8 GHz	-5.87 dB @ 38.8 GHz
	Common-mode	S_{21} < -18.02 dB within 20 ~ 50 GHz and -25.8 dB @38.8 GHz	

4. Summary

We successfully demonstrate the implementation of a millimeter wave balanced bandpass filter in 40 GHz millimeter-wave frequency range. The measured passband centre frequency which is at 38.8 GHz, with a minimum differential-mode insertion loss of 5.87 dB and a 3-dB bandwidth of 9.6 GHz. For common-mode, insertion loss is about 18.02 dB within 20 to 50 GHz and 25.8 dB in 38.8 GHz center frequency.

Acknowledgements

The authors would like to thank Taiwan semiconductor research institute (TSRI) and TSMC for circuit fabrication.

5. References

[1] B. Razavi, A 60-GHz CMOS Receiver Front-End, IEEE Journal of SOLID-STATE CIRCUITS, vol. 41,no. 1, pp. 17-22, Jan. 2006

[2] L.K. Yeh, C.Y. Hsu, C.Y. Chen, and H.R. Chuang, A 24-/60-GHz CMOS On-Chip Dual-Band Bandpass Filter Using Trisection Dual-Behavior Resonators," IEEE EDL, vol. 29, no. 12, pp. 1373-1375, Dec. 2008

[3] C.Y. Hsu, C.Y. Chen, and H.R. Chuang, 70 GHz Folded Loop Dual-Mode Bandpass Filter Fabricated Using 0.18 □m Standard CMOS Technology, IEEE Microwave and Wireless Components Letters, vol. 18, no. 9, pp. 587-590, Sep. 2008

[4] H.C. Kuo, H.L.Yue, Y.W. Ou, C.C. Lin, and H.-R. Chuang, A 60-GHz CMOS Sub-Harmonic RF Receiver With Integrated On-Chip Artificial-Magnetic-Conductor Yagi Antenna and Balun Bandpass Filter for Very-Short-Range Gigabit Communications, IEEE Trans on MTT, vol. 61, no. 4, pp. 1681-1691, Apr. 2013

[5] C.Y. Hsu, C.Y. Chen, and H.R. Chuang, A 60-GHz Millimeter-Wave Bandpass Filter Using 0.18-μm CMOS Technology, IEEE EDL, vol. 29, no. 3, pp. 1681-1691, pp. 246-248, Mar. 2008

[6] S. Sun, J. Shi, L. Zhu, S. C. Rustagi, and K. Mouthaan, Millimeter-Wave Bandpass Filters by Standard 0.18-μm CMOS Technology, IEEE EDL, vol. 28, no. 3, pp. 220-222, Mar. 2007

[7] M. Miao and C. Nguyen, A Novel Multilayer Aperture-coupled Cavity Resonator for Millimeter-wave CMOS RFICs, IEEE Trans. MTT, vol. 55, no. 4, pp. 783–787, Apr. 2007

[8] C.H. Wu, C.H. Wang, and C.H. Chen, Balanced coupled-resonator bandpass filters using multisection resonators for common-mode suppression and stopband extension, IEEE Transactions on Microwave Theory and Techniques, vol. 55, no. 8, pp. 1756-1763, Aug. 2007

[9] L.K. Yeung and K.L. Wu, An LTCC balanced-to-unbalanced extracted-pole bandpass filter with complex load, IEEE Transactions on Microwave Theory and Techniques, vol. 54 no. 4, pp. 1512-1518, Apr. 2006.

[10] D M. Pozar, Microwave Engineering, 3rd edition, John Wiley & Sons, Inc., 2004

[11] C.H. Wu, C.H. Wang, and C.H. Chen, Novel balanced coupled-line bandpass filters with common-mode noise suppression, IEEE Transactions on Microwave Theory and Techniques, vol. 55, no. 2, pp. 287-295, Feb. 2007

[12] C.H. Wu, C.H. Wang, and C. H. Chen, Stopband-extended balanced bandpass filter using coupled stepped-impedance resonators, IEEE Microwave and Wireless Components Letters, vol. 17, no. 7, pp. 507-509, July 2007

IOP Conf. Series: Materials Science and Engineering **644** (2019) 012026 doi:10.1088/1757-899X/644/1/012026

Low Noise Amplifier Design for IoT Wireless Communication Systems

Yeong-Lin Lai[1], Yeong-Kang Lai[2], Guo-Jang Lee[1], Chun-Yi Zheng[1], Ping-Wen Huang[1], Ming-Ho Lee[1], and Yun Wei Chiang[1]

1 Department of Mechatronics Engineering, National Changhua University of Education, 50007 Changhua, Taiwan
2 Department of Electrical Engineering, National Chung Hsing University, 40227 Taichung, Taiwan
Email: yllai@cc.ncue.edu.tw

Abstract. In the era of Internet of Things (IoT), the Internet has evolved from a simple Internet function of web information access to intelligent functions of identification, position, monitoring, and management of things. Devices in the IoT must transmit data between the devices and equipment connecting to cloud. As a fog computing architecture is established proximally at the local ends of the IoT, the data transmission volume and transmission delay can be effectively reduced. IoT wireless communication is one of the essential items for complete data transmission between the devices and on-line equipment. This paper proposes the low noise amplifier (LNA) design that can be applied to the RF front-end receiver of a 2.45-GHz wireless communication system for IoT applications. The LNA is required to have characteristics of low noise factor and high signal gain in order to amplify weak signals received by the antenna. In this study, the design of 2.45-GHz LNA adopts an architecture of power-constrained simultaneous noise and input matching on the basis of the 0.18-μm CMOS process technology in order to achieve simultaneous noise and input matching at low power conditions. Both the architectures of push-pull and forward substrate bias are also utilized. The LNA demonstrates the characteristics of low noise factor, high gain, and good 1-dB gain compression. The LNA shows good potential for IoT wireless communication system applications.

1. Introduction

The concept of a network of devices was discussed as early as 1982. A modified Coke vending machine at Carnegie Mellon University became the first Internet-connected appliance [1]. It was able to report its inventory and whether the newly loading drinks were cold or not [2]. Mark Weiser's paper on ubiquitous computing, "The Computer of the 21st century" published in 1991, and academic venues suchlike UbiComp and PerCom brought out a contemporary vision for the Internet of Things (IoT) [3]. In 1999, the concept of "Internet of Things" became more popular through Auto-ID Center of Massachusetts institute of technology (MIT) [4]. Kevin Ashton, one of the founders of the Auto-ID Center, considered radio-frequency identification (RFID) [5-10] to be an essential element of the IoT [11]. RFID tags can be utilized to mark the unique identity of objects in daily life so that individual things are managed and inventoried.

In 2005, the International telecommunication union (ITU) published "ITU Internet Report 2005: IoT" at the world summit on the information society (WSIS), which pointed out the era of "IoT" coming. The ITU proposed to establish an application product architecture with RFID as the concept of the IoT [12]. The report also presented that objects can communicate and exchange information to server in the network environment with Internet-connected.

—

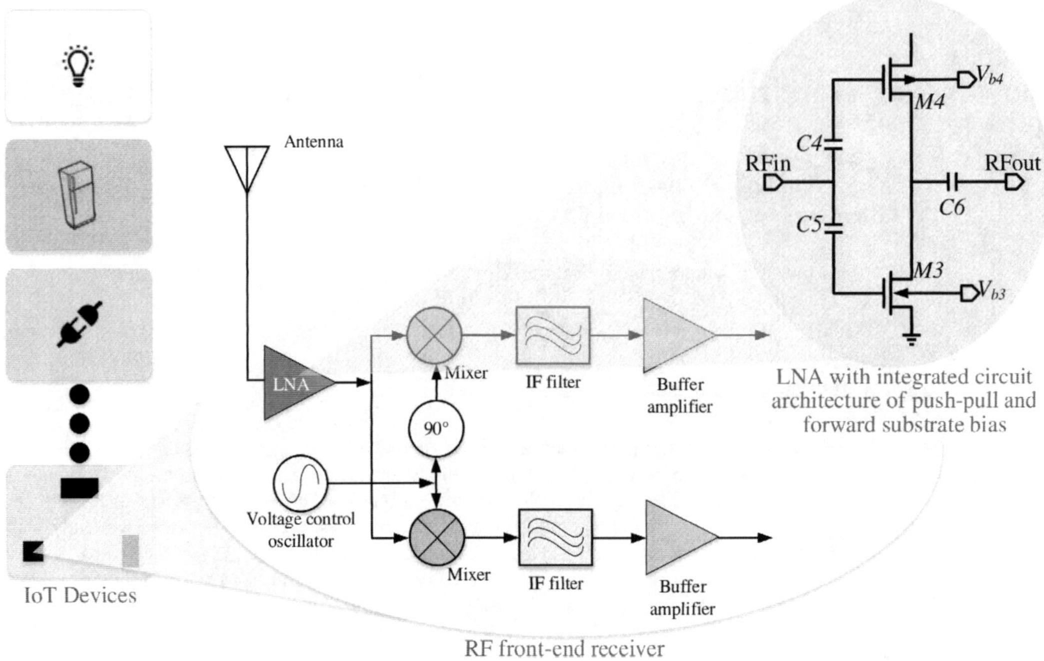

Figure 1. Block diagram of RF front-end receiver for IoT wireless communication systems.

The concept of the IoT is an integrated sensing technology that can transmit sensed human bio-signals or actual physical signals in the environment to the Internet via wireless or wired communication networks for storage in the cloud. The IoT is a network of physical objects including but not limited to vehicles, buildings, equipment and health monitoring devices. This network utilizes electronic devices including sensors and actuators. In this way, the IoT can access the physical data of the objects in the physical world through the Internet, and thus can be monitored and operated remotely to improve the system optimization performance. The IoT may be further combined with the big data analysis in the cloud and even artificial intelligence, which can effectively transform data into services of multiple fields such as individuals, families, communities, cities, and factories, and realize applications such as smart health care, smart homes, smart cities, and smart factories [13-16].

In IoT applications, devices must transfer data between the devices and equipment connecting to the cloud. A fog computing architecture established at local ends of the IoT can effectively reduce the data transmission volume and delay. IoT wireless communication is one of the essential items for complete data transmission between the devices and on-line equipment. IoT wireless communication is one of the essential items for complete data transmission between the devices and on-line equipment. A low noise amplifier (LNA) [17] plays a significant role in the front-end circuit of the receiver. This is the first-stage active amplifier of the front-end circuit of the receiver. Fig. 1 shows the block diagram of the RF front-end receiver for IoT wireless communication systems. The LNA is required in the RF front-end receiver to amplify weak signals received by the antenna. Therefore, the LNA must have the characteristics of low noise factor and high signal gain.

With the advancement of semiconductor process technology, low power feasibility has increased. However, the LNA circuit architecture most commonly used is dominated by a cascode configuration of common source and common gate architecture. However, such a cascode configuration increases the difficulty of the implementation of low power circuits. Thus, some papers proposed other architectures, such as a folded architecture, to achieve circuit characteristics of low power consumption.

This paper proposes the LNA design that can be applied to the RF front-end receiver of a 2.45-GHz wireless communication systems for IoT applications. The LNA designed utilizes the integrated push-pull and forward substrate bias circuit architecture in this paper. The LNA circuit exhibits low DC power consumption and reasonable gain circuit characteristics.

IOP Conf. Series: Materials Science and Engineering **644** (2019) 012026 doi:10.1088/1757-899X/644/1/012026

2. Circuit architecture

2.1. Power-constrained simultaneous noise and input matching

In this paper, the LNA design employed a matching method of LNAs at low power. The matching method is called power-constrained simultaneous noise and input matching [18] that can obtain simultaneous noise and input matching at low power conditions. The circuit architecture is shown in Fig. 2. The input impedance can be expressed as follows:

$$Z_{in3} = sL_{s3} + \frac{1}{sC_{t3}} + \frac{g_{m3}L_{s3}}{C_{t3}} \tag{1}$$

where C_{t3} is the sum of the parasitic capacitor C_{gs3} and the additional capacitor C_{ex3}, g_{m3} is the transconductance of M_3, and L_{s3} is the source degenerate inductor. Simultaneous noise and input matching occurs by tuning L_{g3}, C_{ex3}, L_{s3}, and other parameters until that Z_{s3} is equal to Z_{opt} and Z_{in3} is equal to the conjugate of Z_{opt}, i.e., Z_{s3} is the conjugate of Z_{in3}.

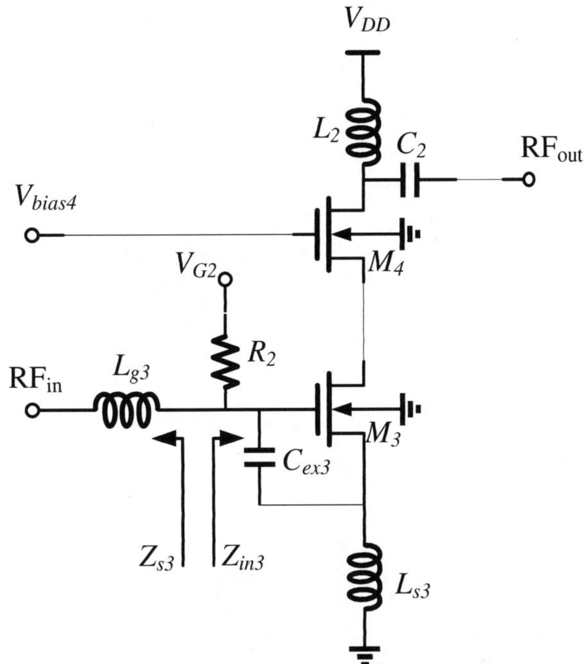

Figure 2. Power-constrained simultaneous noise and input matching architecture.

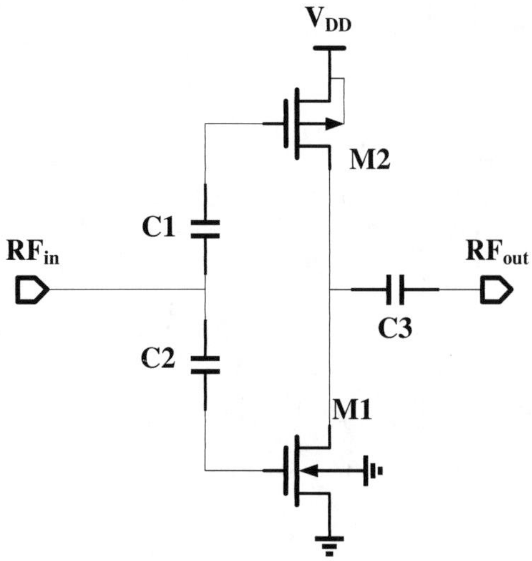

Figure 3. Current-reuse push-pull architecture.

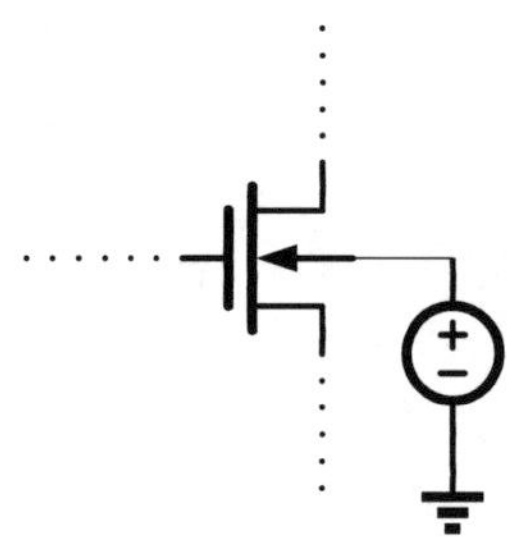

Figure 4. Diagram of forward substrate bias.

2.2. Push-pull architecture

Fig. 3 shows a current-reuse push-pull architecture. Transistors *M1* and *M2* are NMOS and PMOS, respectively. Capacitor *C1*, *C2*, and *C3* are all used for DC-Blocking capacitors. With the same signal gain, the traditional push-pull architecture consumes only a half of the current consumption compared to the common source or common gate architecture. Because *M1* and *M2* are in the same loop, it is the technology of current reuse [19]. And the equivalent transconductance of the overall circuit can be expressed by the equation

$$G_{m(total)} = G_{m(nmos)} + G_{m(pmos)} \quad (2)$$

Therefore, the push-pull circuit architecture can still maintain the same gain at a lower current consumption, thereby contribute to the transconduction of the transistors *M1* and *M2*.

2.3. Forward substrate bias architecture

Fig. 4 is the circuit architecture of the forward substrate bias. The threshold voltage V_{th} is adjustable according to

$$V_{th} = V_{th0} + \gamma \ (\sqrt{2\varphi_f - V_{bs}} - \sqrt{2\varphi_f}) \quad (3)$$

V_{bs}, the voltage of the substrate to the source of the transistor, determines the threshold voltage of the MOS transistor, V_{th}. Usually, the substrate and the source of transistor are either connected or reverse biased. In this study, the substrate and source of the transistor is applied with a forward bias in order to reduce the V_{th} for the improvement of performance. It should be noted that, the first, the forward bias applied to substrate and source has to be biased below P-N junction turn-on voltage of the transistor. And the second, the CMOS process technology must include the deep N-well technology to reduce the noise transmitted through the substrate [20]. TSMC 0.18-μm CMOS process technology is utilized to meet the requirements. The architecture of the forward substrate bias is proposed to reduce the bias voltage and power consumption of the circuit.

2.4. Integrated push-pull and forward substrate bias circuit architecture
Fig. 5 shows the integrated push-pull and forward substrate bias circuit architecture utilized, which leads to the characteristics of low power consumption and reasonable gain. M3 and M4 are NMOS and PMOS, respectively. And C4, C5, and C6 are used for DC Blocking. The PMOS in the circuit reduces the V_{th} in the same way as the NMOS does. This integrated push-pull and forward substrate bias circuit architecture reduces the bias voltage of the LNA to 1.2 V. At such a bias voltage, the LNA operating at 2.3 mA consumes a power of 2.76 mW. The results show that this circuit architecture can achieve low power consumption for the LNA.

3. Circuit characteristics
Agilent ADS and TSMC 0.18-μm CMOS process technology are used for the low power LNA design. Fig. 6 shows the gain characteristics of the designed LNA. At 2.45 GHz, the LNA has a gain of 11.46 dB. Fig. 7 shows the noise factor characteristics of the LNA. At 2.45 GHz, the LNA shows a noise figure of 2.78 dB. In the 2.4–2.485 GHz band, both the input reflection coefficient and the output reflection coefficient are below -10 dB.

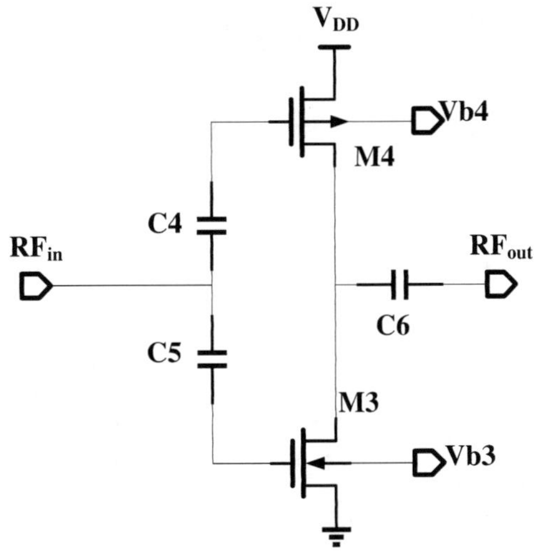

Figure 5. Integrated push-pull and forward substrate bias circuit architecture.

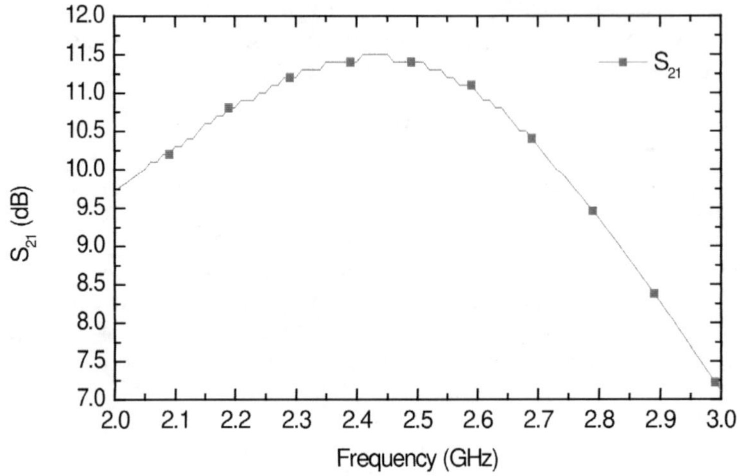

Figure 6. Characteristics of gain (S_{21}) of LNA.

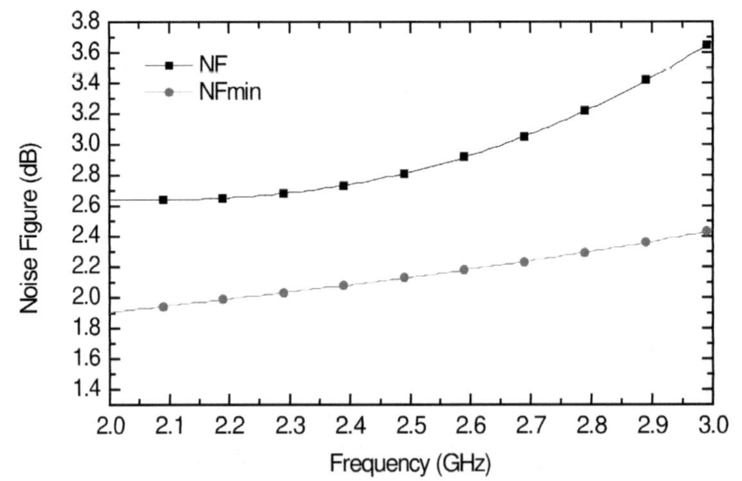

Figure 7. Characteristics of noise figure of LNA.

4. Conclusion

This paper proposes an LNA circuit design that employs a power-constrained simultaneous noise and input matching method and an integrated push-pull and forward substrate bias circuit architecture for IoT wireless communication systems. The LNA operating at a DC bias of 1.2 V consumes a low power consumption and achieve good RF performance. At 2.45 GHz, the LNA shows the characteristics of 11.46-dB signal gain and 2.78-dB noise figure. The designed LNA demonstrates good performance for IoT wireless communication systems.

Acknowledgment

This work was supported in part by the Ministry of Science and Technology of Taiwan, R.O.C. under Contracts MOST 107-2218-E-018-002, MOST 106-2218-E-018-003, and MOST 106-2622-E-018-003-CC3. The authors thank Taiwan Semiconductor Research Institute of National Applied Research Laboratories and Taiwan Semiconductor Manufacturing Company Ltd. for their assistance in chip design.

5. References

[1] The 'Only' coke machine on the internet, Carnegie Mellon University. [Online]. Available: https://www.cs.cmu.edu/~coke/history_long.txt. [Accessed: 25-May-2019]

[2] Internet of Things done wrong stifles innovation - InformationWeek. [Online]. Available: https://www.informationweek.com/strategic-cio/executive-insights-and-innovation/internet-of-things-done-wrong-stifles-innovation/a/d-id/1279157. [Accessed: 25-May-2019]

[3] M. Weiser, The computer for the 21st century, Sci. Am., 265, 94-104 (1991)

[4] K. Ashton, That 'Internet of Things' Thing, RFID J., 1, (2009)

[5] Y.L. Lai, Y.H. Chou, and L.C. Chang, An intelligent IoT emergency vehicle warning system using RFID and Wi-Fi technologies for emergency medical services, Technol. and Health Care, 26, 43-55 (2018)

[6] C.L. Chen, Y.L. Lai, C.C. Chen, C.Y. Zheng, and L.C. Chang, A non-repudiated and intelligent RFID medication safety management system, Intell. Automat. & Soft Comput., 1-7 (2016)

[7] Y.L. Lai, L.C. Chang, T.H. Liu, Y.C. Sung, and C.L. Yin, A novel RFID sensing system using enhanced surface wave technology for battery exchange stations, Mathemat Probl. in Engineeri., 2014, (2014) Art. ID 869489

[8] Y.L. Lai and J. Cheng, A cloud-storage RFID location tracking system, IEEE Trans. Magn., 50 (2014) Art. ID 3501004

[9] C.H. Chang, Y.L. Lai, and C.C. Chen, Implement the RFID position based system of automatic tablets packaging machine for patient safety, J. Med. Syst., 36, 3463-3471 (2012)

[10] Y.L. Lai and C.C. Chen, An intelligent RFID fall notification system, Int. J. Innovative Computing, Information and Control, 7, 3133-3145(2011)

[11] P. Magrassi, Why a Universal RFID Infrastructure Would Be a Good Thing, Gart. Res. Rep., G00106518 (2002)

[12] International telecommunication union, ITU Internet Reports 2005: The Internet of Things, (2005)

[13] M. A. Khan, K.Salah, IoT security: Review, blockchain solutions, and open challenges, Futur. Gener. Comput. Syst., 82, 395-411(2018)

[14] F. Firouzi, A. M. Rahmani, K. Mankodiya, M. Badaroglu, G.V. Merrett, P. Wong, B. Farahani, Internet-of-Things and big data for smarter healthcare: From device to architecture, applications and analytics, Futur. Gener. Comput. Syst., 78, 583–586 (2018)

[15] M.S. Mahdavinejad, M. Rezvan, M. Barekatain, P. Adibi, P. Barnaghi, A.P. Sheth, Machine learning for internet of things data analysis: a survey, Digit. Commun. Networks, 4, 161-175 (2018)

[16] A. Riahi Sfar, E. Natalizio, Y. Challal, Z. Chtourou, A roadmap for security challenges in the Internet of Things, Digit. Commun. Networks, 4, 118-137 (2018)

[17] D.J. Allstot, X. Li, and S. Shekhar, Design considerations for CMOS low-noise amplifiers, in IEEE Radio Frequency Integrated Circuits Symp. Dig., Fort Worth, TX, pp. 97-100, June 2004

[18] T.K. Nguyen, Y.M. Su, S.G. Lee: Proc. Int. Symp. on Circuits and Systems, (2004) IV-281

[19] A.N. Karanicolas, A 2.7V 900MHz CMOS LNA and mixer, in Int. IEEE Solid-State Circuits Conf. Dig., San Francisco, CA, pp. 50–51, Feb. 1996

[20] W.K. Yeh, S.M. Chen, and Y.K. Fang, Substrate noise-coupling characterization and efficient suppression in CMOS technology, IEEE Trans. Electron Devices, 51, 817-819 (2004)

IOP Conf. Series: Materials Science and Engineering **644** (2019) 012027 doi:10.1088/1757-899X/644/1/012027

Design and Simulation of a Porous Rectangular Board-Foldable Trolley

Yau-Ren Shiau[1], Jiunn Fang[2], and Yu-Lan Chen[3]
1 Department of Industrial Engineering and Systems Management, Feng Chia University, Taichung 407, Taiwan
2 Department of Aerospace and Systems Engineering, Feng Chia University, Taichung 407, Taiwan
3 Ph.D. Program of Mechanical and Aeronautical Engineering, Feng Chia University, Taichung 407, Taiwan
Email: wearefamily999@gmail.com

Abstract. To reduce the structural weight and volume and satisfy an optimal balance of performance constraints, a study using a porous rectangular foldable board and structurally improved trolley model was conducted, and motion simulations were performed to derive the static and transient structural properties. The simulations revealed that the fatigue lifetimes of both the main structure and the porous boards were in a high-standard-value range, whereas the transient analysis indicated that both the directional deformation and equivalent stress exhibited minimal local values. For porous boards simply supported only on two sides, the trolley had a much smaller collapsed volume than other types of trolley. The static analysis indicated that this trolley prototype had a good safety coefficient and long fatigue lifetime. The transient analysis showed that the porous area had a higher strength and bending load than the non-porous area. The overall transient response was close to that of a steady-state, showing a negligible effect on the structure.tc.

1. Introduction

Lightweight design is a major requirement in the development of many products. Foldable trolleys, for example, have been widely used at various sites by a range of users. Their main structures are primarily made of steel and aluminium, whereas the load-boards are usually non-porous and made of polyethylene or metals. Generally, after the trolley is folded, the volume range of the width of the board still occupies the storage space. Therefore, to satisfactorily optimise designs, in terms of the weight and performance constraints, the volume and lightness of the product's minimum compliance system are gradually becoming a concern [1].

When a laminated composite plate was subjected to a dynamic load-induced over vibration, high translocation, and acceleration, its structural properties were severely degraded [2]. Therefore, a simulation was conducted of the static, modal, and transient structures, based on the finite-element method (FEM), to determine the loading status. With some improvements to the trolley structure, a side-suspended trolley, suspended on two girders, was developed. Based on the FEM, it was found that the improved trolley satisfied the strength and rigidity standards [3]. Finite element analysis (FEA) was used to determine the static and dynamic response of the components under different load conditions [4]. According to the structure and characteristics of the material, the stress and deformation were analyzed and the maximum deformation was obtained [5]. Through the modeling and simulation of a bicycle frame, it is known from the maximum stress and displacement that, after the structural optimization, the rigidity is improved and the total deformation is reduced [6]. Motion simulation was used to study the structural properties of a trolley by using the software ANSYS to investigate the stress and fatigue of a trolley subjected to a given loading. Application-related parameters, such as the total deformation, stress ratio, and strain, were obtained to analyze the fatigue lifetime, which was shown by the simulations to be far longer than standard, thereby indicating an optimal design [7].

Significant positive correlations can, generally, be observed between material and structure. To reduce the weight of a structure, designers usually fabricate holes at specific sites on the plates in the structure, as long as the structural support capability is not affected. Under a strength constraint on the plates, most of the plates that were produced had a high porosity, specific surface area, and an open cell rate that allowed the products to exhibit the best performance [8]. The perforated board withstood changes in load and support conditions, greatly reducing local buckling and strengthening the fatigue resistance of the material [9]. Supported by a lower plate with a porous surface, the pressure distribution and load carrying capacity were also increased when the surface roughness of the upper plate was increased. [10].

Based on the consideration of a lighter body structure, and with sufficient load-carrying capacity, changing the solid gear to a lattice structure observed not only a reduction in the weight but also reduced the vibrations significantly [11]. A loading analysis of a honeycomb-like, bidirectionally hollow thin plate using ANSYS was performed, and it was found that such a plate had a higher loading capability and better plastic deformation characteristics than a solid plate [12]. The structural optimisations that were performed on this basis significantly increased the stiffness of the bicycle frame, thus reducing its deformations. The buckling response of square composite plates with elliptical holes in the centre was studied, and it was found that large elliptical holes generated a pressure under the thinnest plate, indicating that, to prevent buckling that may be generated under a relatively low stress, large elliptical holes should be avoided on composite laminated plates [13].

This study is based on a reported improvement of the trolley structure, in which the trolleys are foldable on two sides to reduce the collapsed volume [14–18]. Briefly, porous foldable boards were built, and a structurally improved trolley was designed. Then, an FEM-based engineering simulation approach was employed to simulate the static structural properties and transient structural motions of the trolley.

2. Materials and methods

In this study, a static structure simulation was performed to analyze the structural performance and fatigue strength of the entire trolley structure. In order to understand the reaction of the load (object) when it is applied to the structure instantaneously, a transient structure analysis was conducted to analyze the bending behaviour of the load boards.

2.1. Development of the geometric model

Numerous porous foldable boards and a structurally-improved trolley prototype, in which the fixed framework structure was improved to an adjustable one to reduce the collapsed volume of the trolley, were built. The folding operation is as shown in Figure 1.

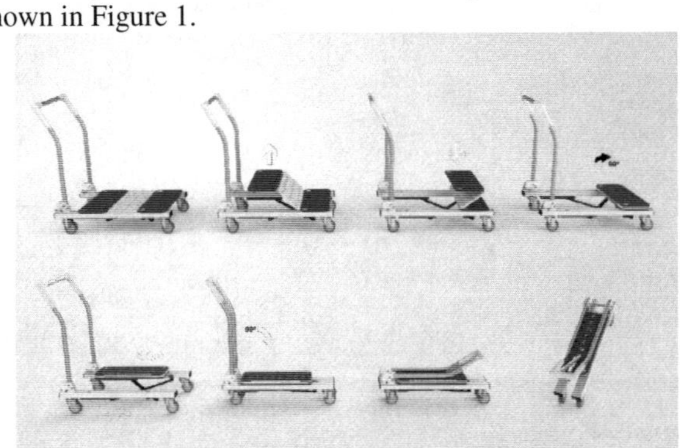

Figure 1. Trolley folding operation.

Three rectangular load boards were designed to have multiple symmetrically-placed holes, each with a diameter of 30 mm. Each board was 450 (length) × 180 (width) × 11 (height) mm when folded, to decrease the weight and reduce the space occupied. When expanded, the trolley had dimensions of 791.49 (length) × 455.2 (width) × 802.64 (height) mm, whereas the collapsed trolley had dimensions of 700 (length) × 180 (width) × 280 (height) mm.

Most of the components are connected by a set of screws, respectively, included the board B is also connected by two groups of hinges between the boards of A and C, as shown in Figure 2. After modelling, the computer aided engineering (CAE) simulation can be simplified as needed.

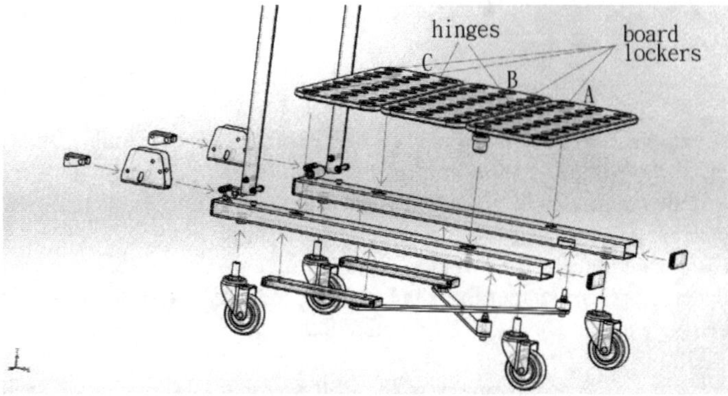

Figure 2. Trolley folding operation.

2.2. *Setting simulation conditions*

For the framework and load boards of the foldable trolley, the main material was an aluminum alloy; a type of material widely used in engineered structures due to its high strength and low weight. We used commercially-available metal parts for several of the hardware fittings of the trolley. Knowledge about the material coefficients of the metal parts was beyond the scope of this study and, therefore, we simply used the aluminum alloy coefficients to represent them. The relevant parameters of geometry and materials for the trolley are listed in Table 1. The element Control was program-controlled, the solver target used Mechanical APDL (ANSYS Parametric Design Language).

Table 1. Model analysis.

Item	Model Geometry	
Bounding Box	Length X	791.49 mm
	Length Y	802.64 mm
	Length Z	455.20 mm
Properties	Volume	4,136,700 mm^3
	Mass	11.459 kg
	Scale Factor	1
Statistics	Nodes	3,383,425
	Elements	2,040,337
	Mesh Metric	Element Quality

When performing a mesh calculation, we set the following parameters: solid, physical preference was set to be mechanical, relevance was 100, size function was adaptive, relevance center was chosen to be coarse, mesh transition was selected to be fast, span angle center was coarse, smoothing was set as medium, mesh metric was set as element quality, nodes were 3,383,425 and elements were 2,040,337, to ensure accurate analysis and simulation (see Figure 3). As shown in Figure 3, the closer the bar of the statistical mesh number is to the right, indicates better quality.

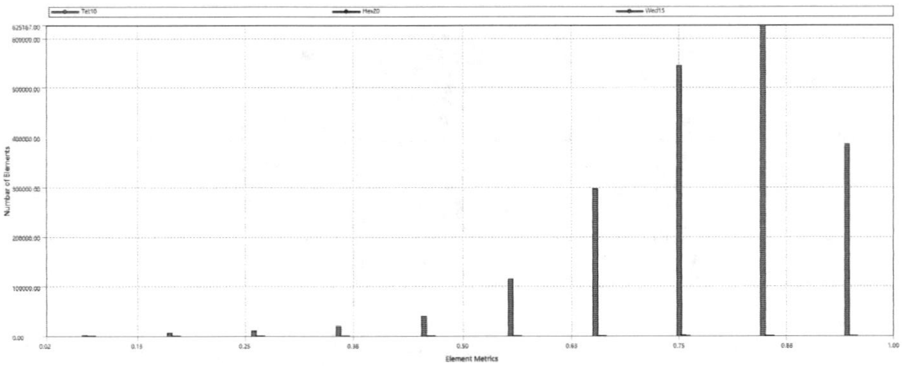

Figure 3. Element Quality.

The constraints for the static analysis were as follows: For the load boards, four board lockers (D in Figure 4) and several hinge screws (J in Figure 4) were used as a fixed support to anchor the three foldable boards (A, B, and C in Figure 4) and the two groups of hinges (E, F, G, and H in Figure 4) that connected two adjacent foldable boards. This was aimed at placing a constraint on the freedom of Ux, Uy, and Uz. The load was set to be 800 N in magnitude and was exerted in a downward direction. Four wheels were used as displacement support and for constraining the freedom of Uy.

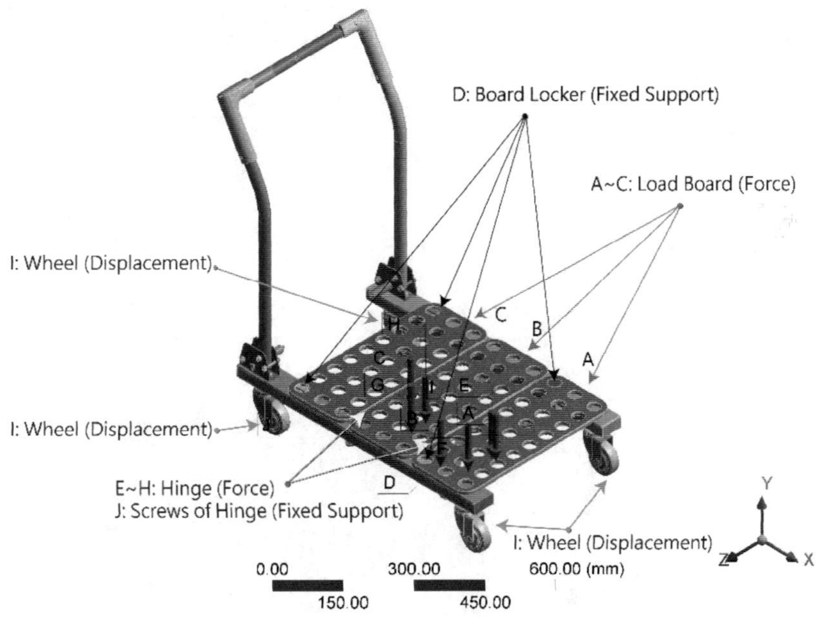

Figure 4. Loads and constraints for the static structural analysis.

Based on the requirement for the loading plate to be supported and fixed, the relevant position is constrained. The constraints for the transient analysis were approximately the same as those for the static analysis, except the displacement support, which was moved to the two sides (D, E, and G–J in Figure 5) of the three load boards. Due to the four board lockers that are connected and fixed between the load board and the fixed support, the freedom of Ux and Uy was constrained. There was no support in the Uz direction and, so, Uz was free.

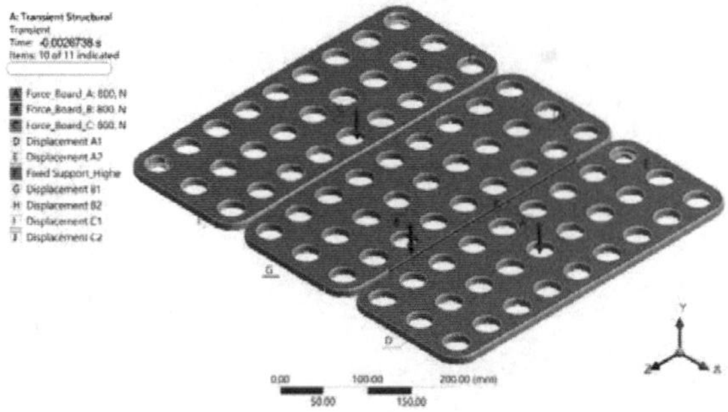

Figure 5. Loads and constraints for the transient structural analysis.

3. Results

After setting the parameters for the materials, meshes, loads, and constraints, we found that, when an 800 N force was exerted on the static structure, the structure underwent a deformation and the force and load board formed a mutual vertical stress. The static and fatigue simulation results are shown in Table 2.

Table 2. Structural statics and fatigue analysis.

Object	Force: 800 N		
	Min.	**Max.**	**Occurs on**
Total Deformation	0 mm	0.12581 mm	Board A (Max)
Equivalent Elastic Strain	0 mm/mm	0.0015746 mm/mm	Screw / Board A (Max)
Equivalent Stress	0 MPa	110.96 MPa	Screw / Board A (Max)
Safety Factor	2.5235	15	Screw / Board A (Min)
Life	4,878,500 cycles	1e+32cycles	Screw / Board A (Min)

3.1. Static analysis.

In this study, an 800 N simulation force was exerted on the boards and hinges of the entire trolley structure (with a total weight of 11.459 kg). After the total deformation analysis, we found that the bending part showed a maximum stress in structural rigidity displacement of 0.12581 mm. It can be seen, in Figure 6, that the maximum displacement occurred at the outer free end of load board A, whereas the main loading regions on the boards did not show any obvious deformation.

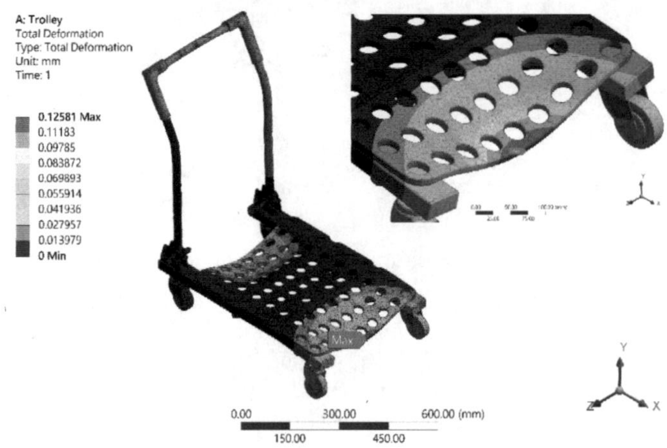

Figure 6. Total deformation.

When the 800 N force was exerted on the load boards, the maximum equivalent elastic strain was 0.0015746 mm/mm, and was located in a small region around the head edge of a screw on the hinge connecting the load board A, as shown in Figure 7.

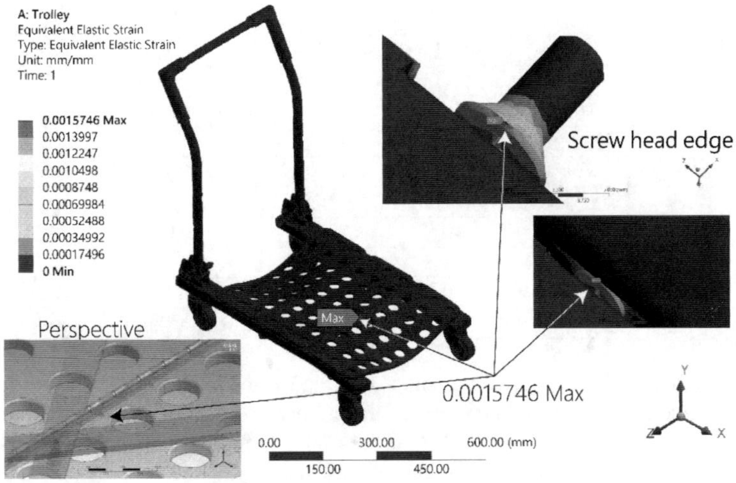

Figure 7. Equivalent elastic strain.

For the purpose of avoiding plastic deformation, which may cause permanent damage to the mechanical materials, the stress should be lower than the yield strength or tensile strength of the materials (i.e., 280 MPa), and a stress below this value does not generate plastic deformation or destructive deformation. The analysis in this study indicated that when the load boards were subjected to the external force, the maximum equivalent stress per unit area inside the boards was 110.96 MPa and was confined to a small region around the head edge of a screw on the hinge connecting load board A, as shown in Figure 8.

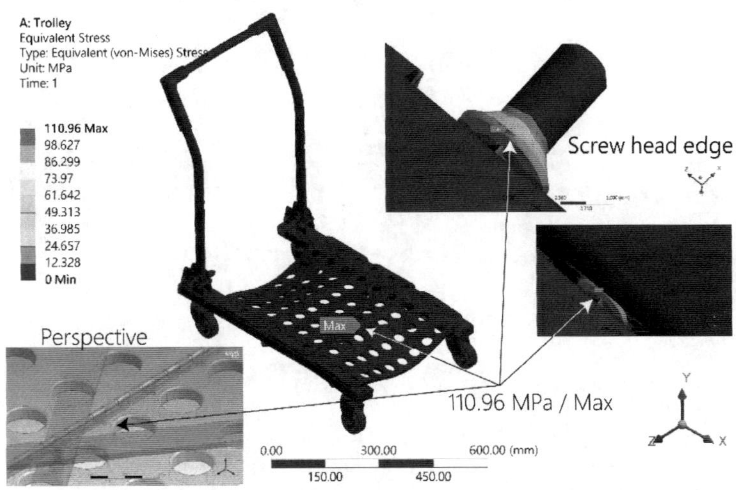

Figure 8. Equivalent stress.

3.2. Fatigue analysis.
A fatigue test was conducted for tension-tension and tension-compression load with a stress ratio, R, of -1, -0.5, 0, and 0.5, and up to the endurance limit of 108 with frequency (Table 3).

Table 3. Alternating stress R-Ratio.

Alternating Stress MPa	Cycles	R–Ratio	Alternating Stress MPa	Cycles	R–Ratio
275.8	1700	-1	77.57	50,000,000	-0.5
241.3	5000	-1	72.39	100,000,000	-0.5
206.8	34,000	-1	144.8	50,000	0
172.4	140,000	-1	120.7	190,000	0
137.9	800,000	-1	103.4	1,300,000	0
117.2	2,400,000	-1	93.08	4,400,000	0
89.63	55,000,000	-1	86.18	12,000,000	0
82.74	100,000,000	-1	72.39	100,000,000	0
170.6	50,000	-0.5	74.12	300,000	0.5
139.6	350,000	-0.5	70.67	1,500,000	0.5
108.6	3,700,000	-0.5	66.36	12,000,000	0.5
87.91	14,000,000	-0.5	62.05	100,000,000	0.5

The safety coefficient of a regular-livelihood product is defined as the ratio of the yield stress to the allowable stress, with a coefficient ≥1.5 indicating the safety of the product. Aluminum is a ductile material. This study found that the minimal safety coefficient of 2.5235 corresponded to a small region around the head edge of a screw on the hinge connecting load board A (Figure 9), whereas the safety coefficients of other parts on the trolley were all greater than that value, and hence were in a safety range.

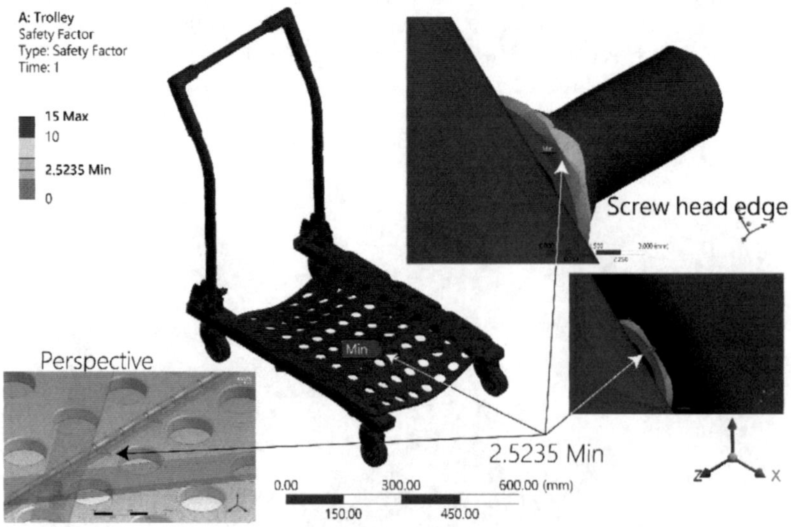

Figure 9. Safety factor.

Based on the mesh distribution and stress intensity factors, we analyzed and predicted the fatigue lifetime, and determined that the minimum value of 4,878,500 cycles was for a small region around the head edge of a screw on the hinge connecting load board A. In comparison, both the structural framework and load boards had a fatigue lifetime greater than 1e+32 cycles, with the fatigue lifetime of the screws lying in between (Figure 10). These results indicated that repeated loading would not compromise the fatigue lifetimes of the structural framework and load boards.

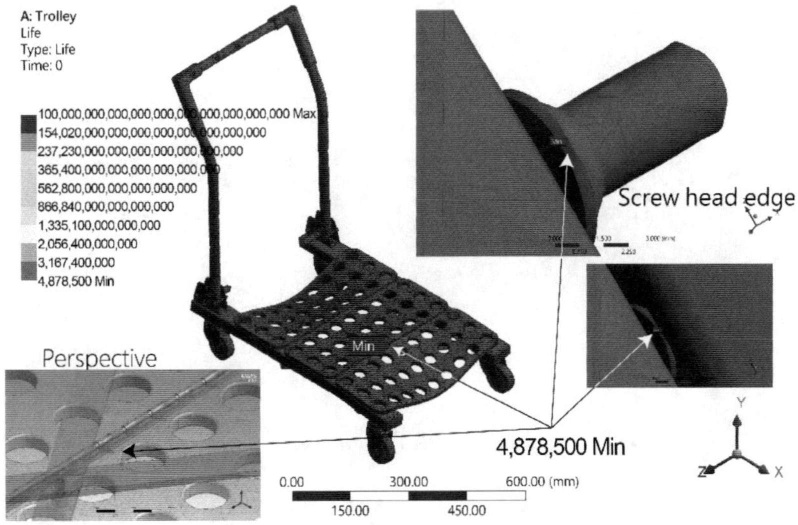

Figure 10. Analysis of fatigue lifetime.

Figure 11 shows a plot of the available lifetime (in cycles) as the fatigue results change, as a function of loading at the critical location (rectangle area) on the trolley, with the loading history expressed as a ratio in the range of 0.5–1.5. With respect to the lifetime, the damage coefficient or safety coefficient varied significantly (with a trend of decline) when the loading history was close to a critical data range, where the sensitivity of the lifetime to the loading history could be estimated.

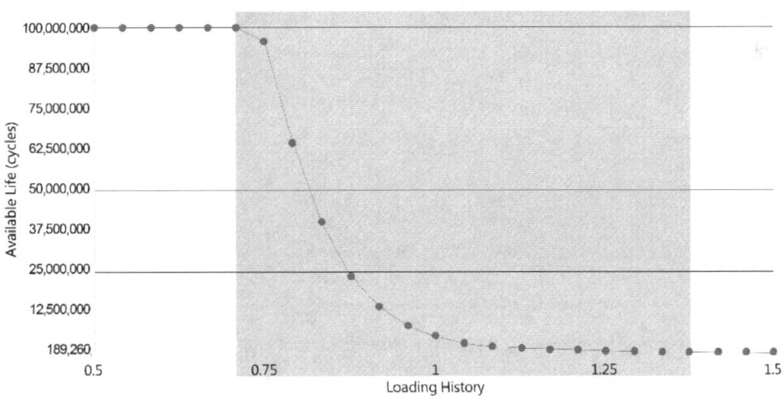

Figure 11. Fatigue sensitivity.

By using the material coefficients of alum–Si cast and extruded aluminum [19] as the foundation for fatigue data calculation and by setting the multiplication factor as 1, our calculation showed that the equivalent stress dropped from its maximum value of 110.96 MPa to the minimum value of 0 MPa within one cycle, as displayed by the stress–strain curve—a plot of local stress versus corrected strain response—in Figure 11. Our calculations also predicted a stable periodicity over repeated stress cycles (Figure 12).

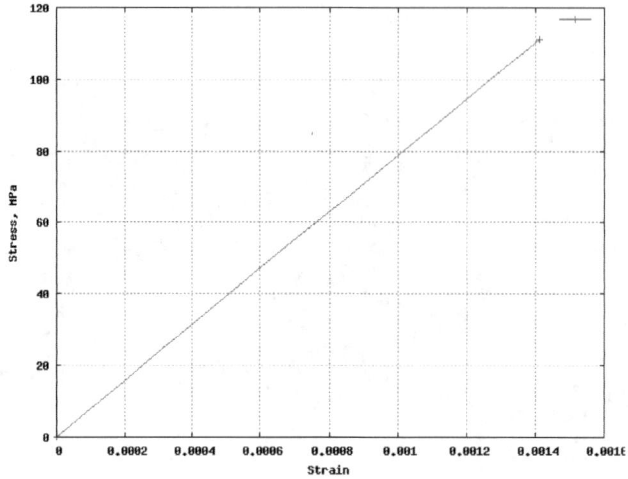

Fig. 12 Local stress and strain response.

3.3. *Transient analysis.*

To determine the characteristics of the transient response, we conducted an analysis of the vibration displacement of the porous board after the force is applied. The transient integration parameters used were: an amplitude decay factor for the second-order transient parameters, transient integration parameters (Gamma=0.1000, Alpha=0.3025, Delta=0.6000), Hilbert-Huang Transform (HHT) time integration (Alphaf=0.1000, Alpham=0.000), and center of mass (X, Y, Z) = (56.899, 35.492, -0.19929). The curve is shown in Figure 13.

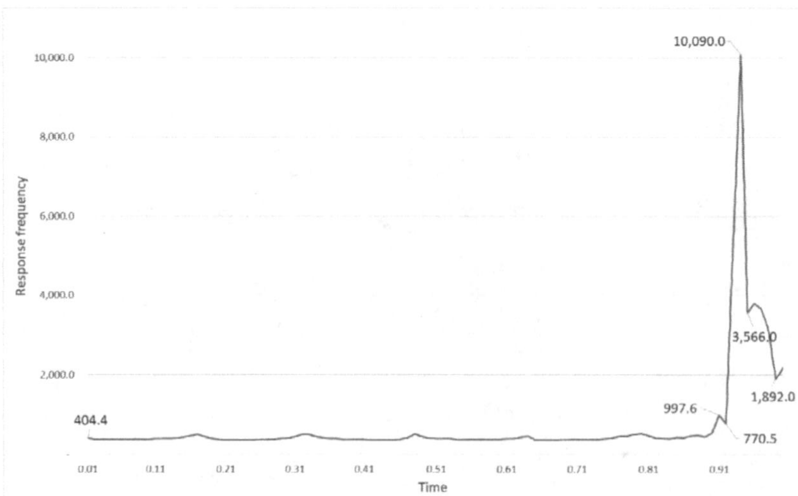

Figure 13. Digitally fitted curve.

The load boards were placed on the two sides of the supporting framework. With such a simple physical structure, the entire load was perpendicular to the load surface. To further understand the strength of the load boards and their bending load, we performed a transient analysis and found that, for directional deformation, the transient response attained a stable status at 0.38 s after the maximum value and at 0.43 s after the minimum value as the transient response evolved gradually from the initial to the stable status, as shown in Figure 14.

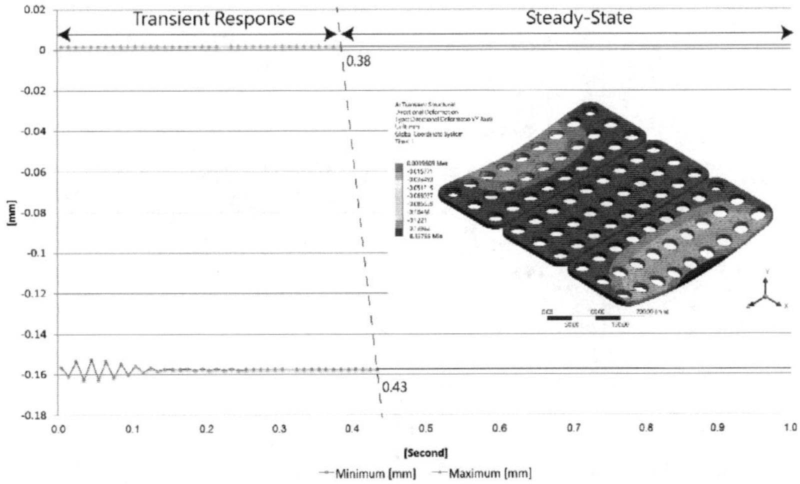

Figure 14. Transient response of directional deformation from the transient structural analysis.

We conducted directional deformation analysis along the Z-axis direction in the porous and non-porous areas of the load boards, and found that the non-porous area of board A exhibited the maximum deformation of -0.08436 mm, followed by the porous area of board A (-0.05746 mm), the porous area of board C (-0.03534), the non-porous area of board C (-0.02146 mm), the porous area of board B (-0.00594 mm), and the non-porous area of board B (-0.00474 mm), as shown in Figure 15.

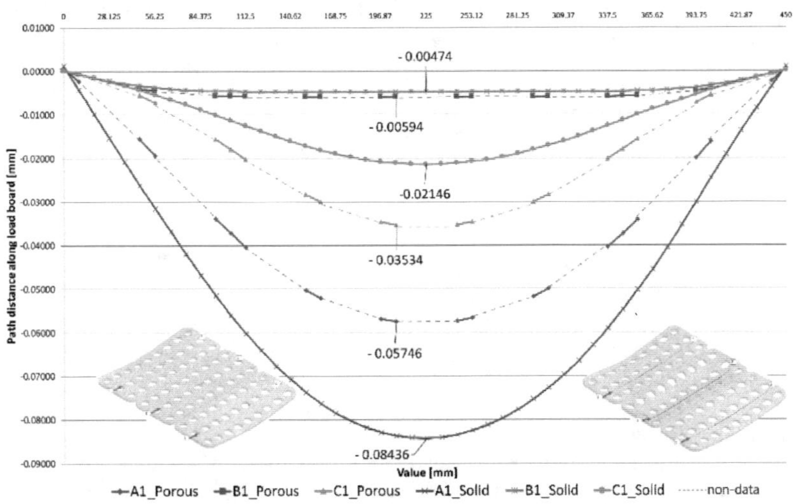

Figure 15. Directional deformation from the transient structural analysis.

In regards to the equivalent stress, the transient response reached a stable status 0.41 s after the maximum value and 0.45 s after the minimum value during the gradual evolution of the response from the initial to the stable status, as shown in Figure 16.

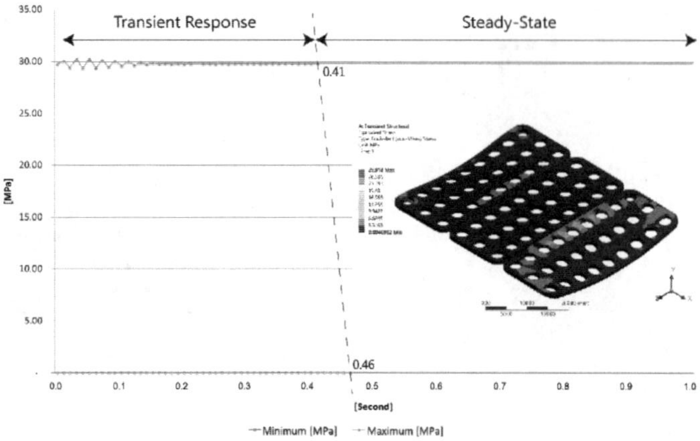

Figure 16. Transient response of equivalent stress from the transient structural analysis.

By analyzing the equivalent stress along the Z-axis direction, we found that the non-porous area of board A had a maximum equivalent stress of 4.826 MPa, followed by the porous area of board A (3.2205 MPa), the porous area of board C (2.9598 MPa), the non-porous area of board C (2.1358 MPa), the porous area of board B (1.7672 MPa), and the non-porous area of board B (0.7127 MPa), as shown in Figure 17. These values were much lower than the yield strength of 280 MPa for the materials.

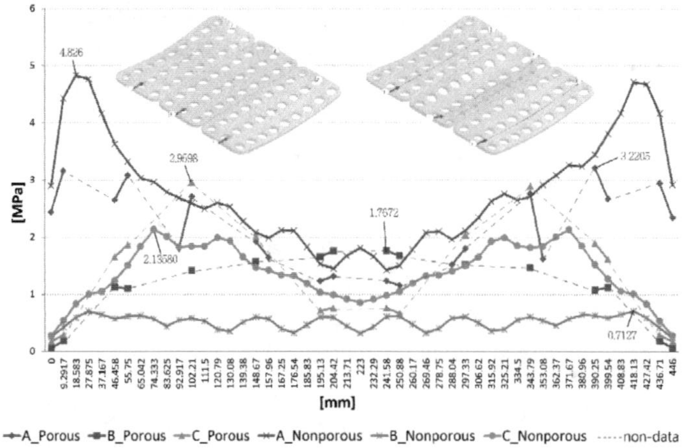

Figure 17. Equivalent stress from the transient structural analysis.

4. Discussion and conclusions

In this study, for the static analysis, we simulated an 800 N force being exerted on the load boards to determine the most probable site at which the trolley materials would show deformation and fatigue, and we found that the maximum displacement of 0.12581 mm—though a very small value—occurred at the free outer end of the load boards, whereas the regions subjected to most of the loading did not show deformation and, consequently, did not undergo the associated plastic deformation.

The maximum equivalent elastic strain of the load boards subjected to loading and the maximum equivalent stress of plastic deformation, as predicted by our simulations, were both present in small regions around the head edge of a screw on the hinge connecting load board A. However, the small values of these two parameters indicated that the plastic deformation characteristics of the structure were not affected.

Regarding the safety coefficient, our analysis revealed that it was always greater than 1.5 and, thus, the safety of the entire structure was only impacted to an extremely low extent. The fatigue lifetime was predicted to be at least 4,878,500 cycles, and, when loaded repeatedly, the trolley did not show a change in the fatigue lifetime, neither in the main framework nor in the load boards.

The screws, which had an almost-negligible effect on the entire performance of the trolley, may be replaced with other high-strength materials, such as low-carbon alloy steels or middle-carbon steels, so that the product quality may be further optimized.
Transient analysis showed that the maximum deformation (in board A) of the porous area had a higher strength and higher bending load than the non-porous area. The overall transient response was close to the steady-state, with a negligible effect on the structure, and the porous plate will not cause a large vibration displacement due to loading with a vertical force under 800N.
The results of the study indicate that replacing regular load boards (that have supports on all the four sides) with rectangular, porous foldable load boards (that have supports on only two sides), significantly reduces the collapsed volume, the weight of the trolley, and the storage space after folding, to achieve an optimal design by providing structural improvements.
Foldable porous board can be applied to tables, chairs, ladders, and so on, to meet the requirements of lightweight and storage-minimum specifications. In terms of materials, materials with high stability and impact strength, such as ABS, can be used.

Acknowledgments

The work was supported by the Ministry of Science and Technology of the Republic of China (MOST106-2221-E- 035-071).

5. References

[1] D. Li, and L.Y. Kim, Multi-material topology optimization for practical lightweight design, Struct. Multidiscip. Optim (2018) 1
[2] K. Maithry and B.C. Rao, M. Static, Modal and transient dynamic analysis of laminated composite plates with holes, I-Manag. J. Civ. Eng., 6 (2016) 40–47
[3] L. Qi, Z.D. Zhang, H. Wen, and H. Gang, Structure finite element analysis of side-suspended trolley based on cosmos works. Hoist, Hoist. Conv. Mach (2011) 30–33
[4] F. Leonardo, D. Giampiero, D. Karim, L. Alfredo and C. Gianni, 2018 Appl. Sci., 8, 2635
[5] S. Chundi, C. Hang, E. Chen, Z. You, R. Tian, R. Zhang and J. Gao, Dynamic response analysis of rutting resistance performance of high modulus asphalt concrete pavement, Appl. Sci. 8 (2018) 2701
[6] C.C. Lin, S.J. Huang, and C.C. Liu, Structural analysis and optimization of bicycle frame designs, Adv. Mech. Eng. (2017) 9
[7] Y.L. Chen, J. Fang, Y.R. Shiau and W.C. Hong, A study on static structural of non-rotor unmanned aerial vehicle, MATEC Web Conf., 119 (2017) 01053
[8] P. Liu, T. Li, C. Fu and M. L., Applications of Porous Metal Materials, J. Funct. Mater. 32 (2001) 12–15
[9] B. Harl, J. Predan, N. Gubeljak, and M. Kefl, On configuration-based optimal design of load-carrying lightweight parts, Int. J. Simul. Model. 16 (2017) 219–228
[10] R.B. Kudenatti, S.M. Patil, P.A. Dinesh, and C.V. Vinay, Numerical study of surface roughness and magnetic field between rough and porous rectangular plates, Math. Probl. Eng (2013)
[11] R. Ramadani, A. Belsak, M. Kegl, J. Predan, and S. Pehan, Topology optimization based design of lightweight and low vibration gear bodies, Int. J. Simul. Model. 17 (2018) 92–104
[12] Y.B. Cheng, W.R. Cheng and J. Dang, Experiments of cast-in-situ concrete cellular voided two-way plates with edge support, J. Southeast. Univ. 37 (2007) 217–221
[13] M. Aydin Komur, F. Sen, A. Atas, and N. Arslan, Buckling analysis of laminated composite plates with an elliptical/circular cutout using FEM, Adv. Eng. Softw. 41 (2010) 161–164
[14] Y.L. Chen, R.O.C. Patent TWM479256 (2014)
[15] Y.L. Chen, W.S. Chen, Y.P. K, and Y.R. Shiau, GTEA Conference, Taichung, Taiwan, (2014) p. GT1006
[16] Y.R. Shiau, Y.L. Chen, W.S. Chen, and Y.P. Kang, Innovative Design for a Foldable Trolley, J. Technol. 30 (2015) 9–14
[17] Y.L. Chen, R.O.C. Patent TWM496587 (2015)
[18] Y.L. Chen, CN Patent ZL2015200625972 (CN204415465U) (2015)

[19] A.S. Waterloo, Cast and extruded aluminum, University of Waterloo Fatigue Design + Evaluation Committee of the SAE Experimental. Available online: http://fde.uwaterloo.ca/Fde/Materials/Alumcast/Al_Si/alumSi.html (accessed on 15 June 2018).